Dr. John Chung's

SAT MATH

Fourth Edition

Good Luck!

Copyright by Dr. John Chung

Made in the USA

Dear Beloved Students,

With these new editions, I would like to thank all the students who sent me emails to encourage me to revise my books. As I said, while creating this series of math tests has brought great pleasure to my career, my only wish is that these books will help the many students who are preparing for college entrance. I have had the honor and the pleasure of working with numerous students and realized the need for prep books that can clearly explain the fundamentals of mathematics. Most importantly, the questions in these books focus on building a solid understanding of basic mathematical concepts. Without these solid foundations, it will be difficult to score well on the exams. These books emphasize that any difficult math question can be completely solved with a solid understanding of basic concepts.

As the old proverb says, "Where there is a will, there is a way." I still remember vividly a fifth-grader who was last in his class who eventually ended up at Harvard University seven years later. I cannot stress enough how such perseverance in the endless quest to master mathematical concepts and problems will yield fruitful results.

You may sometimes find that the explanations in these books might not be sufficient. In such a case, you can email me at drjcmath@gmail.com and I will do my best to provide a more detailed explanation. Additionally, as you work with these books, please notify me if you encounter any grammatical or typographical errors so that I can provide an updated version.

It is my great wish that all students who work with these books can reach their ultimate goals and enter the college of their dreams.

Thank you.

Sincerely,

Dr. John Chung

CONTENTS

60 TIPS

CONTENTS

CONTENTS

Practice Tests

Dr. John Chung's SAT Math

60 TIPS

TIPS

Tip 01 | Linear Function

The functions are called "linear" because they are precisely the functions whose graph in the xy-plane is a straight line.

Such a function can be written as

1) Slope-intercept form

$f(x) = mx + b$, where m is the slope and b is the y-intercept.

2) Point-slope form

$y - y_1 = m(x - x_1)$, where (x_1, y_1) is the known point on the line.

3) General form

$ax + by + c = 0$

4) Standard form

$Ax + By = C$

The slope between any two points on the line is constant.

$$m = \frac{\triangle y}{\triangle x} = \frac{y_2 - y_1}{x_2 - x_1}$$

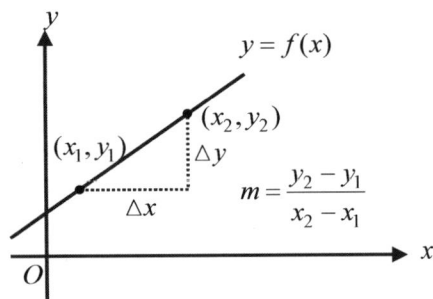

SAT Practice

1. For a linear function f, $f(0) = 2$ and $f(3) = 5$. If $k = f(5)$, what is the value of k?

   A) 5
   B) 6
   C) 7
   D) 8

x	$f(x)$
0	a
1	12
2	b

2.   The table above shows some values for the function f. If f is a linear function, what is the value of $a+b$?

A) 24
B) 36
C) 48
D) 60

3.   A linear function is given by $ax+by+c=0$ and $a>0$, $b<0$, and $c>0$. Which of the following graphs best represents the graph of the function?

A)

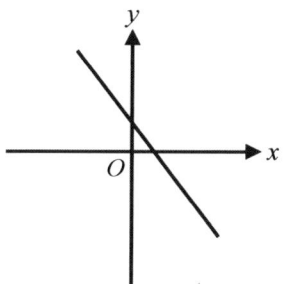

B)

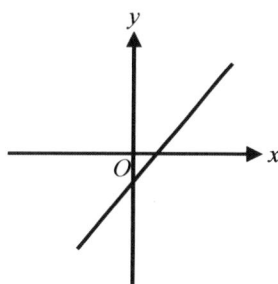

C)

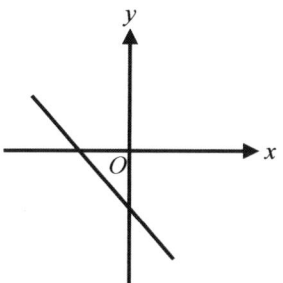

D)

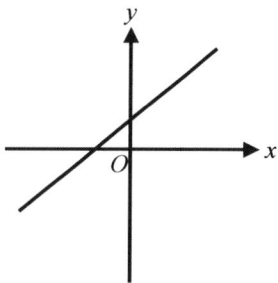

4. If f is a linear function and $f(3) = 2$ and $f(5) = 6$, what is the y-intercept of the graph of f?

A) 4
B) 2
C) −2
D) −4

---

5. If f is a linear function and $f(3) = -2$ and $f(4) = -4$, what is the x-intercept of the graph of f?

A) 3
B) 2.5
C) 2
D) 0

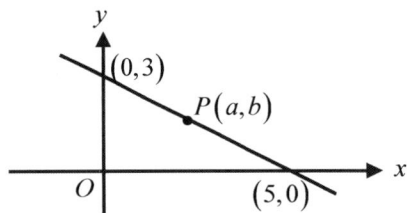

6. The graph of a function f is shown in the xy-plane above. If $b = 2a$, what is the value of a?

A) $\dfrac{5}{2}$ B) $\dfrac{5}{4}$ C) $\dfrac{15}{13}$ D) $\dfrac{16}{15}$

---

x	$f(x)$
−1	6
0	4
1	2
2	0

7. The table above shows some values of the linear function f for selected values of x. Which of the following represents the function f?

A) $f(x) = 4 - x$
B) $f(x) = 4 - 2x$
C) $f(x) = 4 + 2x$
D) $f(x) = 4 + x$

$$F = \frac{9}{5}C + 32$$

8. Fahrenheit (F) and Celsius (C) are related by the equation above. If Fahrenheit temperature increased by 27 degrees, what is the degree increase in Celsius?

A) 15
B) 20
C) 32
D) 81

9. In the formula $P = \frac{7}{12}K + 60$, if P is increased by 35, what is the increase in K?

A) 35
B) 60
C) 80
D) 140

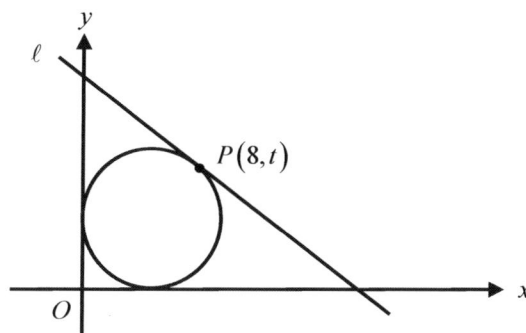

10. In the xy-plane above, a circle is tangent to line ℓ, the x-axis, and the y-axis. If the radius of the circle is 5, what is the value of t?

A) 7
B) 8
C) 9
D) 10

Tip 02 — Slope of a Line

One of the most important properties of a straight line is its angle from the horizontal. This concept is called "**slope**". To find the slope, we need two points from the line.

1) From two points (x_1, y_1) and (x_2, y_2) $\rightarrow$ Slope $m = \dfrac{y_2 - y_1}{x_2 - x_1}$

2) From slope-intercept form of a line $y = mx + b$ $\rightarrow$ $m = $ slope and $b = y$-intercept

3) The slope between any two points on the line is constant.

SAT Practice

1. If f is a linear function and $f(3) = 6$ and $f(5) = 12$, what is the slope of the graph of f?

   A) 2
   B) 3
   C) 4
   D) 5

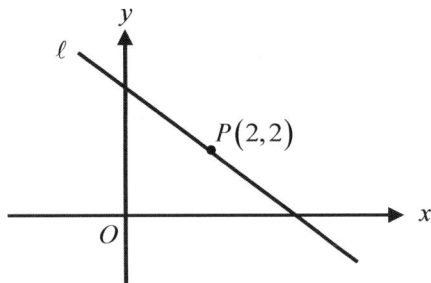

2. In the xy-plane above, line ℓ passes through point P and has a slope of $-\dfrac{1}{2}$. What is the x-intercept of line ℓ?

   A) $(4, 0)$
   B) $(5, 0)$
   C) $(6, 0)$
   D) $(7, 0)$

x	$f(x)$
2	5
4	a
8	23
a	b

3. The table above shows values of the linear function f for selected values of x. What is the value of b?

A) 11
B) 22
C) 32
D) 42

---

x	$f(x)$
2	a
5	6
8	b

4. The table above gives values of the linear function f for selected values of x. What is the value of $a+b$?

A) 8
B) 10
C) 12
D) 18

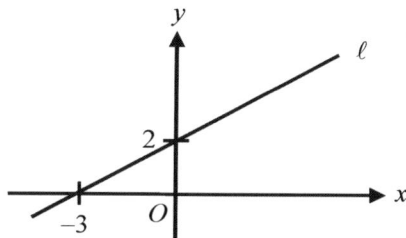

5. In the xy-plane above, point $P(42, m)$ lies on line ℓ. What is the value of m?

A) 24
B) 30
C) 36
D) 42

Tip 03 Average Rate of Change

The slope is the average rate of change of a line.

1) For a line, the average rate of change is the slope of the line and the slope is constant no matter what two points you calculated it for on the line.
2) For a curve, the average rate of change is the slope of the secant line that passes through the two points on the curve.

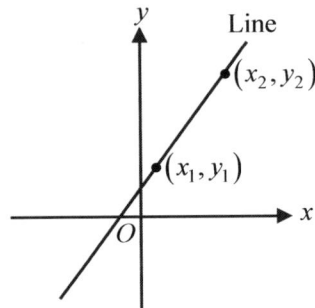

Line

(x_2, y_2)

(x_1, y_1)

Average rate of change $= \dfrac{y_2 - y_1}{x_2 - x_1}$

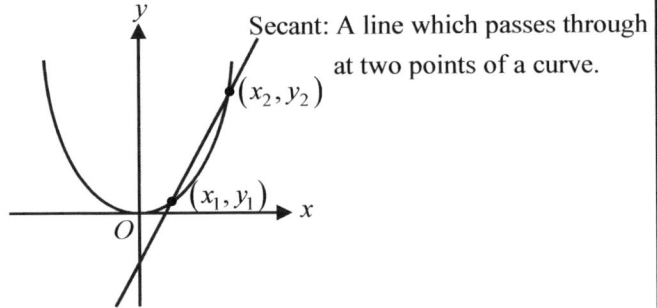

Secant: A line which passes through at two points of a curve.

(x_2, y_2)

(x_1, y_1)

Average rate of change $= \dfrac{y_2 - y_1}{x_2 - x_1}$

SAT Practice

1. What is the average rate of change of $f(x) = \dfrac{1}{2}x^2 - 4$ as x changes from 0 to 4?

A) 2   B) 3   C) 4   D) 5

---

2. If an object is dropped from a cliff, then the distance, in meters, it has fallen after t seconds is given by the function $d(t) = 4.9t^2$. What is its average speed (average rate of change) over the interval between $t = 1$ and $t = 3$?

A) 4.9 meters per second
B) 9.8 meters per second
C) 19.6 meters per second
D) 39.2 meters per second

Questions 3 and 4 refer to the following information.

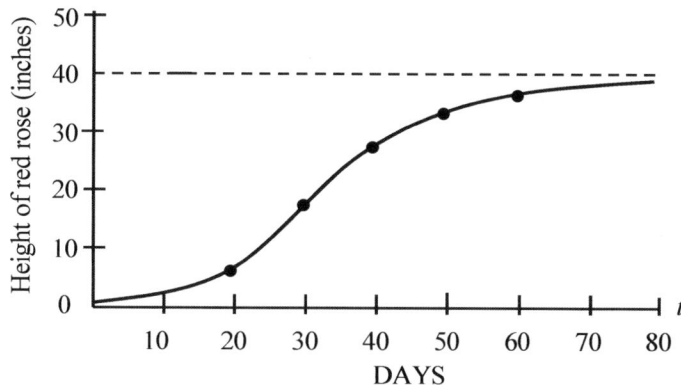

t	$h(t)$
0	2
10	3.5
20	7.4
30	19.8
40	29.8
50	33.4
60	37.5
70	39.2
80	39.4

A flower plant is measured every day t. The height, $h(t)$ inches, of the plant can be modeled by a function which is shown in the graph and table above.

3.   Find the average rate of change in height from 10 to 70 days?

4.   Which of the following intervals has the greatest average rate of change?

A) From 10 to 20 days

B) From 20 to 30 days

C) From 30 to 40 days

D) From 40 to 50 days

TIPS

Tip 04 — Area enclosed by Lines

In order to find the area enclosed by lines, mostly we need to find x-intercept, y-intercept,
And points of intersection of the lines.

SAT Practice

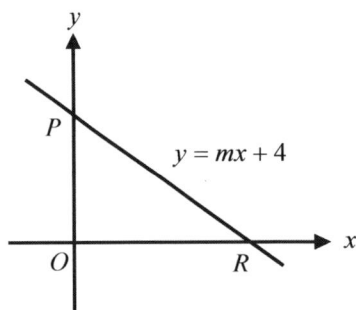

1. The graph of $y = mx + 4$ is shown in the xy-plane above. If the area of triangle POR is 6, what is the value of m?

   A) -2

   B) $-\dfrac{4}{3}$

   C) $-\dfrac{3}{4}$

   D) $-\dfrac{1}{4}$

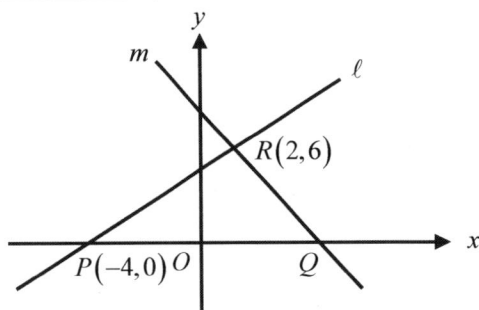

2. In the xy-plane above, line m and line ℓ are perpendicular and intersect at point $R(2,6)$. What is the area of triangle PQR?

   A) 18
   B) 24
   C) 32
   D) 36

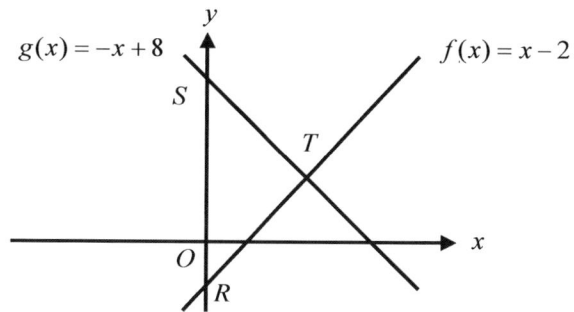

$g(x) = -x + 8$
$f(x) = x - 2$

3. The graphs of the functions f and g are shown in the xy-plane above. What is the area of $\triangle RST$?

A) 25
B) 50
C) 75
D) 100

TIPS

Tip 05 — Midpoint and Distance between Two Points

The midpoint of a line segment: Each coordinate of the midpoint of a line segment is equal to the average of the corresponding coordinates of the endpoints of the line segment. Given the two end points (x_1, y_1) and (x_2, y_2), the coordinates of the midpoint of the line segment are

$$\left(\frac{x_1 + x_2}{2}, \frac{y_1 + y_2}{2} \right).$$

The distance between two points: The distance d between two points (x_1, y_1) and (x_2, y_2) is given by the formula

$$d = \sqrt{\left(x_2 - x_1\right)^2 + \left(y_2 - y_1\right)^2}$$

SAT Practice

1. In the xy-plane, the midpoint of $\overline{AB}$ is $(10, 4)$. If the coordinates of point A are $(5, 1)$, what are the coordinates of point B?

   A) $(5, 3)$ B) $(6, 4)$ C) $(15, 7)$ D) $(20, 10)$

2. If point $M(5, -3)$ is the midpoint of the line segment connecting point $A(2a, b)$ and point $B(b, a)$, what is the value of a?

   A) 8   B) 12   C) 16   D) 20

3. In triangle ABC in the xy-plane, the coordinates of point A are $(-4, 4)$ and the coordinates of point B are $(4, 4)$. If the area of $\triangle ABC$ is 24, which of the following could be the coordinates of point C?

   A) $(3, 8)$
   B) $(2, 10)$
   C) $(2, -5)$
   D) $(-6, -4)$

4. If the distance between $(a, 3)$ and $(b, 8)$ is 13, what is the value of $|a - b|$?

   A) 4
   B) 8
   C) 12
   D) 16

TIPS

Tip 06 | Line Reflection

Reflecting across the x-axis : When we reflect a point (x, y) across the x-axis, the x-coordinate remains the same, but the y-coordinate is transformed into its opposite as follows.

Reflecting across the x-axis: $P(x, y) \rightarrow P'(x, -y)$

Reflecting across the y-axis: $P(x, y) \rightarrow P'(-x, y)$

Reflecting across the $y = x$: $P(x, y) \rightarrow P'(y, x)$

Reflecting across the $y = -x$: $P(x, y) \rightarrow P'(-y, -x)$

Reflecting across the origin: $P(x, y) \rightarrow P'(-x, -y)$

SAT Practice

1.   In the xy-plane, line ℓ is the reflection of line m across the x-axis. If the equation of line m is $y = \dfrac{1}{5}x - 6$, what is the slope of line ℓ?

A) -5 B) $-\dfrac{1}{5}$ C) $\dfrac{1}{5}$ D) 5

2.   In the xy-plane, line ℓ is the reflection of line m across the y-axis. If these two lines intersect at point (a, b), which of the following must be true?

A) $a = -2$ B) $a = 0$ C) $a = 2$ D) $a > 0$

3.   If the graph of $2x - 3y = 6$ is reflected across the x-axis, which of the following represents the equation of the reflected graph?

A) $2x + 3y = -6$ B) $2x + 3y = 6$ C) $2x - 3y = -6$ D) $-2x - 3y = 6$

TIPS

Tip 07 — Parallel and Perpendicular Lines

1. Two non-vertical lines are parallel if and only if their slopes are equal.
2. Two non-vertical lines are perpendicular if and only if the product of their slopes is -1.
   (Negative reciprocal each other)

SAT Practice

1. Which of the following is an equation for the line passing through the point $(-4, 1)$ that is parallel to $4x - 2y = 3$?

   A) $y = 2x - 9$
   B) $y = 2x + 9$
   C) $y = -2x - 9$
   D) $y = -2x + 9$

2. Which of the following is an equation for the line passing through the point $(-4, 1)$ that is perpendicular to $4x - 2y = 3$?

   A) $y = -\dfrac{1}{2}x - 1$
   B) $y = -\dfrac{1}{2}x + 1$
   C) $y = \dfrac{1}{2}x - 1$
   D) $y = \dfrac{1}{2}x + 1$

Tip 08 — System of Linear Equations

A system of linear equations means two or more linear equations. If two linear equations intersect, that point of intersection is called the solution to the system of equations.

1) **The system has exactly one solution.**
   When two lines have different slopes, the system has only one and only one solution.

2) **The system has no solution.**
   When two lines are parallel and have different y-intercept, the system has no solution.

3) **The system has infinitely many solutions.**
   When two lines are parallel and the lines have the same y-intercept.

From the standard form for the system of equations,

$$a_1 x + b_1 y = c_1 \quad \text{and} \quad a_2 x + b_2 y = c_2$$

1) If $\dfrac{a_1}{a_2} \neq \dfrac{b_1}{b_2}$ one solution

2) If $\dfrac{a_1}{a_2} = \dfrac{b_1}{b_2} \neq \dfrac{c_1}{c_2}$ no solution

3) If $\dfrac{a_1}{a_2} = \dfrac{b_1}{b_2} = \dfrac{c_1}{c_2}$ infinitely many solutions

From the slope-intercept form for the system of equations

$$y = m_1 x + b_1 \quad \text{and} \quad y = m_2 x + b_2$$

1) If $m_1 \neq m_2$ one solution

2) If $m_1 = m_2$ and $b_1 \neq b_2$ no solution

3) If $m_1 = m_2$ and $b_1 = b_2$ infinitely many solutions

SAT Practice

$$2x - 5y = 8$$
$$4x + ky = 17$$

1. For which of the following values of k, will the system of equations above have no solution?

A) 10   B) 5   C) -5 D) -10

$$5x - 2y = 3$$
$$ax + by = 6$$

2. In the system of equations above, a and b are constants. If the system has infinitely many solutions, what is the value of $a + b$?

A) 6   B) 4   C) 0   D) −4

$$3x + by = 3$$
$$ax - 4y = 6$$

3. In the system of equations above, a and b are constants. For which of the following values of $\{a, b\}$ will the system have no solution?

A) $\{-1, 2\}$ B) $\{1, 1\}$ C) $\{2, 1\}$ D) $\{3, -4\}$

$$ax + 3y = 6$$
$$(a-1)x + (a-1)y = 2$$

4. In the system of equations above, a is a constant. If the system has no solution, what is the value of a?

A) −3   B) 1   C) 3   D) 5

5. The cost of long distance telephone call is determined by a basic fixed charge for the first 5 minutes and a fixed charge for each additional minute. If a 15-minute call costs $3.50 and a 20-minute call costs $4.75, what is the total cost, in dollars, of a 40-minute call?

A) 8.25   B) 9.50   C) 9.75   D) 10.25

6. The tickets for a movie cost $8.00 for adults and $5.00 for children. If the total of 200 tickets were sold and the total amount of $1360 was collected, how many adult tickets were sold?

Tip 09 | Quadratic Function

In algebra, a quadratic function is a polynomial function in which the highest degree term is of the second degree.

  1) Standard form: $f(x) = ax^2 + bx + c \quad (a \neq 0)$

  2) Vertex form: $f(x) = a(x-h)^2 + k \quad (a \neq 0)$ and vertex (h, k)

  3) Factored from: $f(x) = a(x-x_1)(x-x_2)$, where x_1 and x_2 are the roots of the quadratic function.

Axis of symmetry

  1) Standard form: $x = -\dfrac{b}{2a} \quad \rightarrow \quad \text{Vertex}\left(-\dfrac{b}{2a}, f\left(-\dfrac{b}{2a}\right)\right)$

  2) Vertex form: $x = h \quad \rightarrow \quad \text{Vertex}(h, k)$

  3) Factored form: $x = \dfrac{x_1 + x_2}{2} \quad \rightarrow \quad \text{Vertex}\left(\dfrac{x_1 + x_2}{2}, f\left(\dfrac{x_1 + x_2}{2}\right)\right)$

1) $a > 0$ and $y = ax^2 + bx + c$

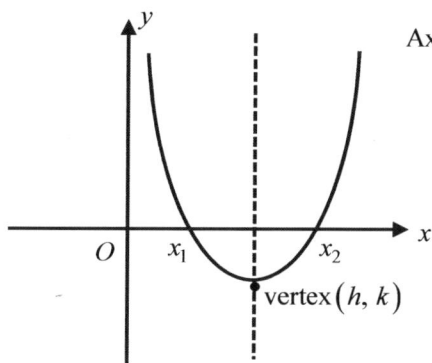

$$\text{Axis of symmetry} = -\frac{b}{2a} \text{ or } x = \frac{x_1 + x_2}{2}$$

2) $a < 0$ and $y = ax^2 + bx + c$

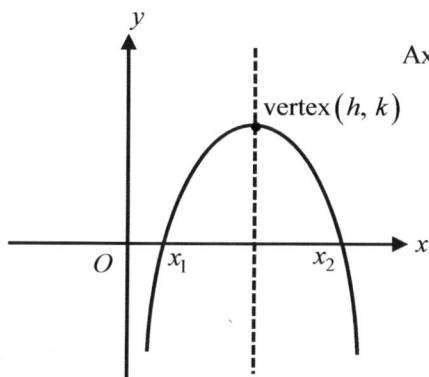

$$\text{Axis of symmetry} = -\frac{b}{2a} \text{ or } x = \frac{x_1 + x_2}{2}$$

Questions 1 and 2 refer to the following information.

$$h = 256t - 16t^2$$

A ball is thrown straight up from the ground with an initial velocity of 256 feet per second. The equation above describes the height the ball can reach in t seconds.

1. If the ball reaches its maximum height in k seconds, what is the value of k?

   A) 8
   B) 12
   C) 16
   D) 24

2. What is the maximum height, in feet, that the ball will reach?

   A) 370
   B) 384
   C) 1024
   D) 1200

$$f(x) = \frac{1}{2}(x+2)(x-10)$$

3. If the function f above has a vertex at point (h, k) in the xy-plane, what is the value of k?

   A) −36
   B) −18
   C) 9
   D) 18

$$y = 3x^2 - 6x + 10$$

4. Which of the following is equivalent to the equation above?

   A) $y = 3x^2 + 10$
   B) $y = 3(x-1)^2 + 7$
   C) $y = 3(x-1)^2 + 10$
   D) $y = 3(x+2)^2 - 2$

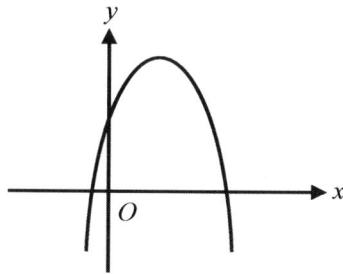

5. The graph of $y = ax^2 + bx + c$ is shown in the xy-plane above, where a, b, and c are constants. Which of the following must be true?

I. $a < 0$ II. $b > 0$ III. $c > 0$

A) I only   B) I and II only   C) I and III only   D) I, II, and III

$$f(x) = a(x - h)^2 + k$$

6. The function f is defined by the equation above, where a, h, and k are constants. If a and k are negative, which of the following CANNOT be true?

A) $f(5) < 0$

B) $f(-5) < 0$

C) $f(1) = k$

D) $f(0) = -k$

Tip 10 — System of Linear and Quadratic Equations

A system of linear and quadratic equations can be solved 1) **graphically** or 2) **using algebra.**

Example

$$y = 2x + 1$$
$$y = x^2 - 5x + 7$$

Solve the system of equations above algebraically.

Solution)

$$x^2 - 5x + 7 = 2x + 1 \rightarrow x^2 - 7x + 6 = 0 \rightarrow$$
$$(x - 1)(x - 6) = 0$$

Therefore, $x = 1$, $x = 6$

At $x = 1$, $y = 2(1) + 1 = 3$.

At $x = 6$, $y = 2(6) + 1 = 13$

$\rightarrow (1, 3), (6, 13)$

Graphically

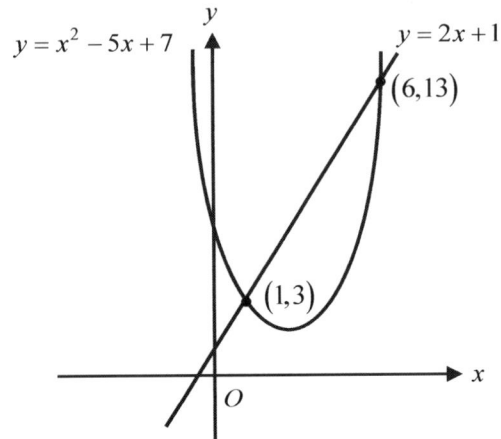

SAT Practice

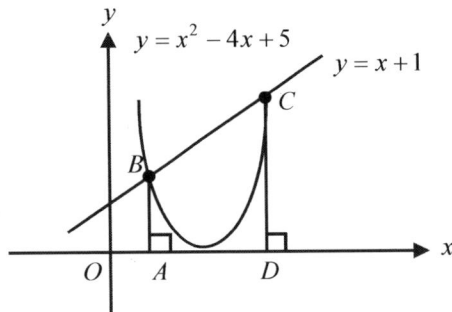

1. In the xy-plane above, the graphs of $y = x^2 - 4x + 5$ and $y = x + 1$ intersect at points B and C.
    What is the area of quadrilateral $ABCD$?

$$y = x^2 - x - 12$$
$$y = 3x$$

2. How many solutions does the system of equations above?

    A) 0   B) 1   C) 2   D) Infinitely many

TIPS

Tip 11 System of Linear Inequalities

The graph of the solution of a system of linear inequalities is the shaded region of the plane containing all points that are common solutions of the inequalities in the system.

Example 1

$$y > x + 4$$
$$x + y \le 4$$

Graph the solution of the system of inequalities above.

Solution)

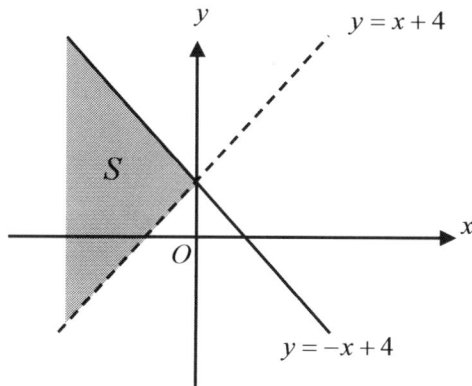

Example 2

$$y + x \le 6$$
$$y - 2x \le 3$$

In the xy-plane, if (a,b) is a solution to the system of inequalities above, what is the maximum possible value of b?

Solution)

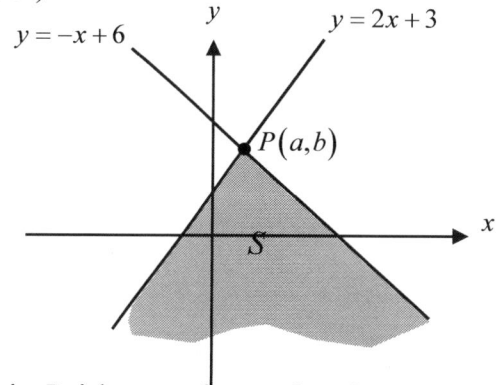

At point P, it has a maximum value of b.
Find the intersection of the graphs.

$$2x + 3 = -x + 6 \quad \rightarrow \quad 3x = 3 \quad \rightarrow \quad x = 1$$
$$y = -1 + 6 = 5$$

Therefore, the maximum value of b is 5.

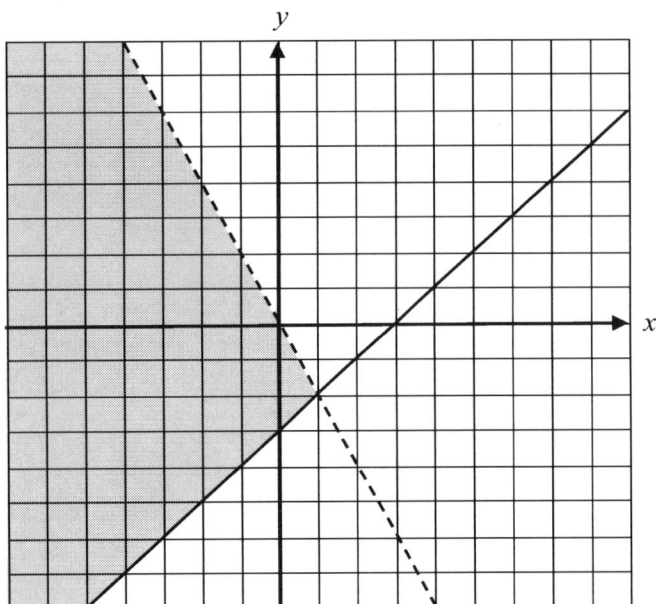

1. Which of the following inequalities describes the shaded region of the graph?

   A) $\begin{cases} 2x - y < 0 \\ x + y \geq 3 \end{cases}$
   B) $\begin{cases} 2x + y < 0 \\ x - y \geq 3 \end{cases}$
   C) $\begin{cases} 2x + y < 0 \\ x - y \leq 3 \end{cases}$
   D) $\begin{cases} x + 2y < 0 \\ y - x \geq 3 \end{cases}$

$$x + 2y \leq 4$$
$$2x - y \leq 0$$

2. In the xy-plane, if (a, b) is a solution to the system of inequalities above, what is the maximum possible value of a?

   A) $\dfrac{2}{5}$
   B) $\dfrac{4}{5}$
   C) $\dfrac{5}{4}$
   D) $\dfrac{5}{2}$

Tip 12   System of Linear and Quadratic Inequalities

$$y \geq x^2$$
$$y \leq x + 6$$

For the system of linear and quadratic inequalities above, the solution is the shaded area between the parabola $y = x^2$ and the line $y = x + 6$.

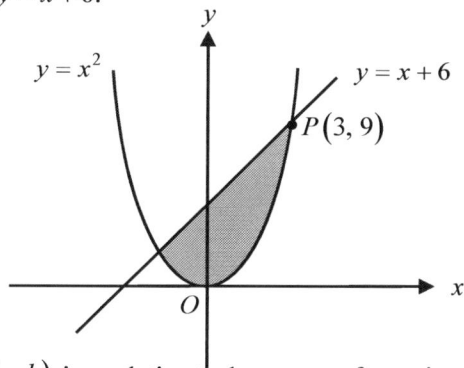

In the xy-plane, if (a,b) is a solution to the system of equations above, what is the maximum possible value of b?

Solution) We can see the maximum value of b at point P. Now solve the system of equations.

$$x^2 = x + 6 \; \rightarrow \; x^2 - x - 6 = 0 \; \rightarrow \; (x-3)(x+2) = 0 \; \rightarrow \; x = 3, -2$$

At $x = 3$, $y = 3 + 6 = 9$.   Therefore, the maximum of $b = 9$.

SAT Practice

$$y \leq -x^2 + 8$$
$$y \geq x + 2$$

1.   In the xy-plane, if (a, b) is a solution to the system of inequalities above, what is the maximum possible value of b?

A) −1

B) 4

C) 8

D) 9

TIPS

Tip 13 | Area enclosed by Curves

In order to find the area enclosed by curves, mostly we need to find x-intercept, y-intercept, and points of intertsection of curves.

SAT Practice

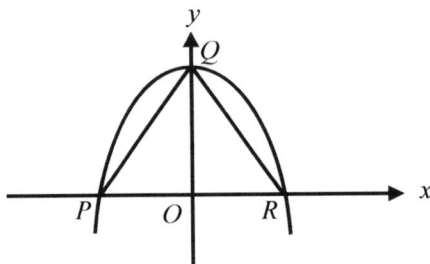

1. The graph of $y = -x^2 + k$ is shown in the xy-plane above, where k is a constant. If the area of $\triangle PQR$ is 64, what is the value of k?

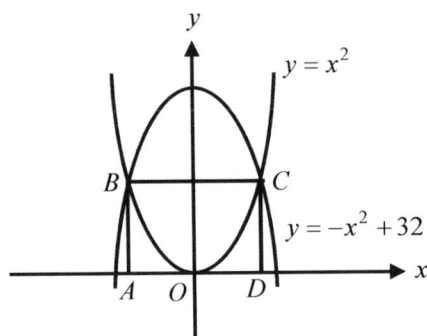

2. The graphs of $y = x^2$ and $y = -x^2 + 32$ are shown in the xy-plane above. What is the area of rectangle $ABCD$?

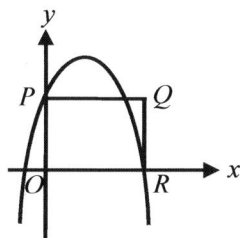

3. The graph of $y = -x^2 + 4x + 5$ is shown in the xy-plane above. What is the area of rectangle $OPQR$?

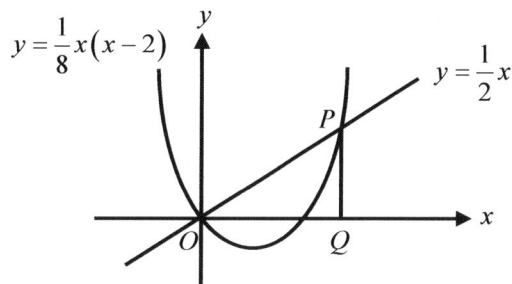

4. The graphs of $y = \frac{1}{8}x(x-2)$ and $y = \frac{1}{2}x$ are shown in the xy-plane above. What is the area of right triangle OPQ?

A) 18
B) 15
C) 9
D) 6

TIPS

Tip 14 Domain and Range

The **domain** of a given function is the complete set of "input" values for which the function is defined. In the xy-plane, the domain is represented on the x-axis (or abscissa).

Note: 1) The denominator of a fraction CANNOT be zero.
2) The number inside a square root sign must be positive or zero.

The **range** of a function is the set of all "output" values produced by that function. In the xy-plane, the range is represented on the y-axis (or ordinate).

Example:

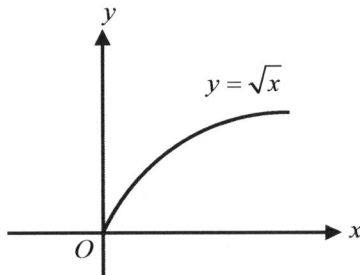

$y = \sqrt{x}$

Domain: $x \geq 0$
Range: $y \geq 0$

SAT Practice

$$f(x) = \frac{\sqrt{x}}{x-3}$$

1. In the function f above, which of the following represents its domain?

   A) $x \geq 0$
   B) $x \neq 3$
   C) $x \geq 3$
   D) $x \geq 0$ and $x \neq 3$

$$g(x) = \sqrt{x-2} - 5$$

2. Which of the following represents the range of the function f above?

   A) $y \geq 0$
   B) $y \geq 2$
   C) $y \geq -2$
   D) $y \geq -5$

TIPS

Tip 15 | Composition of Function

Composition of functions is applying one function to the result of the first function.(Combining functions)

The result of $f(x)$ is sent through $g(\ )$. The composition is written in the form

$$g(f(x)) \quad \text{or} \quad (g \circ f)(x)$$

Example 1:

Given the functions $f(x) = 3x$ and $g(x) = x - 4$, then $g(f(3)) =$

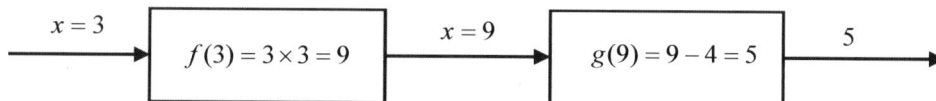

$$(g \circ f)(x) = 3x - 4 \ \rightarrow \ (g \circ f)(3) = 3 \times 3 - 4 = 5$$

Example 2:

For the given functions $f(x) = 3x - 4$ and $g(x) = 2f(x) + 3$, if $g(k) = 0$, what is the value of k?

Since $g(x) = 2(3x - 4) + 3 = 6x - 5$, $g(k) = 6k - 5 = 0 \ \rightarrow \ k = \dfrac{5}{6}$.

SAT Practice

1. The functions f and g are defined by $f(x) = 5x + 3$ and $g(x) = 3f(x) - k$. If $g(2) = 25$, what is the value of k?

    A) 18
    B) 14
    C) 12
    D) 10

$$f(x) = ax + b$$
$$g(x) = 2f(x) - 3$$

2.   In the functions f and g above, if $g(1) = 3$ and $g(3) = 5$, what is the value of b?

A) $\dfrac{1}{2}$ B) $\dfrac{3}{2}$ C) $\dfrac{5}{2}$ D) $\dfrac{7}{2}$

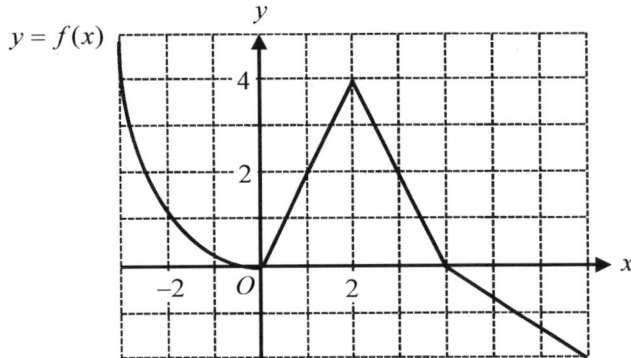

3.   The graph of $y = f(x)$ is shown in the xy-plane above for $-3 \leq x \leq 7$. The function g is defined by $g(x) = 3f(x) - 1$. For how many values of k are there such that $g(k) = 8$?

A) One

B) Two

C) Three

D) Four

TIPS

Tip 16 — Function Undefined

A rational function is a quotient of polynomials in which the denominator is not identically 0.

$$\text{Rational function } R(x) = \frac{P(x)}{Q(x)} \text{ where } Q(x) \neq 0$$

If $Q(x) = 0$, R is undefined.

Example 1

$$h(x) = \frac{1}{(x-3)^2 - 4(x-3) - 12}$$

For what value of x is the function h above undefined?

Sol) Let $X = x - 3$.

$$\frac{1}{(x-3)^2 - 4(x-3) - 12} = \frac{1}{X^2 - 4X - 12}$$

If $Q(X) = X^2 - 4X - 12 = 0$, h is undefined.

$X^2 - 4X - 12 = (X-6)(X+2) = 0 \rightarrow X = 6$ or -2

$x - 3 = 6,\ x - 3 = -2$

Therefore, h is undefined at $x = 9$ and $x = 1$.

Example 2

$$g(x) = \frac{x^2 - 1}{x^2 + x - 2} \text{ and } x > 0$$

For what value of x is the function g above undefined?

Solution) If $x^2 + x - 2 = 0$, g is undefined.

$x^2 + x - 2 = (x+2)(x-1) = 0 \rightarrow x = -2, 1$

Therefore, g is undefined at $x = 1$.

Remember: Do not cross out the common factors.

$$g(x) = \frac{x^2 - 1}{x^2 + x - 2} = \frac{(x-1)(x+1)}{(x-1)(x+2)}$$

SAT Practice

$$h(x) = \frac{x-1}{4(x-1)^2 - 12(x-1) + 9}$$

1. For what value of x is the function h above undefined?

$$f(x) = \frac{1}{x^2 - ax + b}$$

2. For $x = 1$ and $x = 3$, the function f is undefined. Which of the following is the value of b?

A) -4 B) -3 C) 3   D) 4

Tip 17 — Identical Equation

The two expressions, left hand side and right hand side, are always equal for all values we give to the variable. The equations that are true for all values of the variable are called identical equations.

1) $5x - 10 = 0$ is an algebraic equation because the equation is true only for $x = 2$.

2) $5x - 10 = 5x - 10$ is an identical equation because the equation is true for all values of x.
- The equation has infinitely many solutions.
- The expressions of both sides must be equal.

Example 1:
If $2x + 5 = ax + b$ for all values of x, what are values of a and b?

Solution) The expressions are equal. $a = 2$ and $b = 5$

Example 2:
If $ax^2 + bx + c = 0$ is true for all values of x, what are the values of a, b, and c?

Solution) Since $ax^2 + bx + c = 0x^2 + cx + 0$, $a = 0$, $b = 0$, and $c = 0$.

SAT Practice

1.   If $x(k - 2) = 0$ for all values of x, what is the value of k?

A) 0
B) 2
C) 4
D) 6

2.   If $ax^2 + bx + c = 0$ for all values of x, what is the value of $a + b + c$?

A) 0
B) 1
C) 2
D) It cannot be determined from the information given.

$$(k+1)x+5 = ax+k$$

3. In the equation above, k and a are constants. If the equation above is true for all values of x, what is the value of a?

    A) 6
    B) 5
    C) 2
    D) 0

$$a(x+1)+b(x-1) = 2x+4$$

4. In the equation above, a and b are constants. If the equation above is true for all values of x, what is the value of a?

    A) 1
    B) 2
    C) 3
    D) 4

TIPS

Tip 18 | Even and Odd Functions

Even function :
The graph of an even function is symmetric with respect to the y-axis.

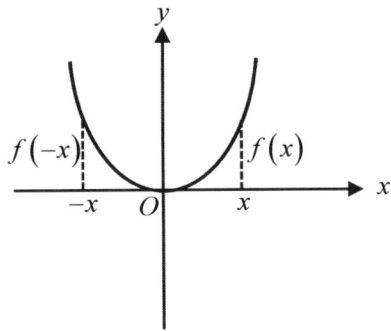

A function is even if $f(x) = f(-x)$.

Odd function:
The graph of an odd function is symmetric with respect to the origin.

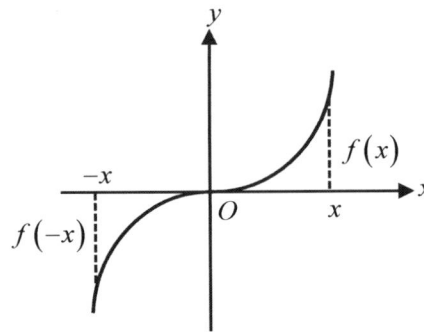

A function is odd if $f(x) = -f(-x)$.

1. Which of the following is an odd function?

   A) $f(x) = x$

   B) $g(x) = x^2$

   C) $h(x) = |x|$

   D) $k(x) = 10$

2. Which of the following graphs is an even function?

   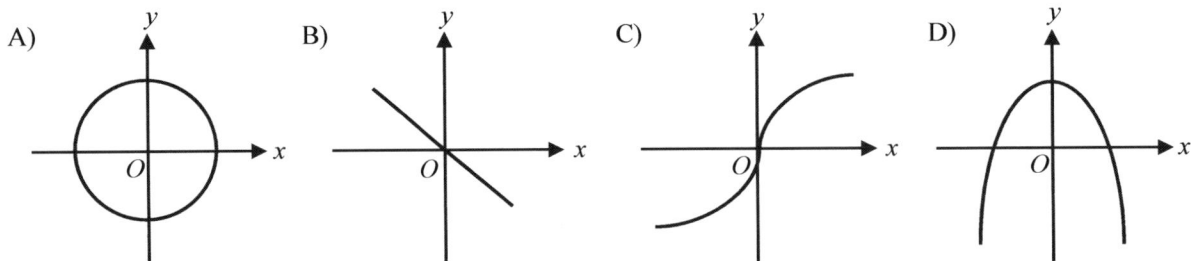

Tip 19 | Factoring

Factoring is to write an expression as a product of factors.

1) $a^2 + 2ab + b^2 = (a+b)^2$

2) $a^2 - 2ab + b^2 = (a-b)^2$

3) $a^2 - b^2 = (a+b)(a-b)$

4) $a^2 - 2a - 3 = (a-3)(a+1)$

5) $6a^2 + a + 1 = (3a-1)(2a+1)$

6) $12a^2 + 2a - 2 = 2(6a^2 + a - 1) = 2(3a+1)(2a-1)$: Complete factoring

SAT Practice

1. If $(x-3)(x+3) = a$, then $(2x-6)(x+3) =$

   A) $2a$
   B) $3a$
   C) $4a$
   D) $6a$

2. Which of the following is equivalent to $\left(n - \dfrac{1}{n}\right)^2 + 4$?

   A) 4

   B) $\left(n + \dfrac{1}{n}\right)^2$

   C) $n^2 + \dfrac{1}{n^2}$

   D) $n^2 - \dfrac{1}{n^2}$

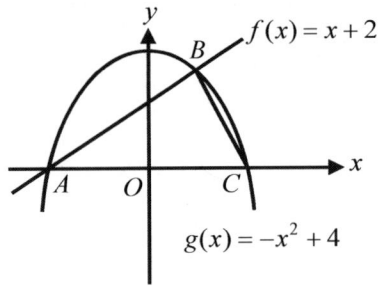

$$f(x) = x + 2$$
$$g(x) = -x^2 + 4$$

3. The graphs of the functions f and g are shown in the xy-plane above. What is the area of $\triangle ABC$?

A) 18

B) 12

C) 9

D) 6

---

4. If $x^2 - y^2 = 24$, where x and y are positive integers $(x > y)$, what is one possible value of x?

TIPS

Tip 20 | Direct Variation (Direct Proportion)

When two variables are related in such a way that $y = kx$, the two variables are said to be in direct variation.

Expression of direct variation:

1) $y = kx$

2) $\dfrac{y}{x} = k$ or $\dfrac{y_1}{x_1} = \dfrac{y_2}{x_2} = \dfrac{y_3}{x_3} = \cdots = k(\text{constant})$

Geometric interpretation:

$y = kx$ is a special linear equation, where the y-intercept is $(0, 0)$. In the xy-plane, k is the slope of the graph.

SAT Practice

1. The value of y varies directly proportional to the value of x. If $y = 15$ when $x = 5$, what is the value of y when $x = 12.5$?

---

2. A group of workers can harvest all the grapes from 10 square meters of a vineyard in 20 minutes. At this rate, how many minutes will the group need to harvest all the grapes from 300 square meters of this vineyard?

A) 60
B) 200
C) 400
D) 600

---

3. To make an orange dye, 5 parts of red dye are mixed with 3 parts of yellow dye. To make a green dye, 4 parts of blue dye are mixed with 2 parts of yellow dye. If equal amount of green and orange dye are mixed, what fraction of the new mixture is yellow dye?

A) $\dfrac{1}{3}$

B) $\dfrac{17}{48}$

C) $\dfrac{9}{24}$

D) $\dfrac{1}{2}$

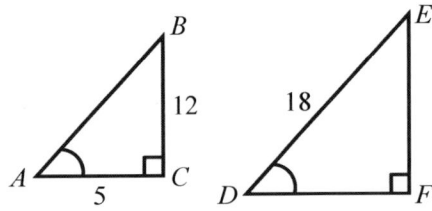

4. In the figure above, $\triangle ABC$ is similar to $\triangle DEF$. What is the length of $\overline{DF}$?

A) 6

B) $\dfrac{82}{13}$

C) $\dfrac{90}{13}$

D) $\dfrac{25}{3}$

x	y
1	a
a	$5a$

5. In the table above, y is directly proportional to x and $a \neq 0$. Which of the following is the value of a?

A) 0
B) 1
C) 5
D) 10

6.   If y is directly proportional to x, which of the following could represent the graph of $y = f(x)$?

A)   B)   C)   D)

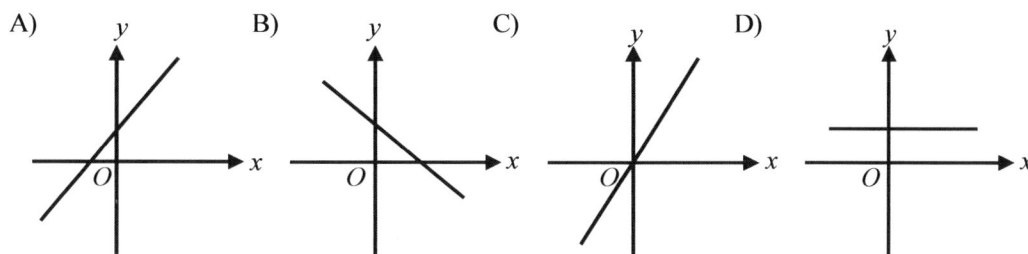

---

7.   If y is directly proportional to x^2, which of the following could be the graph of $y = f(x)$?

A)   B)   C)   D)

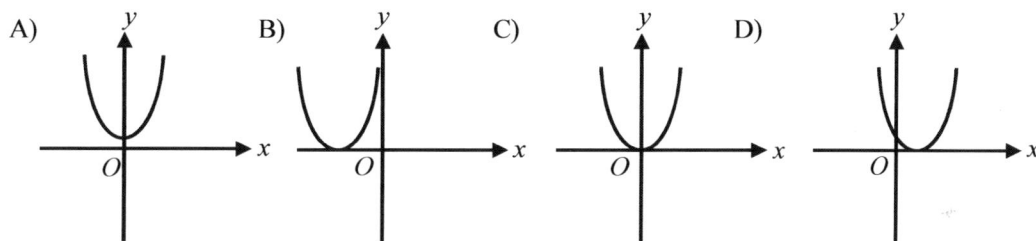

TIPS

Tip 21 — Inverse Variation

When two variables are related in such a way that $xy = k$, the two variables are said to be in inverse variation.

Properties:
1) The values of two variables change in an opposite way, that is, as one variable increases, the other decreases.
2) The product k is unchanged.

In the xy-plane, the graph of $y = f(x)$ is as follows.
1) $k > 0$

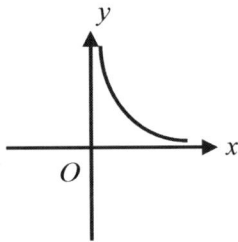

SAT Practice

1. The cost of hiring a bus for a trip to Niagara Falls is $400. If 25 people go on the trip, what is the cost per person in dollars?

---

2. If four typists can complete the typing of a manuscript in 9 days, how long would it take 12 typists to complete the manuscript?

    A) 3 days   B) 4 days   C) 5 days   D) 10 days

---

3. If a man can drive from his home to Albany in 5 hours at 45 miles per hour, how would it take him if he drove at 50 miles per hour?

    A) 4 hours
    B) 4 hours 30 minutes
    C) 5 hours
    D) 5 hours 30 minutes

x	y
2	25
4	a
5	10
8	b

4. In the table above, y varies inversely as x. What is the value of $a + b$?

A) 16   B) 18   C) 18.75   D) 20.25

---

5. If a job can be completed by 2 people in 10 days, then how many people, working at the same rate, are needed to complete the same job in 5 days?

A) 1   B) 2   C) 3   D) 4

---

6. The length of a rectangle varies inversely with the width. If the length is 10 when the width is 20, what is the length when the width is 40?

A) 2   B) 5   C) 10   D) 20

---

7. A certain job can be completed by p persons in h hours. How long would it take n persons, working at the same rate, to complete the same job?

A) $\dfrac{hn}{p}$ B) $\dfrac{n}{hp}$ C) $\dfrac{hp}{n}$ D) $\dfrac{np}{h}$

---

8. If 5 people take d days to install the plumbing for a house, then how many days would it take 2 people to complete one third of the same job? (Assume the people work at the same rate.)

A) $\dfrac{3d}{2}$

B) $\dfrac{5d}{2}$

C) $\dfrac{5d}{6}$

D) $\dfrac{d}{2}$

Tip 22 Sum and Product of the roots of a Quadratic Equation

For a quadratic equation $ax^2 + bx + c = 0$, sum of the roots $= -\dfrac{b}{a}$ and product of the roots $= \dfrac{c}{a}$.

The two roots of $ax^2 + bx + c = 0$ are $x = \dfrac{-b + \sqrt{b^2 - 4ac}}{2a}$ and $x = \dfrac{-b - \sqrt{b^2 - 4ac}}{2a}$.

$$\text{Sum} = \frac{-b + \sqrt{b^2 - 4ac}}{2a} + \frac{-b - \sqrt{b^2 - 4ac}}{2a} = \frac{-2b}{2a} = -\frac{b}{a}$$

$$\text{Product} = \frac{-b + \sqrt{b^2 - 4ac}}{2a} \times \frac{-b - \sqrt{b^2 - 4ac}}{2a} = \frac{(-b)^2 - \left(\sqrt{b^2 - 4ac}\right)^2}{4a^2} = \frac{4ac}{4a^2} = \frac{c}{a}$$

Or, simply, $ax^2 + bx + c = 0 \rightarrow x^2 + \dfrac{b}{a}x + \dfrac{c}{a} = 0$. If α and β are the roots of the equation, then

the equation is $(x - \alpha)(x - \beta) = 0 \rightarrow x^2 - (\alpha + \beta)x + \alpha\beta = 0$.

Since $x^2 + \dfrac{b}{a}x + \dfrac{c}{a} = 0$ and $x^2 - (\alpha + \beta)x + \alpha\beta = 0$ are equivalent. Therefore,

$$\text{Sum: } \alpha + \beta = -\frac{b}{a} \quad \text{and} \quad \text{product: } \alpha\beta = \frac{c}{a}.$$

SAT Practice

1. If one of roots of a quadratic equation $2x^2 + x - k = 0$ is 1, what is the other root of the equation?

A) $-\dfrac{5}{2}$

B) $-\dfrac{3}{2}$

C) $-\dfrac{1}{2}$

D) $\dfrac{3}{2}$

2. If the roots of the equation $x^2 + 4x - 12 = 0$ are α and β, what is the value of $\dfrac{1}{\alpha} + \dfrac{1}{\beta}$?

A) $\dfrac{1}{4}$

B) $\dfrac{1}{3}$

C) $\dfrac{1}{2}$

D) $\dfrac{3}{2}$

TIPS

Tip 23 | Remainder Theorem

When polynomial $f(x)$ is divided by $(x-a)$, the remainder R is equal to $f(a)$.

Polynomial $f(x)$ can be expressed as follows.

$$f(x) = (x-a)Q(x) + R,$$ where $Q(x)$ is the quotient and R is the remainder.

The identical equation above is true for all values of x, especially $x = a$.

Therefore, $f(a) = (a-a)Q(a) + R \rightarrow f(a) = R$.

Examples:

1) Interpretation of $f(2) = 5 \rightarrow$ The remainder is 5 when $f(x)$ is divided by $(x-2)$.

2) Interpretation of $f(-5) = -3 \rightarrow$ The remainder is -3 when $f(x)$ is divided by $(x+5)$.

SAT Practice

1. When $f(x) = x^2 + 3x + k$ is divided by $x-3$, the remainder is 25. What is the value of k?

---

2. What is the remainder when $x^3 - x^2 - 3x - 1$ is divided by $(x+3)$?

   A) -36 B) -28 C) 14   D) 36

---

3. Find the value of k for which the remainder is zero when $x^3 - 5x^2 + x + k$ is divided by $(x-1)$?

Tip 24 | Factor Theorem

If $f(a) = 0$, then $f(x)$ has a factor of $(x-a)$.

$f(x)$ can be expressed with a factor of $(x-a)$ as follows.

$$f(x) = (x-a)Q(x)$$

Therefore, $f(a) = 0$ means that the remainder is 0.

SAT Practice

1. If $(x-3)$ is a factor of $x^3 - 4x + k$, what is the value of the constant k?

   A) -15 B) -10 C) 10 D) 15

2. If a polynomial $P(x) = x^2 + kx - 8$ has a factor of $(x-2)$, what is the value of k?

3. What is the value of k when $P(x) = x^3 - 5x^2 - x + k$ is divisible by $x+1$?

   A) -5
   B) -2
   C) 2
   D) 5

4. What is the value of a when $g(x) = x^3 + ax + b$ is divisible by $(x-1)(x-2)$?

   A) -7
   B) -4
   C) 4
   D) 8

Tip 25 | Circle

A circle is the locus of points equidistant from a given point known as the center.

The standard equation of a circle whose center is at the point (h, k) is

$$(x-h)^2 + (y-k)^2 = r^2, \text{ where } r \text{ is the radius.}$$

SAT Practice

1. What is the area of the circle whose equation is $x^2 - 4x + y^2 + 2y = 11$?

   A) 11π B) 12π C) 16π D) 25π

2. What is the circumference of a circle whose equation is $x^2 + y^2 - 6y = 16$?

   A) 10π B) 20π C) 30π D) 40π

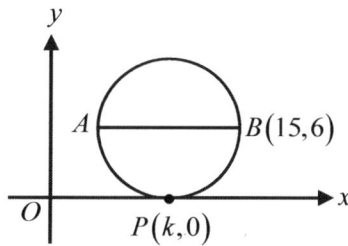

3. In the xy-plane above, $\overline{AB}$ is the diameter of the circle and parallel to the x-axis. What is the value of k?

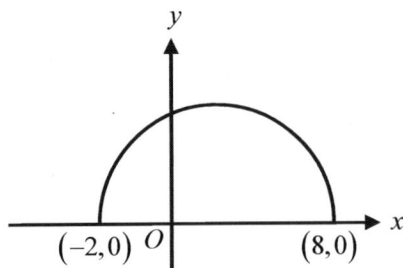

4. The graph of a semicircle is shown in the *xy*-plane above. Which of the following are the *x*-coordinates of two points on this semicircle whose *y*-coordinates are equal?

A) 0 and 7   B) 1 and 6   C) 1 and 5   D) 2 and 3

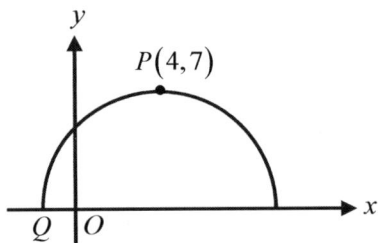

5. In the *xy*-plane above, the semicircle has a maximum at point P. What are the coordinates of point Q?

A) $(-4,0)$
B) $(-3,0)$
C) $(-2,0)$
D) $(-1,0)$

Tip 26 | Average Speed

Average speed is the total distance divided by the total time taken.

$$\text{Average speed} = \frac{\text{Total distance travelled}}{\text{Total time taken}}$$

Example:

Peter travelled from city A to city B at 60 miles per hour, and then he travelled back along the same route at 40 miles per hour. What is his average speed for the entire trip?

Solution)

D = distance between city A and city B

Total distance $= 2D$ and Total time taken $= \dfrac{D}{60} + \dfrac{D}{40} = \dfrac{5D}{120} = \dfrac{D}{24}$

Therefore, average speed $= \dfrac{2D}{\dfrac{D}{24}} = 48\,\text{mph}$

Or

We can use any convenient number for D.

If $D = 120$, then $t_1 = \dfrac{120}{60} = 2$ and $t_2 = \dfrac{120}{40} = 3$.

Therefore, average speed $= \dfrac{\text{Total distance}}{\text{Total time}} = \dfrac{120 + 120}{2 + 3} = 48\,\text{mph}$

SAT Practice

1.   Jason travelled from city A to city B in 4 hours. For the first hour, he drove at a constant speed of 50 miles per hour. Then he increased his speed and kept it at 60 miles per hour for the next 3 hours. What is his average speed, miles per hour, for the trip?

   A) 55.5
   B) 56
   C) 57.5
   D) 58

2. Claire travelled from city C to city D. The first half of the way, she drove at the constant speed of 40 miles per hour. Then she increased her speed and travelled the remaining distance at 50 miles per hour. What is her average speed, miles per hour, for the trip?

A) $44\frac{4}{9}$

B) $44\frac{2}{3}$

C) 45

D) $45\frac{1}{3}$

3. Jackson drove a car from Amherst to Boston at the constant speed of 60 miles per hour. On the way back along the same route, he drive at a constant speed of 40 miles per hour. If he took 6 hours for the entire trip, what is the distance, in miles, from Amherst to Boston?

A) 120
B) 144
C) 160
D) 240

Tip 27   Percentage

- % of increase = $\dfrac{\text{Amount of increase}}{\text{Original amount}} \times 100$
- % of decrease = $\dfrac{\text{Amount of decrease}}{\text{Original amount}} \times 100$

SAT Practice

1. If 20% of 30% of a positive number is equal to 10% of $k\%$ of the same number, what is the number?

    A) 80   B) 60   C) 40   D) 20

---

2. The price of a music CD was first increased by 15 % and then the new price was decreased by 30 %. Which of the following is true about the price after these two changes?

    A) The price decreases by 15%.
    B) The price decreases by 19.5%.
    C) The price decreases by 35%
    D) The price decreases by 45%.

---

3. If $2a + 3b$ is equal to 250 percent of $6b$, what is the value of $\dfrac{a}{b}$?

    A) $\dfrac{1}{3}$

    B) 3

    C) 6

    D) 9

---

4. If 25 percent of m is 50, what is 15 percent of $2m$?

    A) 80
    B) 60
    C) 50
    D) 40

5. The cost of an automobile increases each by 2.5%, and the cost this year is $20,000. If the cost of the automobile is given by $C(n) = 20,000x^n$, what is the value of x?

   A) 1.25
   B) 1.025
   C) 0.25
   D) 0.025

6. Tom's weekly salary was increased from $500 to $1,000 this week. By what percent was his salary increased?

   A) 50%
   B) 100%
   C) 200%
   D) 400%

7. If the price of a stock rises by 6 percent one day falls 5 percent next day, what was the change in the price of the stock after these two days?

   A) The price rose by 0.5%
   B) The price rose by 0.7%
   C) The price rose by 1%
   D) The price rose by 1.5%

8. If a is 25 percent of $2b$, then b is what percent of a?

   A) 50%
   B) 75%
   C) 100%
   D) 200%

Tip 28 | Ratios and Proportion

Ratio

A ratio is a comparison of two numbers. We can write this as 8:12 or as a fraction $\frac{8}{12}$, and we say the ratio is *eight to twelve.*

Proportion

A proportion is an equation with a ratio on each side. It is a statement that two ratios are equal. $\frac{3}{4} = \frac{6}{8}$ is an example of a proportion.

Rate

A rate is a ratio that expresses how long it takes to do something, such as traveling a certain distance. To walk 3 kilometers in one hour is to walk at the rate of 3 km/h. The fraction expressing a rate has units of distance in the numerator and units of time in the denominator.

Example 1:

Juan runs 4 km in 30 minutes. At that rate, how far could he run in 45 minutes?

Solution)

Give the unknown quantity the name n. In this case, n is the number of km Juan could run in 45 minutes at the given rate. We know that running 4 km in 30 minutes is the same as running n km in 45 minutes; that is, the rates are the same. So we have the proportion $\frac{4}{30} = \frac{n}{45}$.

Finding the cross products and setting them equal, we get $30 \times n = 4 \times 45$, or $30n = 180$. Dividing both sides by 30, we find that $n = 180 \div 30 = 6$ and the answer is 6 km.

SAT Practice

1. If the cost of a 6-minute telephone call is $1.20, then at this rate, what is the cost of a 15-minute call?

    A) $2.00
    B) $3.00
    C) $3.25
    D) $3.75

2. In 5 years the ratio of Julie's age to Song's age will be 3:5. In 10 years the ratio of Julie's age to Song's age will be 2:3. What is the sum of their current ages?

A) 15
B) 20
C) 30
D) 35

3. If $\dfrac{x}{y} = \dfrac{2}{3}$ and $2x + 5y = 76$, what is the value of x?

A) 2
B) 4
C) 8
D) 16

4. If a is divisible by 2, b is divisible by 5, and $\dfrac{a}{b} = \dfrac{7}{9}$, where a and b are positive numbers, and $a + b < 400$, what is one possible value of $a + b$?

TIPS

Tip 29 Ratios in Similar Figures

Two polygons are similar if and only if their corresponding angles are congruent and their corresponding sides are in proportion.

Remember:

If the ratio of the corresponding lengths is $a:b$, then the ratio of the areas is $a^2:b^2$ and the ratio of the volumes is $a^3:b^3$.

SAT Practice

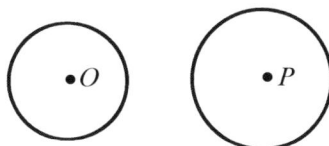

1. In the figure above, the ratio of the circumference of circle O to the circumference of circle P is 2:3. If the area of circle O is 20, what is the area of circle P?

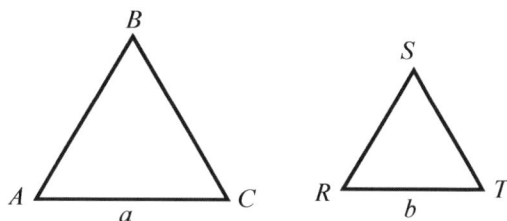

2. The figure above shows two equilateral triangles with a side a and a side b. If $\dfrac{a}{b} = \dfrac{5}{2}$ and the area of $\triangle ABC$ is 30, what is the area of $\triangle RST$?

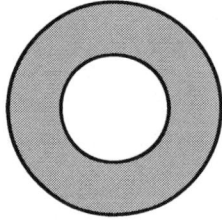

3. In the figure above, the radius of the larger circle is $\dfrac{5}{2}$ times the radius of the smaller circle. If the area of the smaller circle is 28, what is the area of the shaded region?

    A) 70
    B) 112
    C) 147
    D) 175

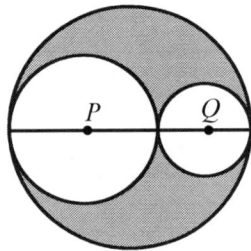

4. In the figure above, circles P and Q are tangent each other and internally tangent to the largest circle. The ratio of the radius of circle P to the radius of circle Q is 3:1. If the area of the shaded region is 96π, what is the radius of circle Q?

    A) 2
    B) 4
    C) 6
    D) 8

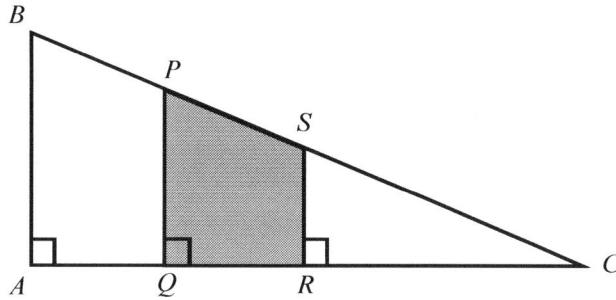

5. In the figure above, the ratio of the lengths $AQ:QR:RC = 2:2:3$. If the area of quadrilateral $PQRS$ is 48, what is the area of $\triangle ABC$?

A) 96
B) 124
C) 147
D) 192

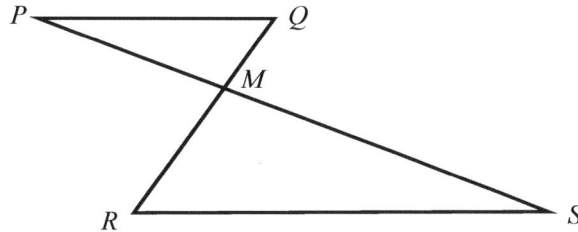

6. In the figure above, $\overline{PQ}$ is parallel to $\overline{RS}$. The ratio of the area of $\triangle PQM$ to the area of $\triangle SRM$ is 4:9. If the perimeter of $\triangle PQM$ is 15, what is the perimeter of $\triangle SRM$?

A) 22.5
B) 33.75
C) 35.5
D) 37.5

Tip 30 — Percent of a Solution (Mixture)

The percent of a solution is expressed as the percentage of solute over the total amount of solution.

$p\%$ of a solution is

$$\frac{\text{Solute}}{\text{Total amount of solution}} = \frac{p}{100}$$

Or

$$\frac{\text{Solute}}{\text{Total amount of solution}} \times 100 = p\%$$

SAT Practice

1. How many gallons of water must be added to 40 gallons of a 10% alcohol solution to produce an 8% alcohol solution?

   A) 5   B) 8   C) 10   D) 12

2. How many gallons of a 20% salt solution must be added to 10 gallons of a 50% salt solution to produce 30% salt solution?

   A) 5 gallons   B) 10 gallons   C) 15 gallons   D) 20 gallons

3. How many quarts of alcohol must be added to 10 quarts of a 25% alcohol solution to produce a 40% alcohol solution?

   A) 2.5 quarts   B) 8 quarts   C) 10 quarts   D) 15quarts

4. How many gallons of acid must be added to G gallons of a $k\%$ acid solution to bring it up to an $m\%$ solution?

   A) $\dfrac{G}{100-m}$

   B) $\dfrac{Gm}{100-m}$

   C) $\dfrac{G(m-k)}{100-m}$

   D) $\dfrac{100-m}{G(m-k)}$

TIPS

Tip 31 | Exponents

The exponent is the number of times the base is used as a factor.

$$5^2 = 25 \begin{cases} 5 = \text{base} \\ 2 = \text{exponent} \\ 25 = \text{power} \end{cases}$$

The mathematical operations of exponents are as follows.

1) $a^m a^n = a^{m+n}$

2) $\left(a^m\right)^n = a^{mn}$

3) $\left(ab\right)^m = a^m b^m$

4) $\left(\dfrac{a}{b}\right)^n = \dfrac{a^n}{b^n}$

5) $a^{-n} = \dfrac{1}{a^n}$

6) $a^{\frac{m}{n}} = \sqrt[n]{a^m}$

SAT Practice

1. If $\left\{(-2)^3 (8)^2\right\}^4 = \left(2^4\right)^n$, what is the positive value of n?

A) 6
B) 7
C) 8
D) 9

---

2. If $4^3 + 4^3 + 4^3 + 4^3 = 2^n$, what is the value of n?

A) 2
B) 4
C) 6
D) 8

3. If m and n are positive and $5m^5n^{-3} = 20m^3n$, what is the value of m in terms of n?

A) $\dfrac{1}{4n}$

B) $\dfrac{4}{n^2}$

C) $\dfrac{4}{n^3}$

D) $2n^2$

---

4. If a and b are positive integers, $\left(a^{-4}b\right)^{-1} = 16$, and $b = a^2$, which of the following could be the value of a?

A) 0
B) 2
C) 4
D) 8

---

5. If $k^{-2} \times 2^3 = 2^7$, what is the value of k?

A) 2   B) 4   C) 8   D) $\dfrac{1}{4}$

---

6. If p and q are positive numbers, $p^{-3} = 2^{-6}$, and $q^{-2} = 4^2$, what is the value of pq?

A) 1
B) 2
C) 3
D) 4

---

7. If a and b are positive integers and $\left(a^6b^4\right)^{\frac{1}{2}} = 675$, what is the value of $a + b$?

A) 2
B) 4
C) 6
D) 8

TIPS

Tip 32 | Defined Operations

The defined operations are mathematical models (symbolic representations/notational systems/sign systems) of certain situations.

Example:

If the operation ▲ is defined by $▲a = a^a$, what is the value of $\dfrac{▲8}{▲4}$?

Solution)

$▲8 = 8^8 = \left(2^3\right)^8 = 2^{24}$ and $▲4 = 4^4 = \left(2^2\right)^4 = 2^8$

Therefore, the answer is $\dfrac{▲8}{▲4} = \dfrac{2^{24}}{2^8} = 2^{24-8} = 2^{16}$.

SAT Practice

1. Let the operation $\odot$ be defined for all numbers by $a \odot b = \dfrac{a+b}{a-b}$. If $p \odot q = 3$, what is the value of $\dfrac{p}{q}$?

   A) $\dfrac{1}{2}$

   B) 1

   C) $\dfrac{3}{2}$

   D) 2

---

2. Let the operation $\triangle$ be defined by $a \triangle b = \dfrac{a}{b}$ for all positive numbers. If $4 \triangle (k \triangle 6) = 3$, what is the value of k?

   A) 4   B) 8   C) 12   D) 20

---

3. Let the operation be defined by $n^{▲} = n(n-1)(n-2)(n-3).....(2)(1)$, where n is a positive integer. Which of the following is equivalent to $(n+1)^{▲}$?

   A) $n(n^{▲})$

   B) $(n+1)(n+1)^{▲}$

   C) $n(n-1)^{▲}$

   D) $(n+1)n^{▲}$

Tip 33 | Functions as Models

Functions as Models

A function can serve as a simple kind of mathematical model, or a simple piece of a larger model. Remember that a function is just a rule. We can think of the rule (given in our model as a graph, a formula, or a table of values) as a representation of some natural cause and effect relationship.

SAT Practice

1. The total cost c, in dollars, of repairing shoes is given by the function $c(x) = \dfrac{200x - 400}{x} + k$, where x is the number of repairing shoes and k is a constant. If 50 shoes were repaired at a cost of $300, what is the value of k?

   A) 100   B) 108   C) 126   D) 150

2. The value of a computer decreases each year by 1.2 percent. This year the price of the computer was $1,200. If the price p of the computer n years from now is given by the function $p(n) = 1,200c^n$, what is the value of c?

   A) 0.012   B) 0.88   C) 0.988   D) 1.012

3. Let the function m, average rate of change between a and b in the domain of the function, be defined by $m(x) = \dfrac{f(b) - f(a)}{b - a}$. If $f(x) = x^2$, what is the value of m between -2 and 3 ?

   A) -2 B) -1 C) 0   D) 1

4. The present value p of a certain car that depreciates for a number of years is defined by $p(t) = k\left(1 - \dfrac{r}{100}\right)^t$, where k is the initial value of the car, r is the percent of depreciation per year, and t is the number of years.

   If a person purchases the car for $20,000 and the value of the car depreciates by 10% per year, how much will the value of the car be after three years from the date of purchase?

   A) $18,000
   B) $16,200
   C) $14,580
   D) $14,000

TIPS

Tip 34 — Combined Rate of Work

These problems involve two people (or any machines) working at different rates. The general formula for solving combined work rate work problems are as follows.

1) Work rate $= \dfrac{1}{\text{Time taken}}$

2) Time taken together $= \dfrac{1}{\text{Combined work rate}}$

Let's assume we have two workers, John and Chris.

1) John can finish a job in a hours when walking alone.

2) Chris can finish a job in b hours when working alone.

If two workers are working together, the number of hours they need to complete the job is given by

Worker	Rate	Combined work rate	Time
John	$1/a$	$\dfrac{1}{a}+\dfrac{1}{b}=\dfrac{a+b}{ab}$	$\dfrac{ab}{a+b}$
Chris	$1/b$		

John's work rate $= \dfrac{1}{a}$ Chris' work rate $= \dfrac{1}{b}$

Combined work rate $= \dfrac{1}{a}+\dfrac{1}{b}=\dfrac{a+b}{ab}$

Time taken together $= \dfrac{1}{\text{Combined work rate}}=\dfrac{ab}{a+b}$

SAT Practice

1. Worker A can finish a job in 5 hours. When worker A works together with worker B, they can finish the job in 4 hours. How long does it take for worker B to finish the job if he works alone?

   A) 8 hours
   B) 12 hours
   C) 16 hours
   D) 20 hours

2. Raymond and Peter can paint a house in 20 hours when working together at the same time. If Raymond works twice as fast as Peter, how long would it take Peter to paint the house if he works alone?

   A) 20 hours
   B) 30 hours
   C) 40 hours
   D) 60 hours

3. The swimming pool can be filled by pipe A in 5 hours and by pipe B in 8 hours. How long would it take to fill the pool if both pipes were used?

   A) $3\frac{1}{13}$ hours

   B) $5\frac{2}{3}$ hours

   C) 7 hours

   D) $8\frac{1}{3}$ hours

4. If it takes 5 people 12 hours to paint 3 identical houses, then how many hours will it take 4 people working at the same rate to paint 5 identical houses? (Assume they work at the same rate.)

   A) 18 hours
   B) 19 hours
   C) 20 hours
   D) 25 hours

Tip 35 | Combined Range of Two Intervals

For the interval of x, $\quad a \leq x \leq b$

a = minimum value $\quad$ and $\quad b$ = maximum value

Minimum ≤ Combined Range ≤ Maximum

Example:

$$5 \leq A \leq 10 \quad \text{and} \quad 2 \leq B \leq 5$$

1) $7 \leq A + B \leq 15$ $\qquad\qquad$ 2) $10 \leq A \cdot B \leq 50$

3) $0 \leq A - B \leq 8$ $\qquad\qquad$ 4) $1 \leq \dfrac{A}{B} \leq 5$

SAT Practice

1. Given $2 \leq P \leq 8$ and $1 \leq Q \leq 4$. By how much is the maximum value of $\dfrac{P}{Q}$ greater than the minimum value of $\dfrac{P}{Q}$?

2. If $-2 \leq A \leq 2$, and $-6 \leq B \leq -2$, and $C = (A - B)^2$, what is the smallest value of C?

3. If $1 \leq P \leq 6$, and $3 \leq Q \leq 10$, what is the smallest value of $P \times Q$?

4. If $-2 < x < 4$ and $-3 < y < 2$, what are all possible values of $x - y$?

A) $-4 < x - y < 2$

B) $1 < x - y < 7$

C) $1 < x - y < 4$

D) $-4 < x - y < 7$

Tip 36 | Absolute Value

The absolute value of x, denoted "$|x|$" (and which is read as "the absolute value of x"), is regarded as the distance of x from zero.

Properties of absolute value:

1) If $|x| = a$ and $a > 0$, then $x = a$ or $-a$

2) If $|x| < a$ and $a > 0$, then $-a < x < a$

3) If $|x| > a$ and $a > 0$, then $x > a$ or $x < -a$

4) $|x| < 5 \leftrightarrow x^2 < 25 \leftrightarrow -5 < x < 5$

5) $|x| > 5 \leftrightarrow x^2 > 25 \leftrightarrow x < -5$ or $x > 5$

6) $|x - 10| = |10 - x|$

How do we convert the general interval into an expression using the absolute value?

Example 1:

For $10 \leq x \leq 30$,

Step 1: Find the midpoint of 10 and 30. $\rightarrow \dfrac{10 + 30}{2} = 20$

Step 2: Find the distance from the midpoint to the end point. $\rightarrow 30 - 20 = 10$ or $20 - 10 = 10$

Step 3: Substitute in the form.
$$|x - \text{midpoint}| \leq \text{distance}$$

Therefore, $|x - 20| \leq 10$.

Example2:

For $x \leq 10$ or $x \geq 30$,

   $|x - 20| \geq 10$.

Example 3:

If $-8 < x < 2$, then express the interval using absolute value.

Solution)

Step 1: Find the midpoint between -8 and 2 $\rightarrow \dfrac{-8 + 2}{2} = -3$

Step 2: Find the distance from the midpoint to the endpoint. $\rightarrow 2 - (-3) = 5$

Step 3: From the figure above, the interval can be expressed with absolute value.

   midpoint $= -3$

      distance $= 5$

   $-8 < x < 2 \longleftrightarrow |x - (-3)| < 5 = |x + 3| < 5$

1. An art class of 20 students took a final exam and ten of the students scored between 78 and 86 in the exam. If s is defined as the scores of the ten students, which of the following describes all possible values of s ?

   A) $|s - 82| = 4$

   B) $|s + 82| = 4$

   C) $|s - 82| < 4$

   D) $|s + 82| < 4$

2. At a bottling company, a computerized machine accepts a bottle only if the number of fluid ounces is greater than or equal to $5\frac{3}{7}$, and less than or equal to $6\frac{4}{7}$. If the machine accepts a bottle containing f fluid ounces, which of the following describes all possible values of f ?

   A) $|f - 6| < \frac{4}{7}$

   B) $|f - 6| \le \frac{3}{7}$

   C) $|f + 6| > \frac{4}{7}$

   D) $|6 - f| \le \frac{4}{7}$

3. At the O.K Daily Milk Company, machine X fills a box with milk, and machine Y eliminates milk-box if the weight is less than 450 grams, or greater than 500 grams. If the weight of the box that will be eliminated by machine Y is E, in grams, which of the following describes all possible values of E ?

   A) $|E - 475| < 25$

   B) $|E - 500| > 450$

   C) $|475 - E| = 25$

   D) $|E - 475| > 25$

Tip 37 | Parallel Lines with Transversal

A transversal is a line that intersects parallel lines. When it intersects parallel lines, many angles are congruent.

If a set of parallel lines are cut by a transversal, each of the parallel lines has 4 angles surrounding the intersection as follows.

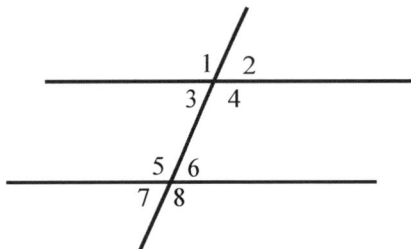

$\angle 1 \cong \angle 4$ and $\angle 2 \cong \angle 3$: Vertical angles

$\angle 3 \cong \angle 6$ and $\angle 4 \cong \angle 5$: Alternate angles

$\angle 2 \cong \angle 6$ and $\angle 4 \cong \angle 8$: Corresponding angles

$\angle 3 + \angle 5 = 180°$ and $\angle 4 + \angle 6 = 180°$:

Sum of the interior angles in the same side is $180°$.

SAT Practice

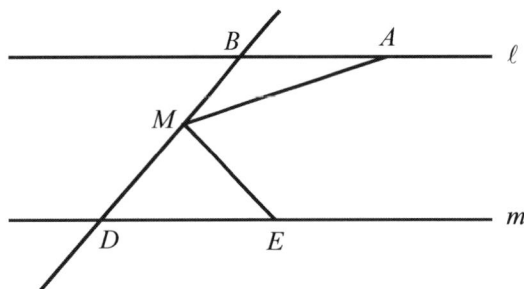

Note: Figure not drawn to scale.

1. In the figure above, If $AB = BM = DM = DE$ and $\ell \parallel m$, what is the measure of $\angle AME$?

A) $50°$

B) $60°$

C) $75°$

D) $90°$

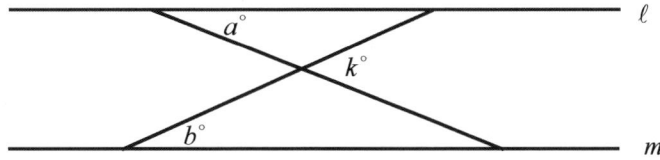

Note: Figure not drawn to scale.

2. In the figure above, $\ell \parallel m$, $a = 65$, and $b = 45$. What is the value of k?

A) 80
B) 90
C) 100
D) 110

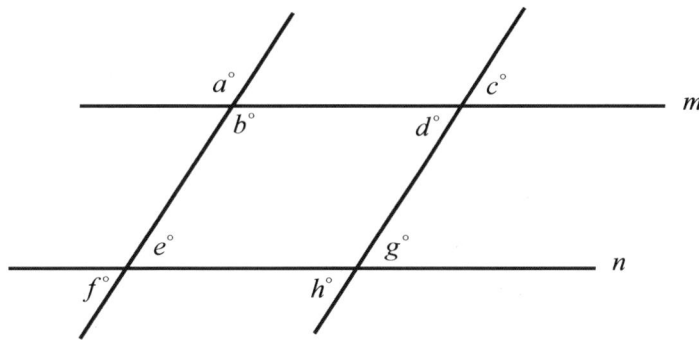

3. In the figure above, line m is parallel to line n. Which of the following must be true?

I. $a = c$
II. $d = g$
III. $b + e = 180$

(A) I only
(B) II only
(C) III only
(D) II and III only

TIPS

Tip 38 — Triangle Inequality

Triangle Inequality Theorem

Theorem 1:
The length of one side of a triangle is less than the sum of the other two sides and is greater than the difference of the other two sides.

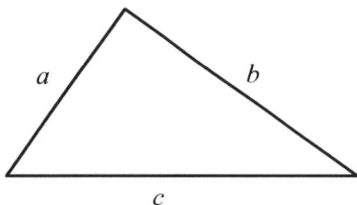

We can say that

$$a \sim b < c < a + b.$$

Theorem 2:

In a triangle, the longest side has the opposite largest angle.

Theorem 3:

The measure of an exterior angle of a triangle is equal to the sum of the measures of its two nonadjacent interior angles.

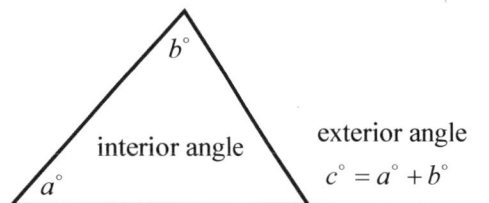

interior angle

exterior angle

$c^\circ = a^\circ + b^\circ$

SAT Practice

1.   If the lengths of the sides of $\triangle ABC$ is 3, $x + 3$, and 9, which of the following could be the value of x?

A) 1
B) 2
C) 3
D) 4

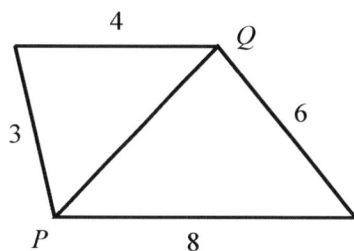

2. In the figure above, which of the following could be the length of $\overline{PQ}$?

    A) 9

    B) 8

    C) 7

    D) 6

---

3. Which of the following CANNOT be possible to construct a triangle with the given side lengths?

    A) 6, 7, 11

    B) 3, 6, 9

    C) 28, 34, 39

    D) 35, 120, 125

TIPS

Tip 39 — Ratio of areas of triangles with the same height

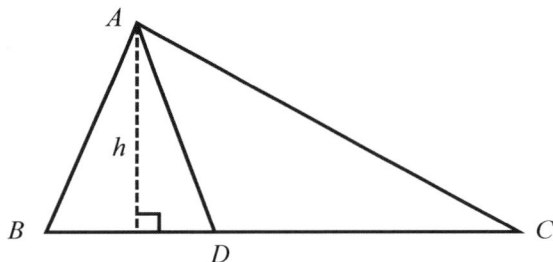

Two triangles with the same height h are shown above. If the ratio of BD to DC is $a:b$, then the ratio of the areas of $\triangle ABD$ to $\triangle ADC$ is also $a:b$.

$$\text{Ratio of areas} \rightarrow \frac{BD \times h}{2} : \frac{DC \times h}{2} = BD : DC = a : b$$

SAT Practice

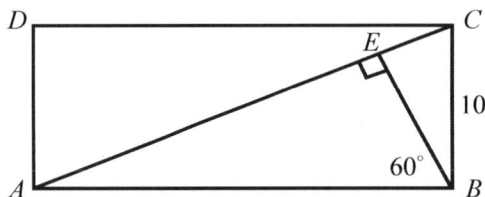

1. In rectangle $ABCD$ above, $\angle ABE = 60°$ and $BC = 10$. what is the area of triangle ABC?

    A) 50
    B) $50\sqrt{3}$
    C) 55
    D) $60\sqrt{3}$

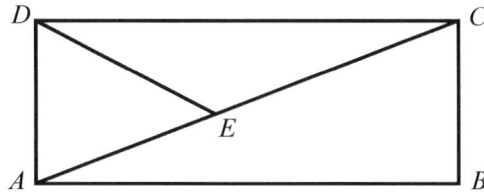

2. In the figure above, the ratio of AE to EC is 3:5. If the area of $\triangle ADE$ is 24, what is the area of rectangle $ABCD$?

A) 64
B) 80
C) 100
D) 128

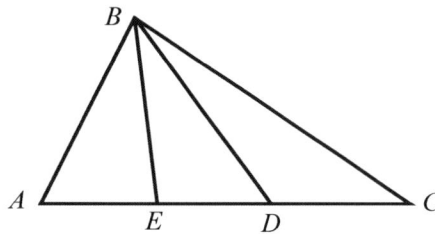

Note: Figure not drawn to scale.

3. In the figure above, the ratio of the lengths AE to ED to DC is 2:3:4. If the area of $\triangle ABD$ is 40, what is the area of $\triangle BDC$?

Tip 40 | Special Right Triangles

Angle-based right triangle:

1) $30° - 60° - 90°$ triangle

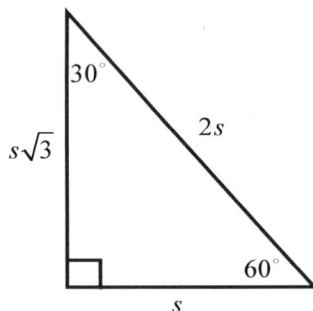

2) $45° - 45° - 90°$ triangle

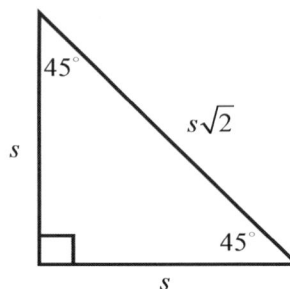

Side-based right triangle:

Right triangles whose sides are Pythagorean triples as follows.

    3:4:5   5:12:13   8:15:17   7:24:25   9:40:41 $\cdots$

SAT Practice

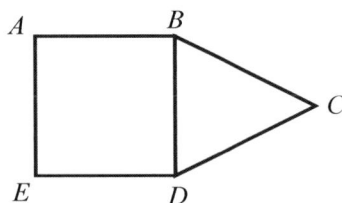

1. The figure above contains square *ABDE* and equilateral triangle *BCD*. If the area of $\triangle BCD$ is $16\sqrt{3}$, what is the area of the square?

A) 32

B) $32\sqrt{3}$

C) 64

D) $64\sqrt{3}$

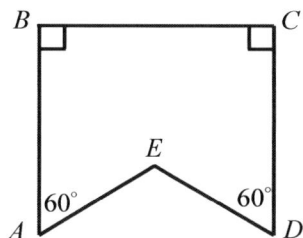

2. In the figure above, $AB = BC = CD = 10$. What is the perimeter of the figure?

A) 50

B) $30 + 10\sqrt{3}$

C) $30 + \dfrac{20\sqrt{3}}{3}$

D) $30 + 20\sqrt{3}$

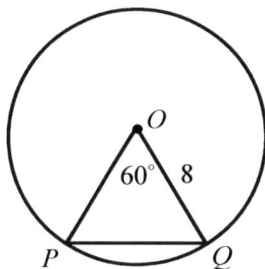

3. In the figure above, the radius of circle O is 8 and measure of $\angle POQ$ is $60°$. What is the area of $\triangle OPQ$?

A) $12\sqrt{3}$
B) $16\sqrt{3}$
C) $20\sqrt{3}$
D) $24\sqrt{3}$

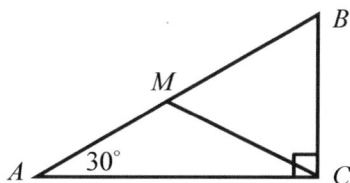

4. In the figure above, M is the midpoint of $\overline{AB}$. If the length of $\overline{BC}$ is 20, what is the area of $\triangle AMC$?

A) $100\sqrt{2}$ B) $100\sqrt{3}$ C) $200\sqrt{2}$ D) $200\sqrt{3}$

TIPS

Proportions in a Right Triangle

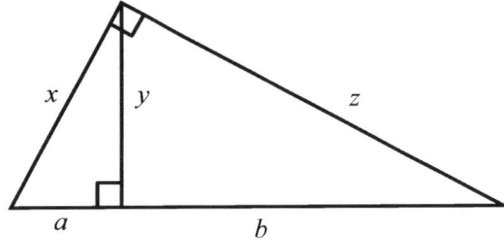

Memorize the formulas.

1) $x^2 = a(a+b)$ 2) $y^2 = a \cdot b$ 3) $z^2 = b(b+a)$ 4) $\dfrac{xz}{2} = \dfrac{(a+b)y}{2}$ (area of the triangle)

SAT Practice

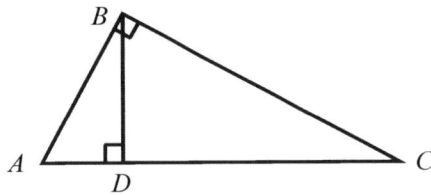

1.   In $\triangle ABC$ above, $AB = 6$ and $AD = 3$. What is the length of $\overline{CD}$?

Questions 2-4 refer to the following information.

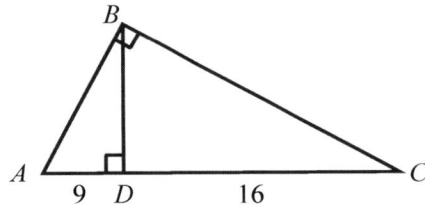

In the figure above, $AD = 9$ and $CD = 16$.

2. What is the length of $\overline{AB}$?

A) 12
B) 14
C) 15
D) 18

---

3. What is the length of $\overline{BC}$?

A) 18
B) 20
C) 25
D) 28

---

4. What is the length of $\overline{BD}$?

A) 10
B) 12
C) 15
D) 18

TIPS

Tip 42 | Pythagorean Theorem

In mathematics, the **Pythagorean Theorem** is a relation in Euclidean geometry among the three sides of a right triangle. The theorem is named after the Greek mathematician Pythagoras, who by tradition is credited with its discovery and proof, although knowledge of the theorem almost certainly predates him. The theorem is as follows:

In any right triangle, the area of the square whose side is the hypotenuse (the side opposite the right angle) is equal to the sum of the areas of the squares whose sides are the two legs (the two sides that meet at a right angle).

$$a^2 + b^2 = c^2 \ \text{(Side)} \qquad A + B = C \ \text{(Area)}$$

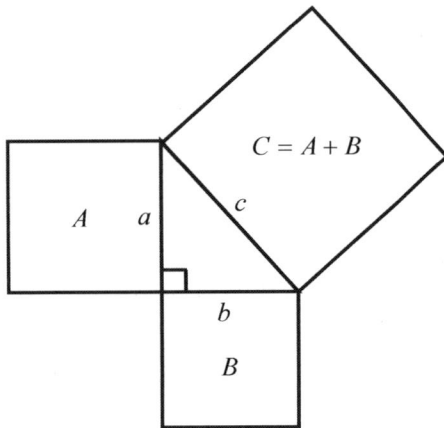

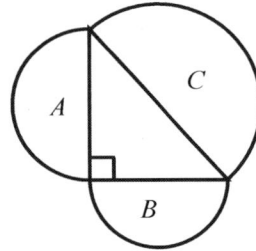

Remember: If the figures attached to the right triangle are similar, $C = A + B$ is always true.

SAT Practice

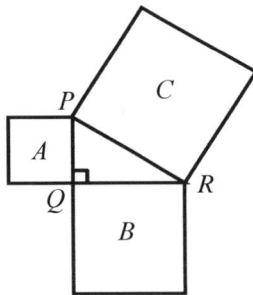

1. In the figure above, the area of square A is 16 and the area of square B is 20. What is the length of $\overline{PR}$?

A) 4   B) 6   C) 25   D) 36

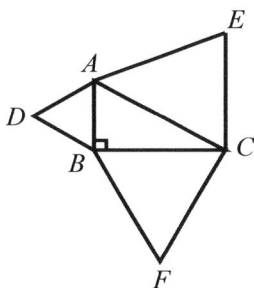

2. In the figure above, $\triangle ABD$, $\triangle ACE$, and $\triangle BCF$ are equilateral triangles, and the ratio of AB to BC is 1:2. If the area of $\triangle ACE$ is 20, what is the area of $\triangle ABD$?

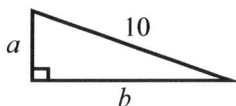

3. In the figure above, which of the following is true about the lengths a and b?

A) $10 < (a+b)^2 < 40$ B) $40 < (a+b)^2 < 80$ C) $80 < (a+b)^2 < 100$ D) $100 < (a+b)^2$

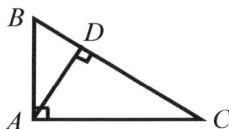

4. In $\triangle ABC$ above, if $AB = 10$ and $AC = 20$, what is the length of $\overline{AD}$?

A) 5   B) $4\sqrt{5}$ C) $6\sqrt{3}$ D) $8\sqrt{3}$

TIPS

Types of transformation in math are:

- Translation: involves "sliding" the object from one position to another.
- Reflection: involves "flipping" the object over a line called the line of reflection.
- Rotation: involves "turning" the object about a point called the center of rotation.
- Dilation: involves a resizing of the object. It could result in an increase in size (enlargement) or a decrease in size (reduction).

Translation of a graph

If the graph of $y = f(x)$ is translated a units horizontally and b units vertically, then the equation of the translated graph is

$$y - b = f(x - a) \quad \text{or} \quad y = f(x - a) + b$$

SAT Practice

1. The graph of the line represented by the equation $y = -2x + 5$ is moved to the left by 3 units and up 1 unit, what is the equation of the graph of the new line?

   A) $y = -2x$
   B) $y = -2x + 4$
   C) $y = -2x - 5$
   D) $y = -2x + 5$

2. How does the graph of $g(x) = x^2 - 1$ compare to the graph of $f(x) = (x - 2)^2 + 1$?

   A) The vertex of the graph of $f(x)$ moved to the right by 2 units and down by 2 units.
   B) The vertex of the graph of $f(x)$ moved to the left by 2 units and down by 2 units.
   C) The vertex of the graph of $f(x)$ moved to the right by 2 units and up by 2 units.
   D) The vertex of the graph of $f(x)$ moved to the left by 2 units and up by 2 units.

Tip 44   Classifying a group in two different ways

Organize the information in a table and use a convenient number.

Example:

In a certain group of only senior and junior students, $\dfrac{3}{5}$ of the students are boys, and the ratio of seniors to juniors is 4:5. If $\dfrac{2}{3}$ of girls are seniors, what fraction of the boys are juniors?

	Boys	Girls	Total
Seniors		$\dfrac{4}{15}$	$\dfrac{4}{9}$
Juniors	A	B	$\dfrac{5}{9}$
Total	$\dfrac{3}{5}$	$\dfrac{2}{5}$	1

$\dfrac{2}{3}$ of girls are seniors $\rightarrow$ $\dfrac{2}{3} \times \dfrac{2}{5} = \dfrac{4}{15}$ $B = \dfrac{2}{5} - \dfrac{4}{15} = \dfrac{2}{15}$

Therefore, $A = \dfrac{5}{9} - \dfrac{2}{15} = \dfrac{25}{45} - \dfrac{6}{45} = \dfrac{19}{45}$.

Question: What **fraction of boys** are juniors?

Final answer is $\dfrac{19/45}{3/5} = \dfrac{19/45}{27/45} = \dfrac{19}{27}$.

Or

Use a convenient number 45 for the number of students in the group.

	Boys	Girls	Total
Seniors	8	12	20
Juniors	19	6	25
Total	27	18	45

SAT Practice

1. Of the 24 company presidents attending a corporate meeting, $\dfrac{3}{4}$ of the presidents are males and $\dfrac{2}{3}$ of the presidents have children. If 2 female presidents do not have children, how many male presidents have children?

   A) 8   B) 10   C) 12   D) 14

Tip 45 | Discriminant

The discriminant $b^2 - 4ac$, which is the expression under the radical sign, describe the nature of the roots of a quadratic equation. By the quadratic formula, the two roots of $ax^2 + bx + c = 0$ are

$$x = \frac{-b \pm \sqrt{b^2 - 4ac}}{2a} \quad \rightarrow \quad D(\text{descriminant}) = b^2 - 4ac$$

Value of discriminant	Nature of roots	Graph of $y = ax^2 + bx + c, \ (a > 0)$
$D > 0$ and $b^2 - 4ac$ is a perfect square	Real, rational, unequal	
$D > 0$ and $b^2 - 4ac$ is not a perfect square	Real, irrational, unequal	
$D = 0$	Real, rational, equal	
$D < 0$	imaginary	

When a line is tangent to a curve, the discriminant of the equation should be 0.

Example:

If he graph of $y = 2x + k$ is tangent to the graph of $y = x^2 - 4$, what is the value of k?

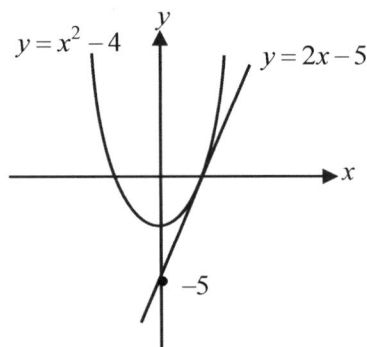

Solution)

$x^2 - 4 = 2x + k \quad \rightarrow \quad x^2 - 2x - 4 - k = 0$,

Since they are tangent, the discriminant D should be 0.

$D = (-2)^2 - 4(1)(-4 - k) = 0 \quad \rightarrow \quad D = 20 + 4k = 0$

Therefore, $k = -5$.

1. Which of the following is true about the graph of $y = 2x^2 - 6x + 3$?

    A) The graph of y is tangent to the x-axis.
    B) The graph of y intersects the x-axis at 2 points.
    C) The graph of y lies entirely above the x-axis.
    D) The graph of y lies entirely below the x-axis.

2. Find the value of k for which of the graph of $y = x^2 + 6x + k$ is tangent to the x-axis ?

    A) 3   B) 6   C) 9   D) 12

3. For what value of k the roots of $x^2 + 2x + k = 0$ are equal?

    A) 0   B) 1   C) 2   D) 3

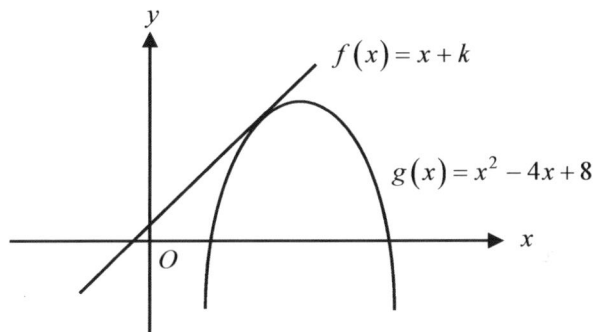

4. In the xy-plane above, the graphs of f and g are tangent each other. Which of the following is the value of k ?

    A) $\dfrac{1}{3}$

    B) $\dfrac{1}{2}$

    C) $\dfrac{3}{2}$

    D) $\dfrac{7}{4}$

TIPS

Tip 46 — Handshakes

If there are five people in a room, and they shake each other's hands once and only once, how many handshakes are there altogether?

(A, B) (A, C) (A, D) (A, E)--------4 handshakes

(B, C) (B, D) (B, E) ----------------3 handshakes

(C, D) (C, E) -----------------------2 handshakes

(D, E) -------------------------------1 handshake

Therefore there will be

$$4+3+2+1= 10 \text{ handshakes}$$

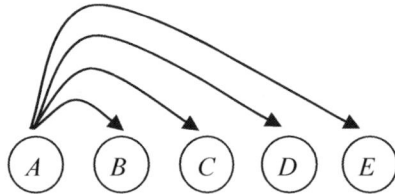

Or, we can use a combination.

The number of handshakes is equal to ways to select two people from 8, because two people make one handshake. Therefore,

$$_5C_2 = \frac{5 \times 4}{2!} = 10$$

SAT Practice

1. If you have 12 people in a group and each person shakes everyone else's hand only once, how many handshakes take place?

    A) 132   B) 112   C) 88   D) 66

2. At a party, everybody shakes hands with each other once. If there are 45 handshakes, how many people are there at the party?

    A) 9   B) 10   C) 11   D) 12

Tip 47 | Consecutive integers I

Integers which follow each other in order, without gaps, from smallest to largest are consecutive integers.

12, 13, 14 and 15 are consecutive integers.
12, 14, 16, 18 are consecutive even integers.
11, 13, 15, 17 are consecutive odd integers.

Properties of consecutive integers:
For consecutive numbers (integers), if the first term is a_1 and the last term is a_n, the average is equal to median.

Average (Arithmetic mean) $= \dfrac{a_1 + a_n}{2}$

Median = Average
Sum of consecutive = median × number of integers

Or, Sum of consecutive = average × number of integers

Example:
For the sequence of consecutive numbers

2, 3, 4, 5, 6, 7, 8, 9, 10

Average $= \dfrac{2+3+4+5+6+7+8+9+10}{9} = 6$

But simply we use the formula   average $= \dfrac{a_1 + a_n}{2}$.

Average $= \dfrac{2+10}{2} = 6$

That is
   Median = Average = 6

Therefore, in the list of consecutive numbers, the average value is equal to the median.

SAT Practice

1. What is the sum of 11 consecutive integers if the middle one is 30?

   A) 60
   B) 120
   C) 330
   D) 660

2. If the median of a list of 99 consecutive integers is 80, what is the greatest integer in the list?

A) 99

B) 128

C) 129

D) 157

---

3. The median of a list of 10 consecutive even integers is 77. What is the sum of the integers?

A) 700

B) 770

C) 780

D) 800

---

4. If the median of a list of 30 consecutive odd integers is 120, what is the greatest integer in the list?

A) 145

B) 147

C) 149

D) 151

Tip 48   Consecutive integers II

Example:

The smallest integer of a set of consecutive integers is -10. If the sum of these integers is 23, how many integers are in this set?

Solution)

$$-10, -9, -8-7, ..., 0, ..., 7, 8, 9, 10, 11, 12$$

sum $= 0$ sum $= 23$

We know that the sum of the consecutive integers from -10 to $+10$ is zero.
Therefore, the number of integers is

$$1, 2, 3, 4, \cdots, 10 = 10 \text{ integers}$$
$$0 = 1 \text{ integer}$$
$$-1, -2, -3, \cdots, -10 = 10 \text{ integers}$$

We need two more integers 11 and 12 to have a sum of 23.

Hence, there are $10 + 1 + 10 + 2 = 23$ integers .

SAT Practice

1. If the sum of the consecutive integers from -30 to x, inclusive, is 96, what is the value of x?

    A) 30
    B) 31
    C) 32
    D) 33

2. The smallest integer of a set of even consecutive integers is -20. If the sum of these integers is 72, how many integers are in the set?

    A) 24
    B) 25
    C) 43
    D) 44

TIPS

Tip 49   Scatterplot and a Line of Best Fit

When data is displayed with a scatterplot, a line of best fit is useful for purpose of **predicting values that may not be displayed on the plot.**

1) A line of best fit is a straight line that best represents the data on a scatterplot. This line may pass through some of the points.
2) The equation of a line of best fit depends upon the points chosen to construct the line.
3) The graphing calculator has the capability of determining which line will actually represent the line-of-best-fit

SAT Practice

Questions 1-5 refer to the following information.

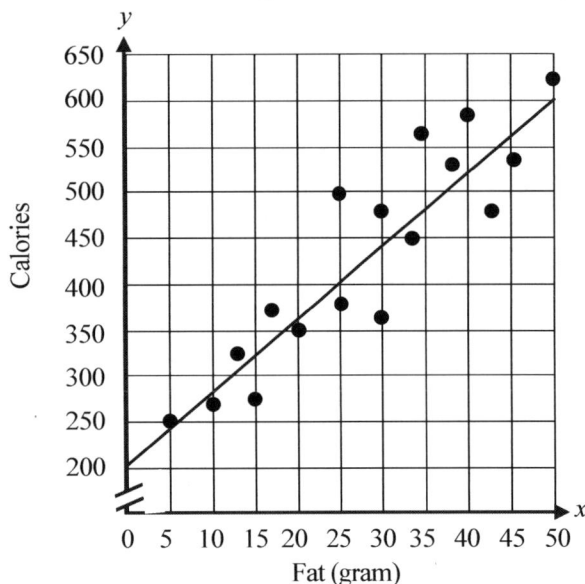

Fat (gram)

The scatterplot above shows the relationship between the fat grams and the calories in a fast food. The line of best fit is also shown above.

1. Which of the following best describes the correlation seen in the graph?

   A) high positive correlation
   B) low positive correlation
   C) high negative correlation
   D) low negative correlation

2. What is the equation of the line of best fit?

A) $y = 8x$

B) $y = 8x + 200$

C) $y = 10x$

D) $y = 10x + 200$

---

3. How many calories would be predicted for 80 fat grams?

A) 600

B) 840

C) 1000

D) 1200

---

4. For a 25 fat grams, which of the following is the percent decrease from the actual calories to the calories predicted by the line of best fit?

A) 10%

B) 15%

C) 20%

D) 25%

---

5. Which of the following is the calories per one fat gram?

A) 8 calories

B) 10 calories

C) 100 calories

D) 200 calories

TIPS

Tip 50 | Must be true or could be true

$$(a+b)^2 = (a-b)^2$$

The questions can be as follows

(1) If the statement above is true, which of the following must also be true (always true)?

or

(2) If the statement above is true, which of the following could be true (possibly true)?

Solution)

Simplify the equation until we can make a conclusion as follows.

$$(a+b)^2 = a^2 + 2ab + b^2 \qquad (a-b)^2 = a^2 - 2ab + b^2$$

Then

$$\cancel{a^2} + 2ab + \cancel{b^2} = \cancel{a^2} - 2ab + \cancel{b^2}$$

$4ab = 0$, or $ab = 0$

$$ab = 0 \begin{cases} a = 0 \\ b = 0 \end{cases} \text{Could be true.}$$

Must be true.

For the question (1), the answer is
$$ab = 0$$
The answer $ab = 0$ implies that

(a) When $a = 0$, b is all real numbers and

(b) When $b = 0$, a is all real numbers.

For the question (2), there are two answers.
$$a = 0, \text{ or } b = 0$$

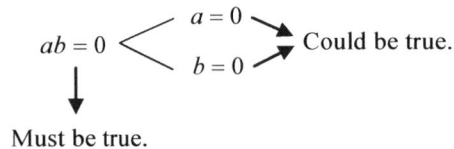

(a) When $a = 0$, b can be any real number.

Or

(b) When $b = 0$, a can be any real number.

SAT Practice

1. If $k(a-b) = a-b$, which of the following could be true?

   I. $k = 1$

   II. $a = 2$ and $b = 2$

   III. $a = b$

   A) II only   B) III only   C) I and III only   D) I, II, and III

---

2. If $a^2 + b^2 = 2ab$, which of the following must be true?

   I. $a = 1$

   II. $a = b$

   III. $a = 0$ and $b = 0$

   A) I only   B) II only   C) III only   D) I and II only

TIPS

Tip 51 | Complex Numbers

Imaginary number: $i = \sqrt{-1}$ and $i^2 = -1$

Example:

$$\sqrt{-4} = i\sqrt{4} = 2i \qquad\qquad \sqrt{-12} = i\sqrt{12} = 2i\sqrt{3}$$

Complex number: $a + bi$, where a and b are real numbers and $i = \sqrt{-1}$.

$a =$ the real part of $a + bi$ and $b =$ imaginary part of $a + bi$

Property:

1) Two complex numbers are equal if and only if their real parts are equal and their imaginary parts are equal.

$$a + bi = c + di \quad \rightarrow \quad a = c \text{ and } b = d$$

2) $|a + bi| = \sqrt{a^2 + b^2}$ (Distance from the origin)

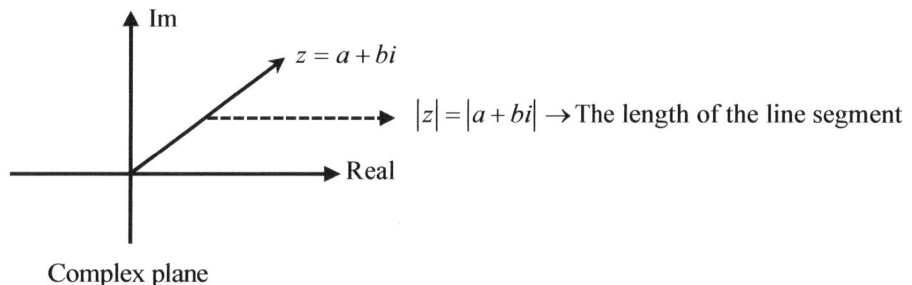

Complex plane

SAT Practice

1. If $a + b + (a - b)i = 6 - 4i$, what is the value of a?

$$\frac{2 + 3i}{1 - 2i} = a + bi$$

2. In the equation above, $i = \sqrt{-1}$. What is the value of b?

A) $-\dfrac{4}{5}$ B) $\dfrac{4}{5}$ C) $\dfrac{7}{5}$ D) $\dfrac{9}{5}$

3. What is the value of $\sqrt{-12} \times \sqrt{-3}$?

    A) -36
    B) -6
    C) 6
    D) 36

$$a = 10 + bi - (a-5)i,$$

4. In the equation above, a and b are real numbers and $i = \sqrt{-1}$. What is the value of b?

---

5. If $i = \sqrt{-1}$, which of the following is equal to i^{126} ?

    A) i B) -1 C) $-i$ D) 1

---

6. Which of the following could be the solution of $x^3 - 2x^2 + 3x = 0$?

    A) 1

    B) 3

    C) $1 + i\sqrt{2}$

    D) $\dfrac{1}{2} + \dfrac{\sqrt{2}}{2}i$

Tip 52 **Circles**

1. Central angle: $\angle AOB$

2. Inscribed angle: $\angle ACB$

3. Major arc: $\overparen{ACB}$

4. Minor arc: $\overparen{AB}$

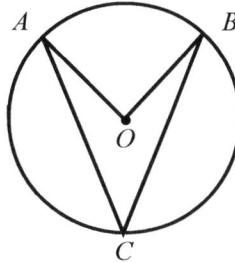

5. Central angle $= 2 \times$ Inscribed angle

6. $\angle ACB = 90°$

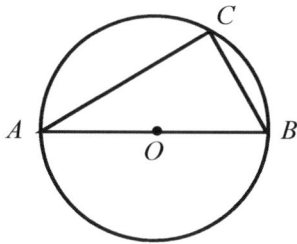

7. $AP = BP$ and $\angle AOB + \angle APB = 180°$, $\overline{AO} \perp \overline{AP}$, $\overline{BO} \perp \overline{BP}$

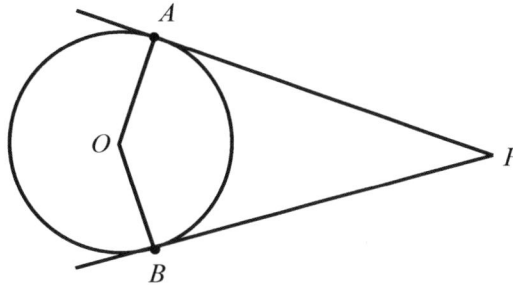

8. $a + c = 180°$ and $b + d = 180°$

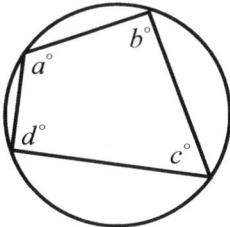

9. $a \times b = c \times d$

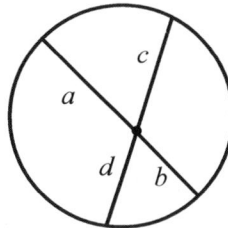

10.

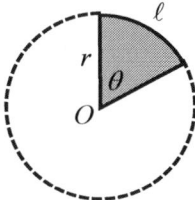

$$\begin{cases} \text{The length of the minor arc: } \ell = r\theta \\ \text{The area of the sector: } A = \dfrac{1}{2}\left(r^2\theta\right) \end{cases}$$, where θ is in radian measure.

11. $\pi(\text{radian}) = 180°$

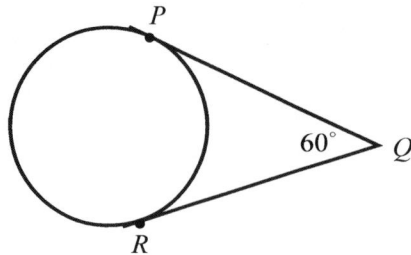

1. In the figure above, $\overline{PQ}$ and $\overline{RQ}$ are tangent to the circle at points P and R. If the radius of the circle is 12, what is the length of minor arc $\overset{\frown}{PR}$?

A) 4π
B) 6π
C) 8π
D) 10π

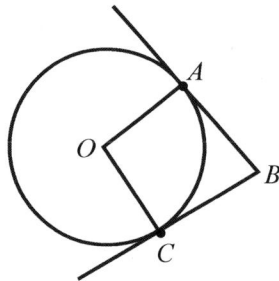

2. In the figure above, point O is the center of the circle, $\overline{BA}$ and $\overline{BC}$ are tangent to the circle at points A and C. If the measure of $\angle AOC$ is $105°$, what is the measure of $\angle ABC$ in degrees?

Questions 3 and 4 refer to the following figure.

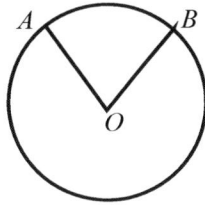

In the figure above, the measure of $\angle AOB$ is $\dfrac{5\pi}{12}$ radians and the radius of the circle is 36.

3.   What is the length of the minor arc $\overarc{AB}$?

A) 10π B) 15π C) 20π D) 36π

4.   What is the area of sector AOB?

A) 135π B) 180π C) 270π D) 360π

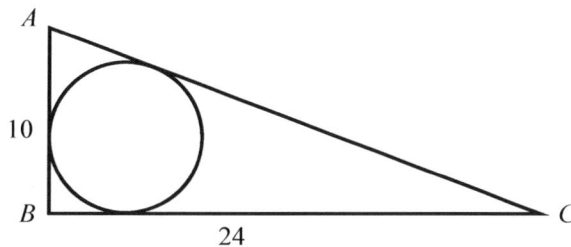

5.   In the figure above, each side of right triangle ABC with $\angle B = 90°$ are tangent to the circle. The length of $\overline{AB}$ is 10 and the length of $\overline{BC}$ is 24. What is the radius of the circle?

| Tip 53 | **Trigonometric Function and Cofunction** |

Trigonometric Function:

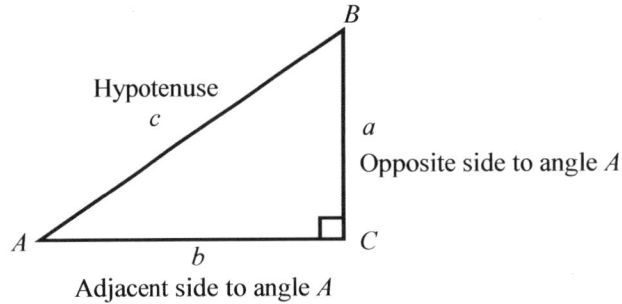

$$\sin\theta = \frac{a}{c} \qquad \cos\theta = \frac{b}{c} \qquad \tan\theta = \frac{a}{b}$$

Cofunction:

$\sin A = \dfrac{a}{c}$ and $\cos B = \dfrac{a}{c}$ $\rightarrow$ $\sin A = \cos B$, because the triangle is a right triangle.

$\left(\angle C = 90° \text{ or } \angle A + \angle B = 90° \right)$

Definition: In a right triangle above,

1) If $\angle A + \angle B = 90°$, then $\sin \angle A = \cos \angle B$. $\rightarrow$ If $\sin \angle A = \cos \angle B$, then $\angle A + \angle B = 90°$

2) If $\angle A + \angle B = 90°$, then $\csc \angle A = \sec \angle B$. $\rightarrow$ If $\csc \angle A = \sec \angle B$, then $\angle A + \angle B = 90°$

3) If $\angle A + \angle B = 90°$, then $\tan \angle A = \cot \angle B$. $\rightarrow$ If $\tan \angle A = \cot \angle B$, then $\angle A + \angle B = 90°$

SAT Practice

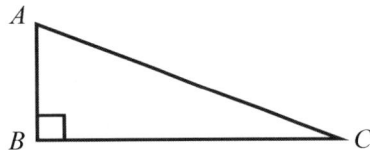

1.   In the figure above, the length of $\overline{AB}$ is 10. If the value of $\sin \angle ACB$ is 0.4, what is the length of $\overline{AC}$?

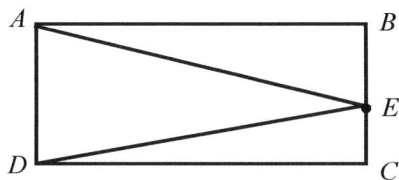

2. In rectangle $ABCD$ above, the value of $\tan \angle BAE = 0.6$ and the value of $\tan \angle EDC$ is 0.25. What is the value of $\dfrac{BE}{EC}$?

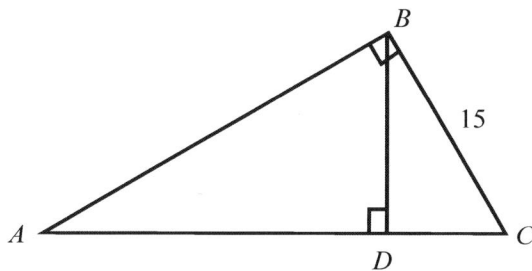

3. In the figure above, the length of $\overline{BC}$ is 15 and the value of $\cos \angle BAC$ is 0.8. What is the length of $\overline{BD}$?

A) 8

B) 10

C) 12

D) 13

4. If $\sin(3x+5)^\circ = \cos(2x-15)^\circ$ in a right triangle, what is the value of x?

Tip 54 | Probability

The probability of an event is the number of ways that the event can occur, divided by the total number of possible outcomes. The symbolic form is

$$P(E) = \frac{n(E)}{n(S)}.$$

1) $P(E)$ represents the probability of event E;

2) $n(E)$ represents the number of ways that event E can occur;

3) $n(S)$ represents the number of possible outcomes in sample space S.

Example 1:
A bag contains 3 red marbles and 4 blue marbles. What is the probability that you select one red marble and one blue, at random, from the bag?

Solution)
There are two different selections.
         RB or BR
Therefore,

$$P(\text{R and B}) = \frac{3}{7} \times \frac{4}{6} = \frac{12}{42}, \ P(\text{B and R}) = \frac{4}{7} \times \frac{3}{6} = \frac{12}{42}$$

The answer is $\frac{12}{42} + \frac{12}{42} = \frac{24}{42} = \frac{4}{7}.$

SAT Practice

1.   A bag contains 3 red marbles and 3 blue marbles. What is the probability that you draw two red marbles without replacement?

   A) $\frac{1}{9}$

   B) $\frac{1}{6}$

   C) $\frac{1}{5}$

   D) $\frac{1}{3}$

TIPS

$$\boxed{3}\ \boxed{5}\ \boxed{9}$$

2. The three cards shown above were taken from a box of ten cards, each with a different integer on it from 1 to 10. What is the probability that the next two cards selected from the box will have both even integer on it?

A) $\dfrac{10}{21}$

B) $\dfrac{12}{23}$

C) $\dfrac{4}{7}$

D) $\dfrac{5}{7}$

3. In a box, there are b blue marbles and g green marbles. If a person selects two marbles, what is the probability that both marbles are blue?

A) $\dfrac{b}{b+g}$

B) $\dfrac{b}{b+g+1}$

C) $\dfrac{b \times b}{(b+g)(b+g-1)}$

D) $\dfrac{b(b-1)}{(b+g)(b+g-1)}$

4. If a number is chosen at random from the set $\{-15,\ -10,\ -5,\ 0,\ 5,\ 10,\ 15,\ 20\}$, what is the probability that it is a member of the solution set of $|x+2|<8$?

(A) 0

(B) $\dfrac{1}{4}$

(C) $\dfrac{3}{8}$

(D) $\dfrac{1}{2}$

Tip 55 | Geometric Probability

"Geometric probability" is the probability dealing with the areas of regions instead of the "number" of outcomes. The equation becomes

$$\text{Probability} = \frac{\text{Favorable region}}{\text{Area of total region}}$$

A typical problem might be this: If you are throwing a dart at the rectangular target below and are equally likely to hit any point on the target, what is the probability that you will hit the small square?

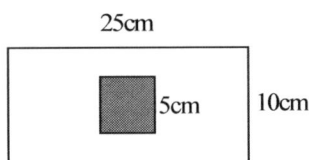

25cm

5cm   10cm

$$\text{Probablity} = \frac{\text{favorable}}{\text{total}} = \frac{25\text{cm}^2}{250\text{cm}^2} = \frac{1}{10}$$

This means that there is a 1 in 10 chance that a dart thrown at the rectangle will hit the small square.

SAT Practice

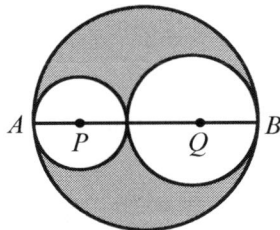

1. In the figure above, the radius of the circle P is 2, the radius of the circle Q is 4, and $\overline{AB}$ is the diameter of the largest circle. If a dart is thrown at the circular target and is equally likely to hit any point on the target, what is the probability that the dart will hit the shaded region?

A) $\dfrac{2}{9}$

B) $\dfrac{1}{3}$

C) $\dfrac{4}{9}$

D) $\dfrac{5}{9}$

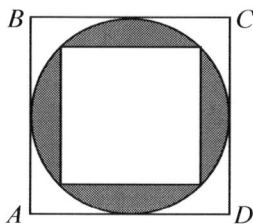

2. In the figure above, a circle is inside of and outside of a square. If a point is chosen at random from square $ABCD$, what is the probability that the point is chosen from the shaded region?

A) $\dfrac{1}{4}$

B) $\dfrac{2\pi - 50}{100}$

C) $\dfrac{\pi - 2}{8}$

D) $\dfrac{\pi - 2}{4}$

Tip 56 | Data Interpretation

Data interpretation problems usually require two basic steps. First, you have to read a chart or graph in order to obtain certain information. Then, second, you have to apply or manipulate the information in order to obtain an answer. Be sure to read all notes related to the data.

GEOMETRY TEST RESULTS
FOR 10 STUDENTS

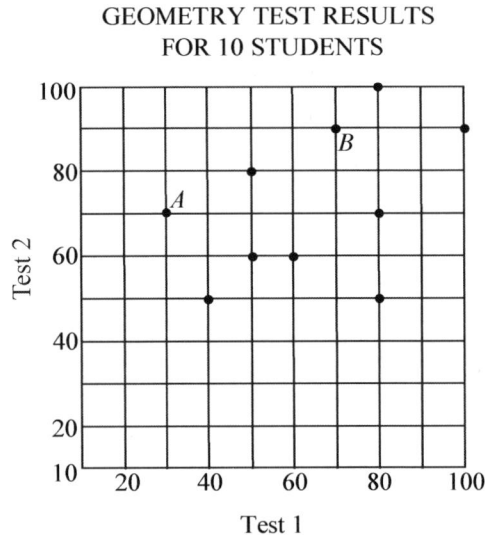

In the scatter plot above, student A got 30 on test 1 and 70 on test 2. Student B got 70 on test 1 and 90 on test 2.

1) What is the median score on test 1 and 2?

Sol) On test 1
30, 40, 50, 50, 60, 70, 80, 80, 80, 100
The median is $\dfrac{60+70}{2} = 65$.

Sol) On test 2
50, 50, 60, 60, 70, 70, 80, 90, 90, 100
The median is 70.

2) What is the average (arithmetic mean) on test 1?

Sol) The average is
$$\dfrac{30+40+50+50+60+70+80+80+80+100}{10} = 64.$$

COST VS. WEIGHT
FOR 10 MEATS

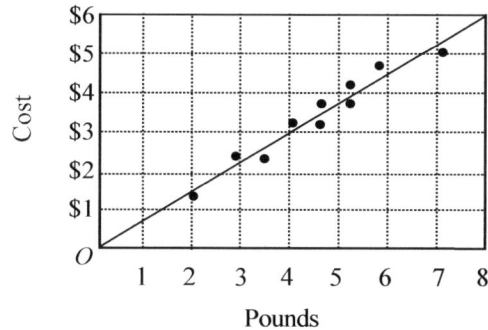

1. For 10 meats of different weights, the cost and weight of each are displayed in the scatter plot above, and the line of best fit for the data is shown. Which of the following is closest to the average (arithmetic mean) cost per pound for the 10 meats?

A) $0.18   B) $0.56   C) $0.62   D) $0.73

ITEMS PURCHASED
BY EACH CUSTOMER

Numbers of Customers	Number of Items
10	10
25	8
45	5
50	Fewer than 5

2. The table above shows the number of items 130 customers purchased from a stationery store during on Sunday. Which of the following can be obtained from the information in the table?

I. The average (arithmetic mean) number
   of items
II. The median number of items.
III. The mode of the number of items.

A) I only
B) II only
C) III only
D) I and II only

Tip 57 | Asymptotes

In analytic geometry, an **asymptote** of a curve is a line such that the distance between the curve and the line approaches zero as they tend to infinity.

Finding asymptotes

1) **Method 1**

For a rational function $f(x) = \dfrac{3}{x-1}$
$\begin{cases} \text{vertical asymptote: Denominator } x-1=0 \ \rightarrow \ x=1 \\ \text{Horizontal asymptote}: \ y = \lim\limits_{x \to \infty} \dfrac{3}{x-1} = 0 \end{cases}$

For a rational function $f(x) = \dfrac{2x+3}{x-1}$
$\begin{cases} \text{vertical asymptote: Denominator } x-1=0 \ \rightarrow \ x=1 \\ \text{Horizontal asymptote}: \ y = \lim\limits_{x \to \infty} \dfrac{2x+3}{x-1} = 2 \end{cases}$

2) **Method 2:** Degree of numerator should be less than degree of denominator

For a rational function $f(x) = \dfrac{3}{x-1} \ \rightarrow \ \boxed{y=0} + \boxed{\dfrac{3}{x-1}}$

Horizontal asymptote: $y = 0$ Vertical asymptote: $x = 1$

Example

1. Find all asymptotes of the graph of $y = \dfrac{x+3}{x-1}$.

Solution) Using long division: $y = \dfrac{x+3}{x-1} \ \rightarrow \ y = 1 + \dfrac{4}{x-1}$

$$\boxed{y = 1} \ + \ \boxed{\dfrac{4}{x-1}}$$

Horizontal asymptote: $y = 1$ Vertical asymptote: $x = 1$

SAT Practice

1. Find all vertical asymptotes of the graph of $y = \dfrac{x-6}{x+1}$.

   A) $x = 1$ B) $x = -1$ C) $x = -1, x = 6$ D) $x = 6$

---

2. Which of the following represents all asymptotes of the graph of $f(x) = \dfrac{x^2 - x - 2}{x^2 + x - 6}$?

   A) $x = 2, y = 1$ B) $x = -3, y = 1$ C) $x = 2, x = -3, y = 1$ D) $x = -2, x = 3, y = 1$

TIPS

Tip 58 | Linear Correlation Coefficient

The linear correlation coefficient, r, measures the **strength** and the **direction** of a linear relationship between two variables.

$$-1 \le r \le 1$$

1) Positive correlation: If variables x and y have a strong positive linear correlation, $0.8 \le r \le 1$.

   For x increases, y also increases.

2) Negative correlation: If variables x and y have a strong negative correlation, $-1 \le r \le -0.8$.

   For x increases, y decreases.

3) No correlation: If there is no linear correlation or a weak linear correlation, r is close to 0.

   Nonlinear relationship between two variables.

4) Perfect correlation: The data points all lie exactly on a straight line. $r = 1$ or $r = -1$.

The strength of the correlation:
1) $-1 \le r \le -0.8$ $\rightarrow$ strong negative correlation
2) $0.8 \le r \le 1$ $\rightarrow$ strong positive correlation
3) $-0.8 < r < 0.8$ $\rightarrow$ weak correlation

SAT Practice

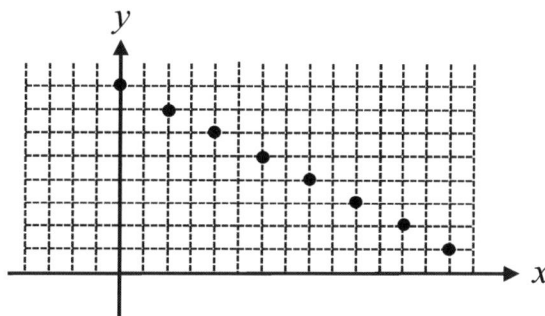

1. In the scatterplot above, which of the following best represents the correlation coefficient between two variables?

    A) -1 B) -0.9 C) 0.9 D) 1

---

2. Which of the following graphs best represents a strong positive association between x and y?

    A)   B)   C)   D)

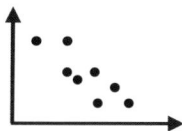 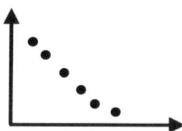 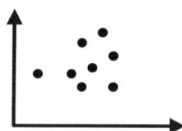 

TIPS

Tip 59 | Standard Deviation

Standard deviation is **a measure of dispersion** of a set of data values from the mean.

Remember: "The greater standard deviation, the greater the spread of data values from the mean"

1) A standard deviation close to 0: The data points tend to be very close to the mean.
2) A high standard deviation: The data points are spread out over a wider range of values.

Example: (Use your graphing calculator and check each standard deviation)

For each data set: $\{10, 10, 10, 10, 10, 10, 10\}$ → Standard deviation is 0.

$\{8, 9, 10, 10, 10, 9, 8\}$ → Standard deviation is 0.83.

$\{7, 8, 9, 10, 9, 8, 7\}$ → Standard deviation is 1.03.

SAT Practice

Questions 1 and 2 refer to the following information.

Data Set 1: $\{1, 1, 1, 3, 3, 3, 3, 5, 6, 8\}$

Data Set 2: $\{2, 2, 3, 3, 3, 4, 4, 4, 5, 5\}$

Data Set 3: $\{3, 3, 3, 3, 3, 4, 4, 4, 4, 4\}$

1.   From the data sets above, which data set appears to have the largest standard deviation?

A) Data set 1
B) Data set 2
C) Data set 3
D) It cannot be determined from the information given.

2.   Which data set appears to have the smallest standard deviation?

A) Data set 1
B) Data set 2
C) Data set 3
D) It cannot be determined from the information given.

Tip 60   Solid

Rectangular solid:

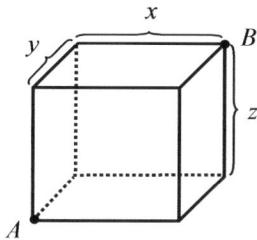

▲ Surface area $= 2(xy + yz + zx)$ ▲ Volume $= xyz$ ▲ Length of diagonal $AB = \sqrt{x^2 + y^2 + z^2}$

Cylinder:

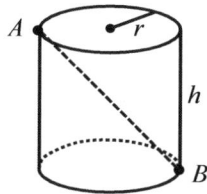

▲ Surface area $= 2\pi r^2 + 2\pi rh = 2\pi r(r + h)$ ▲ Volume $= \pi r^2 h$ ▲ Length of $\overline{AB} = \sqrt{(2r)^2 + h^2}$

Cone:

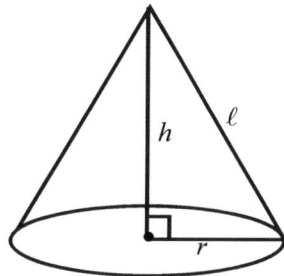

▲ Surface area $= \pi r^2 + \pi r\ell$ ▲ Lateral area $= \pi r\ell$ ▲ volume $= \dfrac{1}{3}\left(\pi r^2 h\right)$

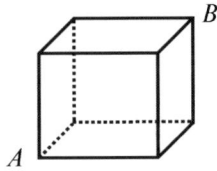

1. In the figure above, if the volume of the cube is 64, what is the length of $\overline{AB}$ (not shown)?

    A) 4   B) $4\sqrt{2}$ C) $4\sqrt{3}$ D) 8

---

2. If the surface area of a cube is 96, what is the volume of the cube?

    A) 8
    B) 27
    C) 64
    D) 81

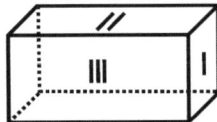

3. In the rectangular solid above, the area of region I (side) is 8, the area of region II (top) is 10, and the area of region III (front) is 20. What is the volume of the solid?

    A) 40
    B) 60
    C) 80
    D) 100

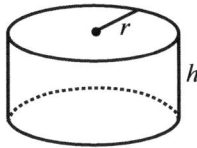

4. The cylinder shown above has a radius of r and a height of h. If $r = h$, what is the surface area of the cylinder?

A) $2\pi r^2$

B) $2\pi r^3$

C) $4\pi r^2$

D) $4\pi r^3$

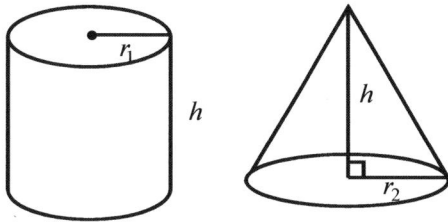

5. In the figure above, the volume of the cylinder is equal to the volume of the cone. What is the value of $\dfrac{r_2}{r_1}$?

A) $\sqrt{2}$

B) $\sqrt{3}$

C) 3

D) 9

No Test Material on This Page

Dr. John Chung's
SAT Math

60 TIPS
Answers and Explanations

Tips Answer

1. C

   Three points $(0, 2), (3, 5), (5, k)$ $\rightarrow$ slope $=\dfrac{5-2}{3-0}=\dfrac{k-5}{5-3}$ $\rightarrow$ $1=\dfrac{k-5}{2}$ $\rightarrow$ $k=7$

2. A

   Slope between two points is constant. $\rightarrow$ $\dfrac{12-a}{1-0}=\dfrac{b-12}{2-1}$ $\rightarrow$ $12-a=b-12$ $\rightarrow$ $a+b=24$

3. D

   Since $ax+by+c=0$ $\rightarrow$ $y=-\dfrac{a}{b}x-\dfrac{c}{b}$. Slope $=-\dfrac{a}{b}>0$ and y-intercept $=-\dfrac{c}{b}>0$

4. D

   Slope $=\dfrac{6-2}{5-3}=2$ $\rightarrow$ $y=2x+b$ $\rightarrow$ putting $(3, 2)$ or $(5,6)$ in the equation $\rightarrow$ $6=2(5)+b$ $\rightarrow$ $b=-4$

5. C

   Method1) $m=\dfrac{-4-(-2)}{4-3}=-2$ $\rightarrow$ $y=-2x+b$ $\rightarrow$ putting $(3,-2)$ in the equation $\rightarrow$ $y=-2x+4$

          For x-intercept, $0=-2x+4$ $\rightarrow$ $x=2$

   Method2) Constant slope: $-2=\dfrac{0-(-2)}{x-3}$ $\rightarrow$ $-2=\dfrac{2}{x-3}$ $\rightarrow$ $x=2$, where $(x,0)$ is the x-intercept.

6. C

   Slope $=\dfrac{3-0}{0-5}=-\dfrac{3}{5}$ $\rightarrow$ slope between $(a,2a)$ and $(0,3)=\dfrac{2a-3}{a-0}=-\dfrac{3}{5}$ $\rightarrow$ $a=\dfrac{15}{13}$

7. B

   Slope is -2 and y-intercept is 4.

8. A

   Since it is a linear equation, slope $=\dfrac{\Delta F}{\Delta C}=\dfrac{9}{5}$ $\rightarrow$ $\dfrac{27}{\Delta C}=\dfrac{9}{5}$ $\rightarrow$ $\Delta C=15$.

9. B

   $\dfrac{\Delta P}{\Delta K}=\dfrac{7}{12}$ $\rightarrow$ $\dfrac{35}{\Delta k}=\dfrac{7}{12}$ $\rightarrow$ $\Delta k=60$

10. C

    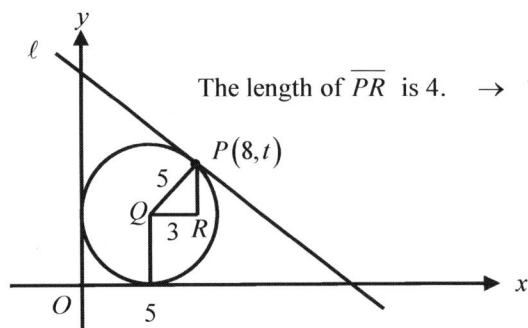

    The length of $\overline{PR}$ is 4. $\rightarrow$ Therefore, $t=5+4=9$.

Tips Answer

Tip 02 — Slope of a Line

1. B

$$\text{Slope} = \frac{12-6}{5-3} = 3$$

2. C

$$(x,0) = x\text{-intercept} \;\rightarrow\; \text{slope} = \frac{0-2}{x-2} = -\frac{1}{2} \;\rightarrow\; x = 6$$

3. C

$$\text{Slope} = \frac{23-5}{8-2} = 3 \;\rightarrow\; \frac{a-5}{4-2} = 3 \;\rightarrow\; a = 11 \;\rightarrow\; \frac{b-5}{11-2} = 3 \;\rightarrow\; b = 32$$

4. C

$$\text{Slope} = \frac{6-a}{5-2} = \frac{b-6}{8-5} \;\rightarrow\; 6-a = b-6 \;\rightarrow\; a+b = 12$$

5. B

$$\text{Slope} = \frac{m-2}{42-0} = \frac{2}{3} \;\rightarrow\; m = 30$$

Tip 03 — Average Rate of Change

1. A

$$\text{Average rate of change} = \frac{f(4)-f(0)}{4-0} \qquad f(4) = 4 \text{ and } f(0) = -4$$

$$\text{Average rate of change} = \frac{4-(-4)}{4-0} = \frac{8}{4} = 2$$

2. C

$$\text{Average speed} = \frac{d(3)-d(1)}{3-1} = \frac{44.1-4.9}{2} = 19.6$$

3. .595

$$\text{Average rate of change} = \frac{h(70)-h(10)}{70-10} = \frac{39.2-3.5}{60} \approx .595 \text{ inches per day}$$

4. B

From the graph you can see that the average rate of change is greatest from 20 to 30 days.

$$\text{From 20 to 30 days:} \quad \frac{19.8-7.4}{10} = 1.24 \text{ inches per day}$$

$$\text{From 30 to 40 days:} \quad \frac{29.8-19.8}{10} = 1 \text{ inches per day}$$

Tip 04 — Area enclosed by Lines

1. B

$$OP = 4 \text{ and } OR = -\frac{4}{m} \;\rightarrow\; \text{area} = \frac{1}{2}(4)\left(-\frac{4}{m}\right) = 6 \;\rightarrow\; m = -\frac{4}{3}$$

Or

$$\text{Area} = \frac{4 \times OR}{2} = 6 \quad \rightarrow \quad OR = 3 \quad \rightarrow \quad \text{slope} \, m = -\frac{4}{3}$$

2. D

$$\text{Slope of line } \ell = \frac{6-0}{2-(-4)} = 1 \quad \rightarrow \quad \text{slope of line } m = -1$$

The equation of line m is $y = -x + b \quad \rightarrow \quad$ putting $(2,6)$ in the equation $\quad \rightarrow \quad y = -x + 8$

x-intercept is $(8,0)$. $PQ = 8-(-4) = 12$ and height $PS = 6 \quad \rightarrow \quad area = \frac{12 \times 6}{2} = 36$

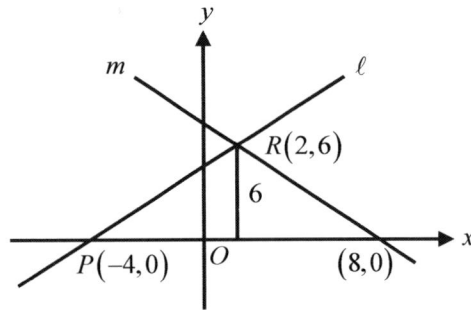

3. A

$$SR = 10 \text{ and } UT = 5 \quad \rightarrow \quad \text{Area} = \frac{10 \times 5}{2} = 25$$

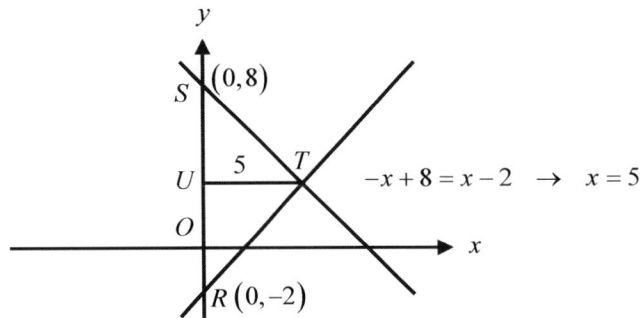

$$-x + 8 = x - 2 \quad \rightarrow \quad x = 5$$

Tip 05	Midpoint and Distance between Two Points

1. C

Let $B(a,b) \quad \rightarrow \quad \left(\frac{a+5}{2}, \frac{b+1}{2}\right) = (10, 4) \quad \rightarrow \quad a = 15 \text{ and } b = 7$

2. C

$\left(\frac{b+2a}{2}, \frac{a+b}{2}\right) = (5, -3) \quad \rightarrow \quad b+2a = 10 \text{ and } a+b = -6 \quad \rightarrow \quad a = 16$

3. B

Since $AB = 4 - (-4) = 8, \quad \rightarrow \quad \frac{8 \times h}{2} = 24 \quad \rightarrow \quad$ The height of the triangle $h = 6$.

Therefore, the y-coordinate of point C must be $4 + 6 = 10$ or $4 - 6 = -2$.

Tips Answer

4.   C
$$\sqrt{(a-b)^2+(3-8)^2}=13 \rightarrow (a-b)^2+25=169 \rightarrow (a-b)^2=144 \rightarrow |a-b|=12$$

Tip 06	Line Reflection

1.   B

Reflection of $y=\dfrac{1}{5}x-6$ is $-y=\dfrac{1}{5}x-6 \rightarrow y=-\dfrac{1}{5}x+6 \rightarrow$ slope$=-\dfrac{1}{5}$

2.   B

The point of intersection must be on the y-axis. Therefore, The x-coordinate is 0.

3.   B

$r_{x\text{-axis}}(x,y)=(x,-y) \rightarrow$ Just replace y with $(-y) \rightarrow 2x-3(-y)=6 \rightarrow 2x+3y=6$

Tip 07	Parallel and Perpendicular Lines

1.   B

From $y=2x-\dfrac{3}{2} \rightarrow$ slope$=2 \rightarrow$ The parallel line must be $y=2x+b \rightarrow$

Putting $(-4,1)$ in the equation $\rightarrow y=2x+9$

2.   A

$y=-\dfrac{1}{2}x+b \rightarrow$ putting $(-4,1)$ in the equation $\rightarrow y=-\dfrac{1}{2}x-1$

Tip 08	System of Linear Equations

1.   D

$\dfrac{2}{4}=\dfrac{-5}{k}\neq\dfrac{8}{17} \rightarrow 2k=-20 \rightarrow k=-10$

2.   A

$\dfrac{5}{a}=\dfrac{-2}{b}=\dfrac{3}{6} \rightarrow a=10$ and $b=-4 \rightarrow a+b=6$

3.   D

$\dfrac{3}{a}=\dfrac{b}{-4}\neq\dfrac{3}{6} \rightarrow ab=-12 \rightarrow \{3,-4\}$

4.   C

$\dfrac{a}{a-1}=\dfrac{3}{a-1}\neq\dfrac{6}{2} \rightarrow \dfrac{a}{a-1}=\dfrac{3}{a-1} \rightarrow a=3 \rightarrow$ when $a=3$, $\dfrac{3}{2}\neq3$.

5.   C

$a+10b=3.5$ and $a+15b=4.75 \rightarrow a=1$ and $b=0.25 \rightarrow$ For 40 minutes call, $1+(40-5)\times0.25=\$9.75$

6.   120

The number of adult ticket $=x \rightarrow 8x+5(200-x)=1360 \rightarrow x=120$

Tips Answer

Tip 09	**Quadratic Function**

1. A

$$t = \frac{-256}{2(-16)} = 8$$

2. C

$$h_{max} = 256(8) - 16(8^2) = 1025$$

3. B

Axis of symmetry $h = \dfrac{-2+10}{2} = 4 \;\rightarrow\; k = f(4) = \dfrac{1}{2}(4+2)(4-10) = -18$

4. B

$$y = 3(x^2 - 2x + 1) + 10 - 3 \;\rightarrow\; y = 3(x-1)^2 + 7$$

5. D

The graph opens downward $\rightarrow a < 0$, axis of symmetry $= \dfrac{-b}{2a} > 0 \;\rightarrow\; b > 0$, y-intercept: $f0) = c > 0$

6. D

Since $a < 0$ and $k < 0$, then $f(x) < 0$ for all values of x. But D) $f(0) = -k > 0$

Tip 10	**System of Linear and Quadratic Equations**

1. 10.5

Find the intersections of the graphs. $x^2 - 4x + 5 = x + 1 \;\rightarrow\; x^2 - 5x + 4 = 0 \;\rightarrow\; (x-1)(x-4) = 0$

$x = 1, 4 \;\rightarrow\; \begin{cases} x = 1 \rightarrow y = 2 \\ x = 4 \rightarrow y = 5 \end{cases}$ Therefore, $AB = 2, CD = 5, AD = 4 - 1 = 3$.

Area of trapezoid $ABCD = \dfrac{(2+5)3}{2} = 10.5$

2. C

Find the discriminant. $x^2 - x - 12 = 3x \;\rightarrow\; x^2 - 4x - 12 = 0 \;\rightarrow\; D = (-4)^2 - 4(1)(-12) = 64$

Since $D = 64 > 0$, the system has two solutions.

Tip 11	**System of Linear and Quadratic Equations**

1. C

Solid line: $y = x - 3$ and dotted line: $y = -2x$

For the shaded region, $y \geq x - 3$ and $y < -2x \;\rightarrow\; x - y \leq 3$ and $2x + y < 0$

2. **B**

$$\begin{cases} x+2y \le 4 \\ 2x-y \le 0 \end{cases} \rightarrow \begin{cases} y \le -\dfrac{1}{2}x+2 \\ y \ge 2x \end{cases}$$

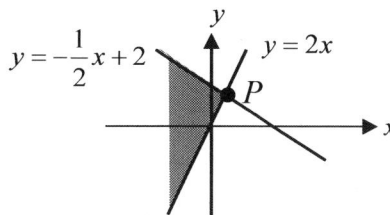

$y = -\dfrac{1}{2}x+2$ $y = 2x$ P

At point P, it has the maximum value of b.

Find the intersection: $2x = -\dfrac{1}{2}x+2 \rightarrow \dfrac{5}{2}x = 2 \rightarrow x = \dfrac{4}{5} \rightarrow$ Therefore, $a = \dfrac{4}{5}$.

Tip 12 — System of Linear and Quadratic Inequalities

1. **C**

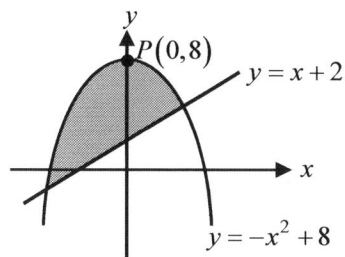

$P(0,8)$ $y = x+2$ $y = -x^2+8$

The shaded region shows the maximum value of b occurs
At point P. therefore, the maximum value of b is 8.

Tip 13 — Area enclosed by Curves

1. **16**

x-intercept: $0 = -x^2+k \rightarrow x = k$ and $-k \rightarrow PR = 2\sqrt{k}$ and $OQ = k$

Area of $\triangle PQR = \dfrac{k\left(2\sqrt{k}\right)}{2} = 64 \rightarrow k\sqrt{k} = 64 \rightarrow \sqrt{k}\sqrt{k}\sqrt{k} = 64 \rightarrow \sqrt{k} = 4 \rightarrow k = 16$

2. **128**

$x^2 = -x^2+32 \rightarrow x = \pm 4 \rightarrow AD = 8$ and $CD = f(4) = 16 \rightarrow$ Area is $8 \times 16 = 128$.

3. **25**

$OP = f(0) = 5 \rightarrow -x^2+4x+5 = 0 \rightarrow (x-5)(x+1) = 0 \rightarrow x = 5, -1 \rightarrow OR = 5$

Therefore, the area of $OPQR = 5 \times 5 = 25$.

4. **C**

The intersection of the two graphs: $\dfrac{1}{8}x(x-2) = \dfrac{1}{2}x \rightarrow x(x-2) = 4x \rightarrow x(x-6) = 0 \rightarrow$

$x = 0$ and $6 \rightarrow OQ = 6$ and $PQ = 3 \rightarrow$ Therefore, the area is $\dfrac{6 \times 3}{2} = 9$.

Tip 14 — Domain and Range

1. **D**
2. **D**

Tips Answer

1. B

$$g(x) = 3f(x) - k = 3(5x + 3) - k \quad \rightarrow \quad g(2) = 39 - k = 25 \quad \rightarrow \quad k = 14$$

2. C

$$g(x) = 2(ax + b) - 3 \quad \rightarrow \quad g(1) = 2a + 2b - 3 = 3 \text{ and } g(3) = 6a + 2b - 3 = 5 \quad \rightarrow \quad a = \frac{1}{2} \text{ and } b = \frac{5}{2}$$

3. C

$$g(k) = 3f(k) - 1 = 8 \quad \rightarrow \quad f(k) = 3 \quad \rightarrow \quad \text{There are three values of } k \text{ whose } y\text{-coordinate is 3.}$$

1. $\dfrac{5}{2}$

Let $X = x - 1$, then denominator $4X^2 - 12X + 9 = 0$. $4X^2 - 12X + 9 = (2X - 3)^2 = 0 \quad \rightarrow \quad X = \dfrac{3}{2}$

Now $x - 1 = \dfrac{3}{2} \quad \rightarrow \quad x = \dfrac{5}{2}$

2. C

Since the function f is undefined at $x = 1$ and $x = 3$, $x^2 - ax + b = (x - 1)(x - 3)$.

$x^2 - ax + b = (x - 1)(x - 3) \quad \rightarrow \quad x^2 - ax + b = x^2 - 4x + 3$ Therefore, $b = 3$.

1. B

$$x \times 0 = 0$$

2. A

Both sides have the same expressions: $ax^2 + bx + c = 0x^2 + 0x + 0 \quad \rightarrow \quad a = 0, b = 0, c = 0$

3. A

$k + 1 = a$ and $k = 5 \quad \rightarrow \quad a = 6$

4. C

$$a(x + 1) + b(x - 1) = 2x + 4 \quad \rightarrow \quad (a + b)x + a - b = 2x + 4 \quad \rightarrow \quad a + b = 2 \text{ and } a - b = 4$$

$a + 3$ and $b = -1$

1. A

Odd function is symmetric with respect to the origin.

2. D

Even function is symmetric with respect to the y-axis.

Tips Answer

1. A
2. B

$$\left(n-\frac{1}{n}\right)^2 + 4 = n^2 - 2 + \frac{1}{n^2} + 4 = n^2 + 2 + \frac{1}{n^2} = \left(n+\frac{1}{n}\right)^2$$

3. D

$$0 = -x^2 + 4 \rightarrow x = 2, -2 \rightarrow AC = 2 - (-2) = 4$$

$$x + 2 = -x^2 + 4 \rightarrow (x+2)(x-1) = 0 \rightarrow x = -2, 1 \rightarrow B(1, 3) \rightarrow \text{The height of } \triangle ABC \text{ is 3.}$$

Therefore, area of $\triangle ABC = \dfrac{3 \times 4}{2} = 6$.

4. 7 or 5

$$x^2 - y^2 = 24 \rightarrow (x+y)(x-y) = 24 \rightarrow x+y=12 \text{ and } x-y=2 \text{ ,or } x+y=6 \text{ and } x-y=4$$

Therefore, from the system of equations $x = 7$ or 5.

1. 37.5

$$\frac{15}{5} = \frac{y}{12.5} \rightarrow y = 37.5$$

2. D

$$\frac{10}{20} = \frac{300}{x} \rightarrow x = 600$$

3. B

$$\begin{cases} \text{Orange dye: } 5R+3Y \\ \text{Green dye: } 4B+2Y \end{cases} \rightarrow \text{equal amount (24 parts)} \rightarrow \begin{cases} 15R+9Y \\ 16B+8Y \end{cases} \rightarrow \frac{17Y}{48\,\text{parts}}$$

4. C

Since $AB = 13$, $\rightarrow \dfrac{13}{18} = \dfrac{5}{x} \rightarrow x = \dfrac{90}{13}$

5. C

$$\frac{1}{a} = \frac{a}{5a} \rightarrow a = 5$$

6. C
7. C

The vertex must be at the origin.

1. 16

$$1 \times 400 = 25x \rightarrow x = 16$$

2. A

$$4 \times 9 = 12 \times x \rightarrow x = 3$$

3. B

$5 \times 45 = 50 \times x \rightarrow x = 4.5$ or 4 hours 30 minutes

4. C

$4a = 8b = 50 \rightarrow a = 12.5$ and $b = 6.25 \rightarrow a + b = 18.75$

5. D

$2 \times 10 = x(5) \rightarrow x = 4$

6. B

$10 \cdot 20 = x \cdot 40 \rightarrow x = 5$

7. C

$ph = nx \rightarrow x = \dfrac{ph}{n}$

8. C

$5d = 2x \rightarrow x = \dfrac{5d}{2} \rightarrow$ For $\dfrac{1}{3}$ of the job, $\dfrac{1}{3} \times \left(\dfrac{5d}{2} \right) = \dfrac{5d}{6}$

Tip 22   Sum and Product of the roots

1. B

Method1): Using sum of the roots $\rightarrow 1 + r = -\dfrac{1}{2} \rightarrow r = -\dfrac{3}{2}$

Method2): Using Factoring $\rightarrow 2 + 1 - k = 0 \rightarrow k = 3 \rightarrow 2x^2 + x - 3 = 0 \rightarrow (2x + 3)(x - 1) = 0$

The other root is $x = -\dfrac{3}{2}$.

2. B

$x^2 + 4x - 12 = 0 \rightarrow (x + 6)(x - 2) = 0 \rightarrow x = -6, x = 2 \rightarrow \dfrac{1}{\alpha} + \dfrac{1}{\beta} = \dfrac{1}{-6} + \dfrac{1}{2} = \dfrac{1}{3}$

Or, using sum and product: $\dfrac{1}{\alpha} + \dfrac{1}{\beta} = \dfrac{\alpha + \beta}{\alpha\beta} = \dfrac{-4}{-12} = \dfrac{1}{3}$, because $\alpha + \beta = -4$ and $\alpha\beta = -12$.

Tip 23   Remainder Theorem

1. 7

$f(3) = 3^2 + 3(3) + k = 25 \rightarrow k = 7$

2. B

$f(-3) = -27 - 9 + 9 - 1 = -28$

3. 3

$f(1) = 1 - 5 + 1 + k = 0 \rightarrow k = 3$

Tip 24   Factor Theorem

1. A

$f(3) = 27 - 12 + k = 0 \rightarrow k = -15$

2. 2

$P(2) = 4 + 2k - 8 = 0 \rightarrow k = 2$

3. D

$$P(-1) = (-1)^3 - 5(-1)^2 - (-1) + k = 0 \rightarrow k = 5$$

4. A

Since $g(x) = x^3 + ax + b = (x-1)(x-2)Q(x)$, $g(1) = 0$ and $g(2) = 0$.

Therefore, $1 + a + b = 0$ and $8 + 2a + b = 0 \rightarrow a = -7$ and $b = 6$.

Tip 25   Circle

1. C

$$x^2 - 4x + y^2 + 2y = 11 \rightarrow (x-2)^2 + (y+1)^2 = 16 \rightarrow \text{Area} = \pi r^2 = 16\pi$$

2. A

$$x^2 + y^2 - 6y = 16 \rightarrow x^2 + (y-3)^2 = 25 \rightarrow r = 5 \rightarrow C = 2\pi r = 10\pi$$

3. 9

$$k = 15 - 6 = 9$$

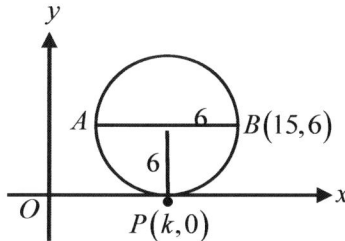

4. C

Axis of symmetry $x = \dfrac{-2+8}{2} = 3 \rightarrow$ C) $\dfrac{1+5}{2} = 3$

5. B

Radius of the circle is 7. Therefore, $4 - 7 = -3$.

Tip 26   Average Speed

1. C

$$d_1 = 50 \times 1 = 50 \text{ and } d_2 = 60 \times 3 = 180 \rightarrow d = d_1 + d_2 = 230 \rightarrow \text{Average speed} = \dfrac{230}{1+3} = 57.5$$

2. A

Use 400 for the distance. $\rightarrow \dfrac{200}{40} = 5$ hours and $\dfrac{200}{5} = 4$ hours $\rightarrow$ Average speed $= \dfrac{400}{5+4} = 44\dfrac{4}{9}$

3. B

$$\dfrac{d}{60} + \dfrac{d}{40} = 6 \rightarrow 2d + 3d = 720 \rightarrow 5d = 729 \rightarrow d = 144$$

Tip 27   Percentage

1. B

$$\left(\frac{20}{100}\right)\left(\frac{30}{100}\right)x = \left(\frac{10}{100}\right)\left(\frac{k}{100}\right)x \;\rightarrow\; k = 60$$

2. B
$$P = (1+0.15)(1-0.3)x = 0.805x = (1-0.195)x \;\rightarrow\; 19.5\% \text{ decreases.}$$

3. C
$$2a + 3b = 2.5(6b) \;\rightarrow\; a = 6b \;\rightarrow\; \frac{a}{b} = \frac{6b}{b} = 6$$

4. B
$$0.25m = 50 \;\rightarrow\; m = 200 \;\rightarrow\; 0.15(2m) = 0.15(400) = 60$$

5. B
$$C = 20,000(1+0.025)^x = 20,000(1.025)^x$$

6. B
$$\frac{1000-500}{500} \times 100 = 100\%$$

7. B
$$P = (1+0.06)(1-0.05)x = 1.007x = (1+0.07)x \;\rightarrow\; 7\% \text{ increases.}$$

8. D
$$a = 0.25(2b) = 0.5b \;\rightarrow\; b = 2a \;\rightarrow\; b \text{ is } 200\% \text{ of } a.$$

Tip 28	Ratios and Proportion

1. B
$$\frac{6}{1.2} = \frac{15}{x} \;\rightarrow\; x = 3$$

2. C

$$\begin{cases} & \text{Now} & 5\rightarrow & 10\rightarrow \\ \text{Julie} & ? & 3k & 3k+5 \\ \text{Song} & ? & 5k & 5k+5 \end{cases} \;\rightarrow\; \frac{3k+5}{5k+5} = \frac{2}{3} \;\rightarrow\; k = 5$$

Therefore, $3k = 15$ and $5k = 25$ $\rightarrow$ Sum of their current ages is $10 + 20 = 30$.

3. C
$$3x = 2y \text{ and } 2x + 5y = 76 \;\rightarrow\; x = 8 \text{ and } y = 12$$

4. 160 or 320

Since a and b must be multiples of 10,
$$\frac{a}{b} = \frac{7}{9} = \frac{70}{90} = \frac{140}{180} \;\rightarrow\; a+b = 70+90 = 160 \text{ or } a+b = 140+180 = 320.$$

Tip 29	Ratios in Similar Figures

1. 45

Since ratio of the corresponding lengths is 2:3, the ratio of their areas is 4:9.
$$4k = 20 \;\rightarrow\; k = 5 \;\rightarrow\; 9k = 45$$

Tips Answer

2. 4.8

Ratio of their area is $\dfrac{25}{4}$. $\rightarrow$ $25k = 30$ $\rightarrow$ $k = 1.2$ $\rightarrow$ $4k = 4(1.2) = 4.8$

3. C

The ratio of their areas is 25:4. $\rightarrow$ $4k = 28$ $\rightarrow$ $k = 7$ $\rightarrow$ Shaded area $= 21k = 21(7) = 147$

4. B

The ratio of their areas is 16:9:1. $\rightarrow 16k, 9k, k$ $\rightarrow$ Shaded area$6 = 6k = 96\pi$ $\rightarrow$ $k = 16\pi$

$\pi r^2 = 16\pi$ $\rightarrow$ $r = 4$

5. C

$\begin{cases} \text{Length} & 7: \ 5: \ 3 \\ \text{Area} & 49: \ 25: \ 9 \end{cases}$ $\rightarrow$ $49k, 25k, 9k$ $\rightarrow$ Shaded region $16k = 48.$ $\rightarrow$ $k = 3$

Therefore, the area of $\triangle ABC = 49k = 49(3) = 147$

6. A

Ratio of the corresponding lengths $= \sqrt{4} : \sqrt{9} = 2:3$ $\rightarrow$ $2k$ and $3k$

Since $2k = 15$, $3k = 22.5$.

| Tip 30 | Percent of a Solution (Mixture) |

1. C

Water $= x$ $\rightarrow$ The amount of solute are equal: $0.1 \times 40 = 0.08 \times (40 + x)$ $\rightarrow$ $x = 10$

2. D

$0.2x + 0.5(10) = 0.3(x + 10)$ $\rightarrow$ $x = 20$

3. A

Alcohol $= x$ $\rightarrow$ $x + 0.25(10) = 0.4(x + 10)$ $\rightarrow$ $x = 2.5$

4. C

Acid $= x$ $\rightarrow$ $x + \dfrac{k}{100}G = \dfrac{m}{100}(G + x)$ $\rightarrow$ $100x + kG = mG + mx$ $\rightarrow$ $100x - mx = mG - kG$

$x(100 - m) = G(m - k)$ $\rightarrow$ $x = \dfrac{G(m-k)}{100 - m}$

| Tip 31 | Exponents |

1. D

$\left\{ (-2)^3 (8)^2 \right\}^4 = \left(2^4 \right)^n$ $\rightarrow$ $(-2)^{12} (8)^8 = 2^{4n}$ $\rightarrow$ $2^{12} \left(2^{24} \right) = 2^{36} = 2^{4n}$ $\rightarrow$ $36 = 4n$ $\rightarrow$ $n = 9$

2. D

$4^3 + 4^3 + 4^3 + 4^3 = 2^n$ $\rightarrow$ $4(4^3) = 2^n$ $\rightarrow$ $2^2 (2^6) = 2^n$ $\rightarrow$ $2^8 = 2^n$ $\rightarrow$ $n = 8$

3. D

$5m^5 n^{-3} = 20m^3 n$ $\rightarrow$ $\dfrac{m^5}{m^3} = \dfrac{20n}{5n^{-3}}$ $\rightarrow$ $m^2 = 4n^4$ $\rightarrow$ $m = 2n^2$

4. C

$\left(a^{-4} b \right)^{-1} = 16$ $\rightarrow$ $a^4 b^{-1} = 16$ $\rightarrow$ $\dfrac{a^4}{b} = \dfrac{a^4}{a^2} = 16$ $\rightarrow$ $a^2 = 16$ $\rightarrow$ $a = 4$

Tips Answer

5.   D

$$k^{-2} \times 2^3 = 2^7 \rightarrow k^{-2} = \frac{2^7}{2^3} = 2^4 \rightarrow k^2 = \frac{1}{2^4} \rightarrow k = \frac{1}{4}$$

6.   A

$$p^{-3} = 2^{-6} \rightarrow \left(p^{-3}\right)^{\left(-\frac{1}{3}\right)} = \left(2^{-6}\right)^{\left(-\frac{1}{3}\right)} \rightarrow p = 2^2 = 4$$

$$q^{-2} = 4^2 \rightarrow \left(q^{-2}\right)^{\left(-\frac{1}{2}\right)} = \left(4^2\right)^{\left(-\frac{1}{2}\right)} \rightarrow q = 4^{-1} = \frac{1}{4} \rightarrow \text{Therefore, } pq = 4\left(\frac{1}{4}\right) = 1.$$

7.   D

$$\left(a^6 b^4\right)^{\frac{1}{2}} = 675 \rightarrow a^3 b^2 = 3^3 \times 5^2 \text{ (prime factorization)} \rightarrow a = 3 \text{ and } b = 5 \rightarrow a + b = 8$$

Tip 32	Defined Operations

1.   D

$$p \odot q = \frac{p+q}{p-q} = 3 \rightarrow p = 2q \rightarrow \frac{p}{q} = 2$$

2.   B

$$4\triangle(k\triangle 6) = 3 \rightarrow 4\triangle\left(\frac{k}{6}\right) = \frac{4}{\frac{k}{6}} = \frac{24}{k} \rightarrow \frac{24}{k} = 8 \rightarrow k = 3$$

3.   D

Tip 33	Functions as Models

1.   B

$$300 = \frac{200(50) - 400}{50} + k \rightarrow 300 = 200 - 8 + k \rightarrow k = 108$$

2.   C

$$p(n) = 1,200\left(1 - 0.012\right)^n = 1,200\left(0.988\right)^n \rightarrow c = 0.988$$

3.   D

$$f(3) = 9 \text{ and } f(-2) = 4 \rightarrow m = \frac{9-4}{3-(-2)} = 1$$

4.   C

$$P = 20,000\left(1 - 0.1\right)^3 = 14,580$$

Tip 34	Combined Rate of Work

1.   D

$$\frac{1}{5} + \frac{1}{x} = \frac{1}{4} \rightarrow \frac{1}{x} = \frac{1}{20} \rightarrow x = 20$$

2. D

$$\begin{cases} \text{Raymond} & x \text{ hours} \\ \text{Peter} & 2x \text{ hours} \end{cases} \rightarrow \frac{1}{x} + \frac{1}{2x} = \frac{1}{20} \rightarrow x = 30 \rightarrow 2x = 60 \text{ hours}$$

3. A

$$\frac{1}{5} + \frac{1}{8} = \frac{1}{x} \rightarrow \frac{1}{x} = \frac{13}{40} \rightarrow x = \frac{40}{13} = 3\frac{1}{13}$$

4. D

$$5 \times 12 = 4 \times x \rightarrow x = 15 \text{ hours for 3 houses} \rightarrow \text{For 5 houses, } 15 \times \frac{5}{3} = 25 \text{ hours.}$$

Tip 35	Combined Range of Two Intervals

1. 7.5

$$\frac{2}{4} \le \frac{P}{Q} \le \frac{8}{1} \rightarrow 8 - 0.5 = 7.5$$

2. 0

$$0 \le A - B \le 8 \rightarrow \text{The smallest value of } C = 0^2 = 0.$$

3. 3

$$3 \le PQ \le 60 \rightarrow \text{Smallest one is 3.}$$

4. D

Tip 36	Absolute Value

1. C

$$\text{midpoint} = \frac{78 + 86}{2} = 82 \text{ and distance} = 86 - 82 = 4 \rightarrow |s - 82| < 4$$

2. D

$$\text{Midpoint} = \frac{1}{2}\left(5\frac{3}{7} + 6\frac{4}{7}\right) = 6 \text{ and distance} = \frac{4}{7}$$

3. D

For $E < 450$ or $E > 500 \rightarrow \text{midpoint} = 475$ and distance $= 25$

Tip 37	Parallel Lines with Transversal

1. D

$$2(a + b) = 180 \rightarrow a + b = 90$$

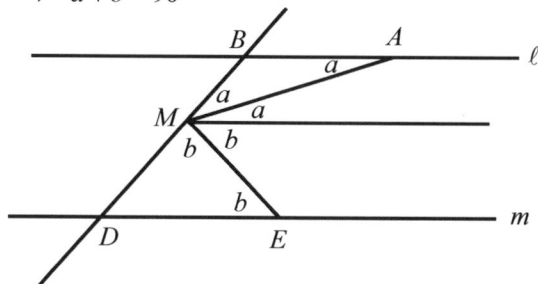

Tips Answer

2. D

$k = 65 + 45 = 110$

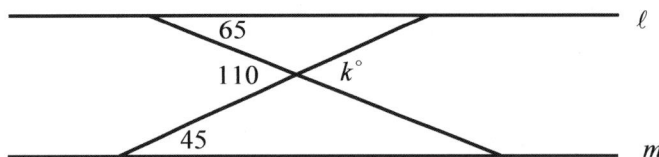

3. D

The other two lines may be not parallel.

Tip 38   Triangle Inequality

1. D

$9 - 3 < x + 3 < 3 + 9 \rightarrow 6 < x + 3 < 12 \rightarrow 3 < x < 9$

2. D

$\begin{cases} 1 < PQ < 7 \text{ and} \\ 2 < PQ < 14 \end{cases} \rightarrow 2 < PQ < 7$

3. B

$3 + 6 = 9$: Triangle inequality theorem fails.

Tip 39   Ratio of Areas of triangles with the same height

1. B

$AC = 20$ and $BE = 5\sqrt{3} \rightarrow$ Area of $\triangle ABC = \dfrac{20 \times 5\sqrt{3}}{2} = 50\sqrt{3}$

2. D

The ratio of areas of $\triangle ADE$ to $\triangle DEC = 3 : 5 \rightarrow 3k$ and $5k \rightarrow 3k = 24 \rightarrow k = 8$

Therefore, the area of $ABCD$ is $16k = 16(8) = 128$.

3. 32

$5k = 40 \rightarrow k = 8$

$4k = 32$

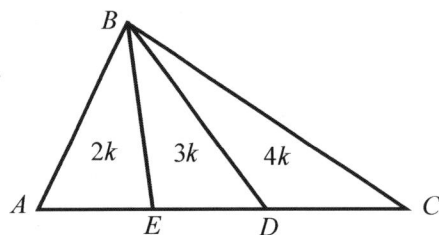

Tip 40   Special right triangles

1. C

The length of a side of the equilateral $\triangle = s \rightarrow$ Area $= \dfrac{s^2\sqrt{3}}{4} = 16\sqrt{3} \rightarrow s = 8$

Therefore, the area of the square is $8 \times 8 = 64$.

2. C

    Perimeter
    $= 30 + \dfrac{20\sqrt{3}}{3}$

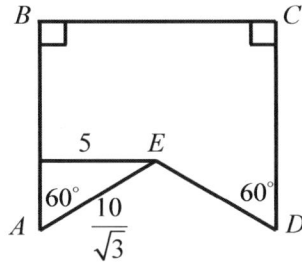

3. B

    Triangle OPQ is equilateral. Area $= \dfrac{s^2\sqrt{3}}{4} = \dfrac{64\sqrt{3}}{4} = 16\sqrt{3}$

4. B

    $AC = 20\sqrt{3}$ and $h = 10$ $\rightarrow$ Area $= \dfrac{20\sqrt{3} \times 10}{2} = 100\sqrt{3}$

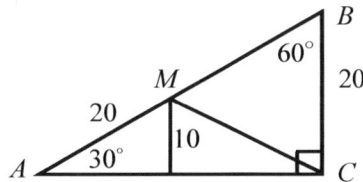

Tip 41	Proportions in a right triangle

1. 9

    $6^2 = 3 \times AD$ $\rightarrow$ $AD = 12$ $\rightarrow$ $CD = 12 - 3 = 9$

2. C

    $AB^2 = 9(9 + 16) = 225$ $\rightarrow$ $AB = 15$

3. B

    $BC^2 = 16(16 + 9)$ $\rightarrow$ $BC = 20$

4. B

    $BD^2 = 9 \times 16 = 144$ $\rightarrow$ $BD = 12$

Tip 42	Pythagorean Theorem

1. B

    $C = 16 + 20 = 36$ $\rightarrow$ $PR^2 = 36$ $\rightarrow$ $PR = 6$

2. 4

    Ratio of the areas of $\triangle ADB$ to $\triangle BCF = 1:4$ $\rightarrow$ $k, 4k$ $\rightarrow$ $5k = 20$ $\rightarrow$ $k = 4$

    The area of $\triangle ADB = k = 4$

3. D

    Since $a^2 + b^2 = 100$, $\left(a + b\right)^2 = a^2 + b^2 + 2ab = 100 + 2ab > 100$

4.   B

$$BC = \sqrt{100 + 400} = 10\sqrt{5} \quad \rightarrow \quad 10 \times 20 = 10\sqrt{5} \times AD \quad \rightarrow \quad AD = 4\sqrt{5}$$

Tip 43	Transformation

1.   A

$$y = -2(x+3) + 5 + 1 \quad \rightarrow \quad y = -2x - 6 + 6 \quad \rightarrow \quad y = -2x$$

2.   B

Tip 44	Classifying a group in two different ways

1.   C

	Males	Females	
Children	12		16
No Children	6	2	8
	18	6	24

Tip 45	Discriminant

1.   B

$$D = (-6)^2 - 4(2)(3) = 12 > 0 \quad \rightarrow \quad \text{Two unequal real roots}$$

2.   C

$$D = 6^2 - 4(1)(k) = 0 \quad \rightarrow \quad k = 9$$

3.   B

$$D = 2^2 - 4 \cdot 1 \cdot k = 0 \quad \rightarrow \quad 4 = 4k \quad \rightarrow \quad k = 1 \quad \text{Discriminant must be 0 for equal roots.}$$

4.   D

$$x^2 - 4x + 8 = x + k \quad \rightarrow \quad x^2 - 5x + 8 - k = 0 \quad \text{must have equal roots since the graphs are tangent each other.}$$

$$D = 25 - 4(8 - k) = 0 \quad \rightarrow \quad k = \frac{7}{4}$$

Tip 46	Handshakes

1.   D

$$_{12}C_2 = \frac{12 \times 11}{2} = 66 \quad \text{Or} \quad 11 + 10 + 9 + 8 + 7 + 6 + 5 + 4 + 3 + 2 + 1 = \frac{(11+1)11}{2} = 66$$

2.   B

$$_nC_2 = \frac{n(n-1)}{2} = 45 \rightarrow n^2 - n - 90 = 0 \rightarrow (n-10)(n+9) = 0 \rightarrow n = 10 \ (n \neq -9)$$

Or add numbers until you get $45.\ 1 + 2 + 3 + 4 + 5 + 6 + 7 + 8 + 9 = 45 \rightarrow$ The number of people $= 9 + 1 = 10$

Tip 47	Consecutive Integers I

1   C

Since the median is the average, sum of the numbers is $30 \times 11 = 330$.

2.   C

$80 + 49 = 129$

3.   B

Since average is 77, sum of the numbers $77 \times 10 = 770$.

4.   C

The 16th term is 121. Therefore, the greatest one is $121 + 2 \times 14 = 149$

Tip 48	Consecutive Integers II

1.   D

Since $-30 + (-29) + \cdots + 0 + \cdots + 29 + 30 = 0$ and $31 + 32 + 33 = 96$, $x = 33$

2.   A

Since $-20 + (-18) + (-16) + \cdots + 0 + 2 + 4 + \cdots + 20 = 0$ and $22 + 24 + 26 = 72$, there are 24 numbers.

Tip 49	Scatterplot and a Line of Best fit

1.   A

The line had a positive slope and the points are very close to the line of best fit.

2.   B

Choose two points on the line. $(25, 400)$ and $(0, 200) \rightarrow$ Slope $= \frac{400 - 200}{25 - 0} = 8$ and the y-intercept

is 200. Therefore, $y = 8x + 200$.

3.   B

$y = 8(80) + 200 = 840$

4.   C

Actual calories is 500 and the calories predicted by the line of best fit is 400.

Therefore, percent decrease is $\frac{|400 - 500|}{50} \times 100 = 20\%$.

5.   A

The slope 8 is the rate.

Tip 50	Must be true or Could be true

1.   D

Tips Answer

$$k(a-b)=a-b \;\rightarrow\; k(a-b)-(a-b)=0 \;\rightarrow\; (a-b)(k-1)=0 \;\rightarrow\; a=b \text{ or } k=1$$

2. B

$$a^2+b^2=2ab \;\rightarrow\; (a-b)^2=0 \;\rightarrow\; a=b$$

C) $a=0$ and $b=0$ is the answer for "could be true."

Tip 51	Complex Numbers

1. 1

$$a+b+(a-b)i=6-4i \;\rightarrow\; a+b=6 \text{ and } a-b=-4 \;\rightarrow\; a=1 \text{ and } b=5$$

2. C

$$\frac{2+3i}{1-2i}=a+bi \;\rightarrow\; \frac{(2+3i)(1+2i)}{(1-2i)(1+2i)}=-\frac{4}{5}+\frac{7}{5}i \;\rightarrow\; b=\frac{7}{5}$$

3. B

$$\sqrt{-12}\times\sqrt{-3}=i\sqrt{12}\left(i\sqrt{3}\right)=i^2\sqrt{36}=6i^2=-6$$

4. 5

$$a=10+bi-(a-5)i \;\rightarrow\; a=10+(b-a+5)i \;\rightarrow\; a=10 \text{ and } b-a+5=0 \;\rightarrow\; b=5$$

5. B

$$i^{126}=i^{24}\times i^2=1\times(-1)=-1$$

6. C

$$x^3-2x^2+3x=0 \;\rightarrow\; x(x^2-2x+3)=0 \;\rightarrow\; \text{Solutions}\rightarrow\; x=0,\; x=\frac{2\pm\sqrt{4-12}}{2}=\frac{2\pm2i\sqrt{2}}{2}=1\pm i\sqrt{2}$$

Tip 52	Circles (Geometry)

1. C

$$24\pi\times\frac{1}{3}=8\pi$$

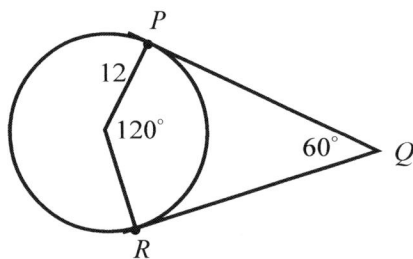

2. 75

$$180-105=75$$

3. B

$$\ell=r\theta=36\times\frac{5\pi}{12}=15\pi$$

4. C

$$\text{Area}=\frac{1}{2}\left(r^2\theta\right)=\frac{1}{2}(36)(36)\left(\frac{5\pi}{12}\right)=270\pi$$

5.   4

$AC = 26$

$10 - r + 24 - r = 26$

$2r = 8 \rightarrow r = 4$

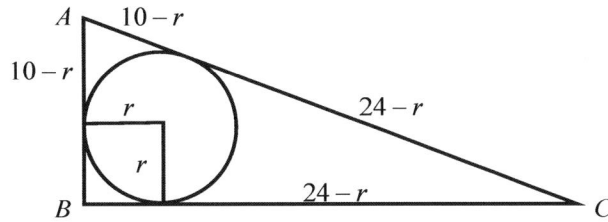

| Tip 53 | Trigonometric Function |

1.   25

$\dfrac{10}{AC} = 0.4 \rightarrow AC = \dfrac{10}{0.4} = 25$

2.   2.4

$AB = DC = x \rightarrow \tan \angle BAE = \dfrac{BE}{x} = 0.6 \rightarrow BE = 0.6x$

$\tan \angle EDC = \dfrac{EC}{x} = 0.25 \rightarrow EC = 0.25x \rightarrow$ Therefore, $\dfrac{BE}{EC} = \dfrac{0.6x}{0.25x} = 2.4$

3.   C

Since $\cos \angle B = \sin \angle C = 0.8$, $\dfrac{BD}{15} = 0.8 \rightarrow BD = 15(0.8) = 12$.

4.   20

Cofunction: $3x + 5 + 2x - 15 = 90 \rightarrow 5x = 100 \rightarrow x = 20$

| Tip 54 | Probability |

1.   C

$P = \dfrac{3}{6} \times \dfrac{2}{5} = \dfrac{6}{30} = \dfrac{1}{5}$ Or $P = \dfrac{{}_3C_2}{{}_6C_2} = \dfrac{3}{15} = \dfrac{1}{5}$

2.   A

Since 3, 5, and 9 were taken already, seven numbers are left as follows.

1, 2, 4, 6, 7, 8, 10 (5 even out of 7) $\rightarrow$ $P(\text{even}, \text{even}) = \dfrac{5}{7} \times \dfrac{4}{6} = \dfrac{10}{21}$

3.   D

Since there are b blue marbles out of $(b + g)$ marbles. The probability of selling a blue one first is $\dfrac{b}{b+g}$. Now

we have $(b-1)$ blue marbles out of $(b + g - 1)$. Therefore, $P = \left(\dfrac{b}{b+g}\right)\left(\dfrac{b-1}{b+g-1}\right)$.

4.   C

$|x + 2| < 8 \rightarrow -8 < x + 2 < 8 \rightarrow -10 < x < 6 \rightarrow$ Only three numbers $-5, 0, 5$ are satisfied.

Therefore, $P = \dfrac{3}{8}$.

Tips Answer

Tip 55 — Geometric Probability

1. C

    The ratio of the lengths of the three circles $=1:2:3$, and the ratio of the areas $=1:4:9=k:4k:9k$.

    The area of the shaded region $=9k-(k+4k)=4k$. Therefore, $P=\dfrac{4k}{9k}=\dfrac{4}{9}$.

2. D

    If the length of a side of the smaller square is a, The area of the smaller circle $=a^2$, the area of the circle

    $=\pi\left(\dfrac{a\sqrt{2}}{2}\right)^2=\dfrac{\pi a^2}{2}$, and the area of the larger square $=\left(a\sqrt{2}\right)^2=2a^2$. Therefore,

    $$P=\dfrac{\text{the area of the shaded region}}{\text{the area of the larger square}}=\dfrac{\dfrac{\pi a^2}{2}-a^2}{2a^2}=\dfrac{\pi-2}{4}.$$

Tip 56 — Data Interpretation

1. D

    The average cost per pound is equal to the slope of the line. Therefore, the slope of the line is

    $\dfrac{6-0}{8-0}\cong 0.75$. The closest number is 0.73.

2. B

    I. No exact information for 50 customers.

    II. The median is 5 items.

    III. Because not enough information for 50 customers. If all 50 customers bought 5 items, then the mode can be 5 items.

Tip 57 — Asymptotes

1. B

    Denominator $x+1=0 \;\rightarrow\; x=-1$

2. B

    $$f(x)=\dfrac{x^2-x-2}{x^2+x-6}=\dfrac{(x-2)(x+1)}{(x-2)(x+3)}=\dfrac{x+1}{x+3} \;\rightarrow\; y=1-\dfrac{2}{x+3}$$

    Therefore, $x=-3$ (vertical asymptote) and $y=1$ (horizontal asymptote).

Tip 58 — Linear Correlation Coefficient

1. A

    All of the points lie on the straight line.(Negatively related)

2. D

Tip 59	Standard Deviation

1.   A

The data are more spread out.

2.   C

The data are close to the mean.

Tip 60	Solid

1.   C

Let x be the length of an edge. Then $x^3 = 64 \rightarrow x = 4$. Therefore, $AB = \sqrt{x^2 + x^2 + x^2} = \sqrt{48} = 4\sqrt{3}$.

2.   C

Let x = the length of an edge. Since $6x^2 = 96$, then $x = 4$. Therefore, volume $= 4^3 = 64$.

3.   A

Volume $= xyz$.

The areas of the faces: $xz = 8$, $xy = 10$, and $yz = 20$. Thus, $(xy)(yz)(xz) = (10)(20)(8)$.

$(xyz)^2 = 1600 \rightarrow xyz = 40$.

4.   C

The area of two circles $= 2\pi r^2$. The lateral area $= 2\pi r \times h = 2\pi r^2$. Therefore, $2\pi r^2 + 2\pi r^2 = 4\pi r^2$.

5.   B

$$\pi r_1^2 h = \frac{\pi r_2^2 h}{3} \rightarrow r_1^2 = \frac{r_2^2}{3} \rightarrow \frac{r_2^2}{r_1^2} = 3 \rightarrow \frac{r_2}{r_1} = \sqrt{3}$$

No Test Material on This Page

PRACTICE TEST 1

Dr. John Chung's SAT Math

Math Test - No Calculator

25 MINUTES, 20 QUESTIONS

Turn to Section 3 of your answer sheet to answer the questions in this section.

DIRECTIONS

For questions 1–15, solve each problem, choose the best answer from the choices provided, and fill in the corresponding circle on your answer sheet. **For questions 16–20,** solve the problem and enter your answer in the grid on your answer sheet. Please refer to the directions before question 16 on how to enter your answers in the grid. You may use any available space in your test booklet for scratch work.

NOTES

1. The use of a calculator **is not permitted**.

2. All variables and expressions used represent real numbers unless otherwise indicated.

3. Figures provided in this test are drawn to scale unless otherwise indicated.

4. All figures lie in a plane unless otherwise indicated.

5. Unless otherwise indicated, the domain of a given function f is the set of all real numbers x for which $f(x)$ is a real number.

REFERENCE

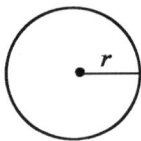

$A = \pi r^2$
$C = 2\pi r$

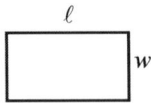

$A = \ell w$

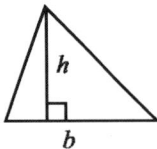

$A = \dfrac{1}{2}bh$

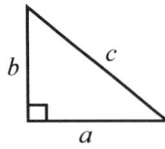

$c^2 = a^2 + b^2$

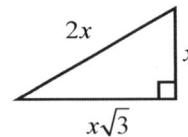

Special Right Triangles

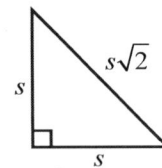

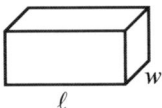

$V = \ell w h$

$V = \pi r^2 h$

$V = \dfrac{4}{3}\pi r^3$

$V = \dfrac{1}{3}\pi r^2 h$

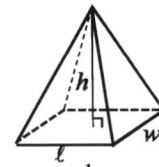

$V = \dfrac{1}{3}\ell w h$

The number of degrees of arc in a circle is 360.

The number of radians of arc in a circle is 2π.

The number of the measures in degrees of the angles of a triangle is 180.

CONTINUE

1

If $\dfrac{2x-3}{2} = k-1$ and $k=5$, what is the value of $2x$?

A) 4

B) 5.5

C) 8

D) 11

2

$$(5+3i)-(8-2i) = a+bi$$

In the equation above, a and b are real numbers. If $i = \sqrt{-1}$, what is the value of b?

A) -1

B) 1

C) -5

D) 5

3

If Claire paid k dollars for a computer that was only 20 dollars more than half the original price, what was the original price, in dollars?

A) $k+20$

B) $k-40$

C) $2k-20$

D) $2k-40$

4

Jenny is on the school swim team and has swim practice m hours in the morning and p hours in the evening each day. The schedule is the same each day. If she swims a total of k hours for five days, which of the following is the expression for m?

A) $\dfrac{k-p}{5}$

B) $\dfrac{k-5p}{5}$

C) $k-5p$

D) $5(k-p)$

5

A certain business is marketing its product and has determined that, when it raised the selling price of its product, its sales went down. The number of units sold, P, is modeled by the equation $P = 1200 - 20s$, where s is the selling price, in dollars. Based on this model, what is the decrease in selling price from 700 units sold to 900 units sold?

A) 5

B) 10

C) 15

D) 20

CONTINUE

6

$$\left(x^2 + y^2\right)^2 - \left(x^2 - y^2\right)^2$$

Which of the following is equivalent to the expression above?

A) $x^4 - y^4$

B) $2\left(x^2 + y^2\right)$

C) $2x^2 y^2$

D) $4x^2 y^2$

7

Kimberly earns k dollars per week. At this rate how many weeks will it take her to earn p dollars?

A) $\dfrac{p}{k}$

B) $\dfrac{k}{p}$

C) kp

D) $\dfrac{10p}{k}$

8

If $\dfrac{2a}{b} = 5$, what is the value of $\dfrac{5b}{a}$?

A) 2

B) 4

C) 10

D) 12.5

9

$$2x + by = 10$$
$$ax + 4y = 15$$

In the system of equations above, a and b are constants and $a = 2b$. If the system has no solution, which of the following could be a possible value of a?

A) -2

B) $\dfrac{1}{2}$

C) 4

D) 8

CONTINUE

10

$$f(x) = ax^2 - 15$$

For the function f defined above, a is a constant and $f(3) = 10$. Which of the following is equal to the value of $f(5)$?

A) $f(0)$

B) $f(3)$

C) $f(-3)$

D) $f(-5)$

11

A certain job can be done in 20 hours by 4 people. How many people are needed to do the same job in 10 hours?

A) 2

B) 4

C) 8

D) 10

12

Which of the following is equivalent to $f(x) = x^2 - 6x + 7$?

A) $f(x) = (x+3)^2 + 5$

B) $f(x) = (x-3)^2 + 2$

C) $f(x) = (x-3)^2 - 2$

D) $f(x) = (x-7)(x+1)$

13

If $24x^2 - kx + 16 = (3x+4)(ax-b)$ for all values of x, where $a, b,$ and k are constants, what is the value of k?

A) -44

B) -12

C) 12

D) 44

CONTINUE

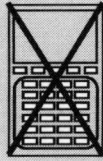

14

In the xy-plane, the equation of line ℓ is $x + 3y = 5$. If line m is perpendicular to line ℓ, what is a possible equation of line m?

A) $y = -\dfrac{1}{3}x + 2$

B) $y = \dfrac{1}{3}x - 1$

C) $y = -3x + 1$

D) $y = 3x + \dfrac{2}{3}$

15

If $a + b = 8$ and $\dfrac{27^a}{3^b} = 81$, what is the value of a?

A) 3

B) 4

C) 5

D) 6

CONTINUE

Answer: $\dfrac{7}{12}$

Answer: 2.5

DIRECTIONS

For questions 16–20, solve the problem and enter your answer in the grid, as described below, on the answer sheet.

Write answer → in boxes.

← Fraction line

← Decimal point

Grid in → result.

1. Although not required, it is suggested that you write your answer in the boxes at the top of the columns to help you fill in the circles accurately. You will receive credit only if the circles are filled in correctly.
2. Mark no more than one circle in any column.
3. No question has a negative answer.
4. Some problems may have more than one answer.
5. **Mixed numbers** such as $3\frac{1}{2}$ must be gridded as 3.5 or 7/2. (If $\boxed{3\,|\,1\,|\,/\,|\,2}$ is entered into the grid, it will be interpreted as $\dfrac{31}{2}$, not $3\frac{1}{2}$.)
6. **Decimal answers:** If you obtain a decimal answer with more digits than the grid can accommodate, it may be either rounded or truncated, but it must fill the entire grid.

Acceptable ways to grid $\dfrac{2}{3}$ are:

Answer: 201
Either position is correct.

Note: You may start your answers in any column, space permitting. Columns you don't need to use should be left blank.

CONTINUE

16

In a right triangle, one of the angles is $x°$. If $\tan x° = \dfrac{5}{12}$, what is the value of $\sin x°$?

17

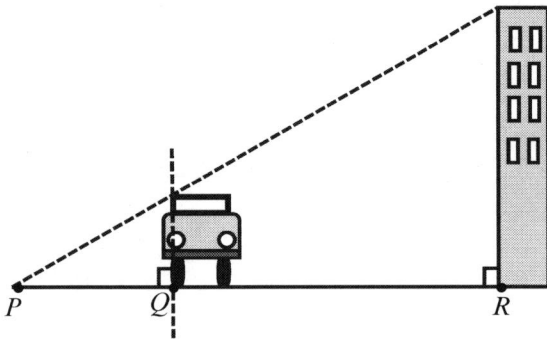

Dawson needs to measure the height of a building near his house. He chooses a point P on the ground where he can visually align the roof of his car with the edge of the building roof. The height of the car is 4 feet and the distance from point P to point Q is 10 feet, as shown in the figure above. If the distance from point Q to point R is 80 feet, and the height of the building is k feet, what is the value of k ?

18

If $a(x+1) + b(x-1) = 7x$ for all real number x, where a and b are constants, what is the value of a ?

19

According to the formula $p = \dfrac{4}{3}k + 81$, if the value of p is increased by 16, by how much does the value of k increase?

20

$$x^2 + y^2 = 56$$
$$y = \sqrt{x}$$

According to the system of equations above, what is the value of x ?

STOP

If you finish before time is called, you may check your work on this section only.
Do not turn to any other section in the test.

No Test Material on This Page

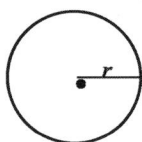
Math Test - Calculator

55 MINUTES, 38 QUESTIONS

Turn to Section 4 of your answer sheet to answer the questions in this section.

DIRECTIONS

For questions 1-30, solve each problem, choose the best answer from the choices provided, and fill in the corresponding circle on your answer sheet. **For questions 31-38**, solve the problem and enter your answer in the grid on your answer sheet. Please refer to the directions before question 31 on how to enter your answers in the grid. You may use any available space in your test booklet for scratch work.

NOTE

1. The use of a calculator **is permitted**.

2. All variables and expressions used represent real numbers unless otherwise indicated.

3. Figures provided in this test are drawn to scale unless otherwise indicated.

4. All figures lie in a plane unless otherwise indicated.

5. Unless otherwise indicated, the domain of a given function f is the set of all real numbers x for which $f(x)$ is a real number.

REFERENC

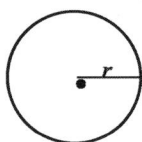 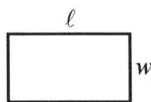 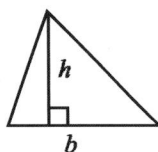 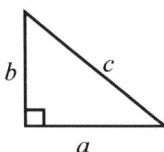 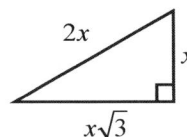 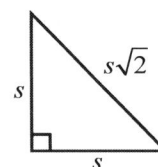

$A = \pi r^2$ $A = \ell w$ $A = \frac{1}{2}bh$ $c^2 = a^2 + b^2$ Special Right Triangles
$C = 2\pi r$

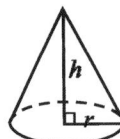 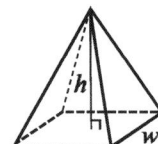

$V = \ell wh$ $V = \pi r^2 h$ $V = \frac{4}{3}\pi r^3$ $V = \frac{1}{3}\pi r^2 h$ $V = \frac{1}{3}\ell wh$

The number of degrees of arc in a circle is 360.

The number of radians of arc in a circle is 2π.

The number of the measures in degrees of the angles of a triangle is 180.

CONTINUE

1

x	$f(x)$
1	6
2	10
3	14
4	18
5	22

The selected values of a function shown in the table above represent a linear function. Which of the following equals $f(10)$?

A) 36

B) 40

C) 42

D) 44

2

If $3(a + 2b - c) = 12,$ what is the value of $a + 2b$ in terms of c?

A) $3c - 4$

B) $c - 12$

C) $c + 4$

D) $c - 4$

3

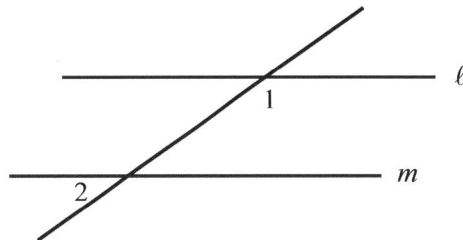

In the figure above, lines ℓ and m are parallel. If the measure of $\angle 1$ is twice the measure of $\angle 2$, what is the measure of $\angle 1$?

A) $100°$

B) $120°$

C) $135°$

D) $145°$

4

If $8^n \times 4^2 = 2^{10},$ what is the value of n?

A) 2

B) 3

C) 4

D) 5

CONTINUE

5

For what value of n is $|n+4|+1$ less than 0?

A) -5

B) -4

C) 3

D) There is no such value of n.

6

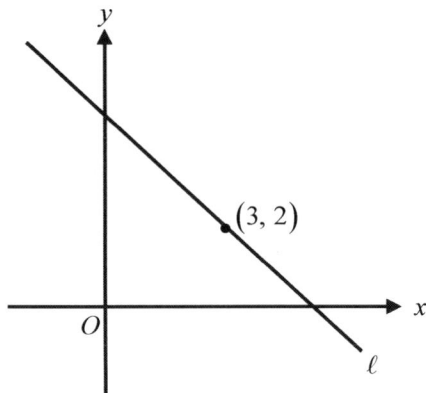

The equation of the graph of line ℓ in the xy-plane above is $y = mx + 6$, where m is a constant. If the line passes through a point $(3, 2)$, what is the value of m?

A) $-\dfrac{4}{3}$

B) $-\dfrac{2}{3}$

C) $-\dfrac{1}{2}$

D) $-\dfrac{1}{4}$

7

In Ms. Lee's class, the number of boys is more than twice the number of girls. There are at least 7 girls and there are no more than 15 boys. How many students are in the class?

A) 19

B) 20

C) 21

D) 22

8

Test Score for a class of **20** Students

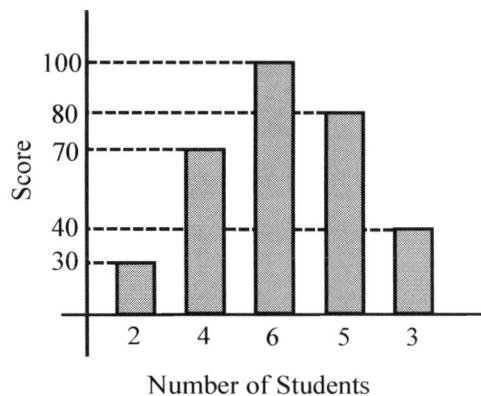

Number of Students

The graph above shows the test scores of 20 students. Based on the histogram above, what is the average (arithmetic mean) score on the test?

A) 70

B) 73

C) 75

D) 78

CONTINUE

Questions 9 and 10 refer to the following information.

$$h(t) = -16t^2 + 128t + 320$$

A science class determined that the motion of a ball launched from the top of a 10-story building could be described by the function above, where t represents the time the ball is in the air in seconds and h, the height in feet of the ball above the ground.

9

What is the number of seconds it takes for the ball to reach its peak?

A) 2

B) 4

C) 8

D) 10

10

At what time will the ball hit the ground?

A) 5

B) 8

C) 10

D) 12

11

The perimeter of a rectangle is 54 cm. If the length is 2 cm more than its width, what is the area of the rectangle?

A) $181.25\,\text{cm}^2$

B) $728\,\text{cm}^2$

C) $800\,\text{cm}^2$

D) $820\,\text{cm}^2$

12

$$3x - y > 0$$
$$2x + y > 1$$

Which of the following is NOT a solution of the system of inequalities above?

A) $(3,0)$

B) $(2,5)$

C) $(0,-3)$

D) $(5,-8)$

CONTINUE

13

x	y
0	2
k	14
k + 2	17

The table above shows the point (x, y) represented on a straight line. If the point $(16, m)$ lies on the same line, what is the value of m?

A) 26

B) 24

C) 22

D) 20

14

James spent $\frac{3}{4}$ of his allowance on a music CD.

He spent $\frac{2}{3}$ of what was left on a hamburger. If this left him P dollars, which of the following was his allowance in dollars?

A) $12P$

B) $14P$

C) $16P$

D) $18P$

Questions 15 and 16 refer to the following information.

Radioactive decay is an exponential function where the amount , y, of radioactive material is reduced by one-half over a certain period of time t. Material M has a half- life of 50 years.

15

If there are 800 grams of radioactive material M, which of the following best represents the decay equation?

A) $y = 800 - 400t$

B) $y = 800\left(\frac{1}{2}\right)^{t}$

C) $y = 800\left(\frac{1}{2}\right)^{\frac{t}{50}}$

D) $y = 800(1 - 0.5t)$

16

If there are 800 grams of radioactive material M, then how much of this material would remain radioactive after 200 years?

A) 25 grams

B) 50 grams

C) 100 grams

D) 200 grams

CONTINUE

17

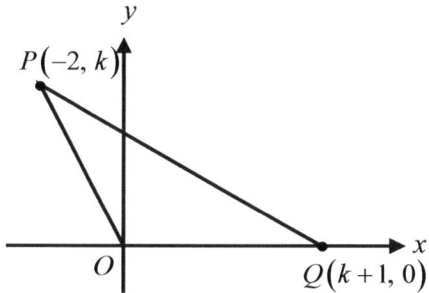

Note: Figure not drawn to scale.

In the xy-plane above, the area of $\triangle OPQ$ is 3. What is the value of k?

A) 2

B) 4

C) 6

D) 8

18

A circle in the xy-plane with center $(4,0)$ passes through point $(7,4)$. Which of the following is the equation of the circle?

A) $(x-4)^2 + y^2 = 9$

B) $(x-4)^2 + y^2 = 25$

C) $(x-4)^2 + y^2 = 5$

D) $(x+4)^2 + y^2 = 5$

19

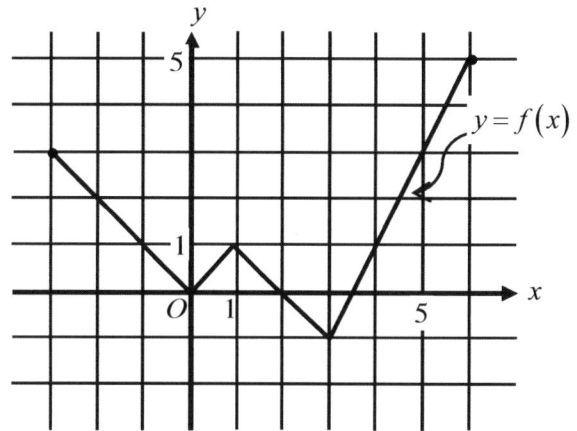

The graph of the function f is shown in the xy-plane above. Which of the following is the average rate of change between $x = -3$ and $x = 6$?

A) $\dfrac{2}{9}$

B) $\dfrac{8}{9}$

C) 2

D) It cannot be determined from the given information.

CONTINUE

20

Emily traveled 60 miles on the highway and 16 miles on the local roads to reach her destination. On the highway, she traveled 30 miles faster than on the local roads. If her speed on local roads is 20 miles per hour, then what was her average speed, in miles per hour, during her entire trip?

A) 24

B) 25

C) 35

D) 38

21

For O.K theater tickets, a ticket for an adult is 5 dollars more than a ticket for a child. If a group of 6 adults and 10 children pay a total of 142 dollars, what is the cost, in dollars, of a ticket for one adult and one child?

A) 19

B) 18

C) 17

D) 16

22

Traveled Distance vs. Hours
for 10 Taxi Drivers

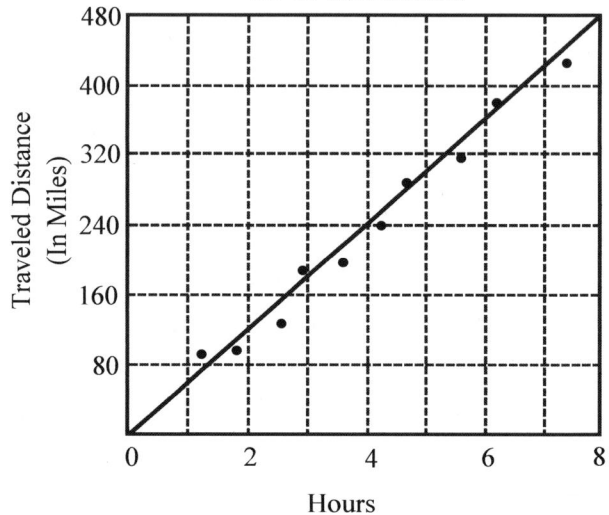

The scatterplot above shows the distance traveled in hours for 10 taxi drivers and the line of best fit for the data. Which of the following is closest to the average speed, in miles per hour, for the drivers?

A) 54

B) 59

C) 65

D) 68

CONTINUE

23

For a polynomial $p(x)$, the value of $p(-5) = 0$.
Which of the following must be true about $p(x)$?

A) $(x - 5)$ is a factor of $p(x)$.

B) $(x + 5)$ is a factor of $p(x)$.

C) x is a factor of $p(x)$.

D) When $p(x)$ is divided by $(x + 5)$, the remainder is -5.

24

$$y \le 3x + \frac{1}{2}$$
$$y \ge \frac{1}{2}x + 3$$

If the system of inequalities above is graphed in the xy-plane, which quadrant contains solutions to the system?

A) Quadrant I

B) Quadrant II

C) Quadrant III

D) Quadrant IV

25

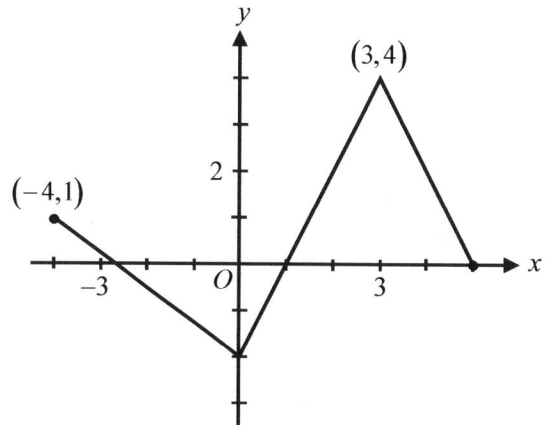

The figure above shows the graph of the piece-wise function f defined for $-4 \le x \le 5$. For which of the following values of x is $f(x) < |f(x)|$?

A) -3

B) -1.3

C) 2.5

D) 3.7

26

If a and b are positive integers and $a^2 - b^2 = 24$, which of the following could be the smallest value of a?

A) 4

B) 5

C) 7

D) 8

CONTINUE

27

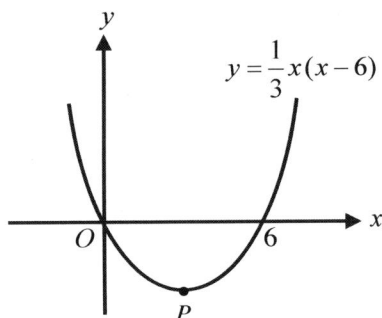

The graph of $y = \dfrac{1}{3}x(x-6)$ is shown in the xy-plane above. Which of the following are the coordinates of vertex P?

A) $(3,-2)$

B) $(2,-4)$

C) $(3,-3)$

D) $(3,-4.5)$

28

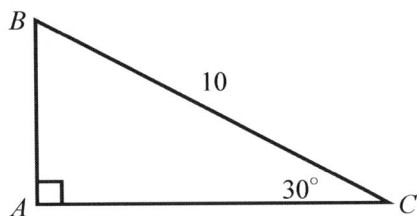

In right triangle ABC above, if $BC = 10$ and $\angle C = 30°$, what is the approximate perimeter of the triangle?

A) 20

B) 23.7

C) 25.8

D) 27.2

29

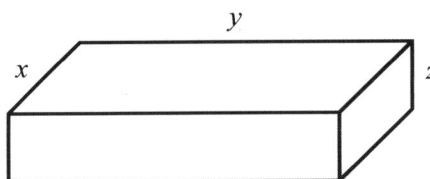

The figure above shows a rectangular solid with width x, length y, and height z. If $xy = 20$, $yz = 10$, and $xz = 18$, what is the volume of the solid?

A) 60

B) 70

C) 80

D) 90

30

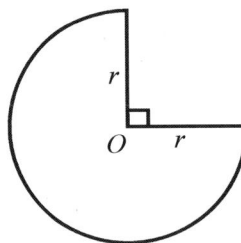

The figure above shows a sector O with radius r. If the area of the sector is 3π, what is the approximate perimeter of the sector?

A) 10

B) 12.5

C) 13.4

D) 15.6

CONTINUE

DIRECTIONS

For questions 31-38, solve the problem and enter your answer in the grid, as described below, on the answer sheet.

1. Although not required, it is suggested that you write your answer in the boxes at the top of the columns to help you fill in the circles accurately. You will receive credit only if the circles are filled in correctly.
2. Mark no more than one circle in any column.
3. No question has a negative answer.
4. Some problems may have more than one answer.
5. **Mixed numbers** such as $3\frac{1}{2}$ must be gridded as 3.5 or 7/2. (If $3\,1\,/\,2$ is entered into the grid, it will be interpreted as $\frac{31}{2}$, not $3\frac{1}{2}$.)
6. **Decimal answers:** If you obtain a decimal answer with more digits than the grid can accommodate, it may be either rounded or truncated, but it must fill the entire grid.

Answer: $\frac{7}{12}$

Write answer in boxes.

Fraction line

Grid in result.

Answer: 2.5

Decimal point

Acceptable ways to grid $\frac{2}{3}$ are:

Answer: 201
Either position is correct.

Note: You may start your answers in any column, space permitting. Columns you don't need to use should be left blank.

CONTINUE

31

Kara needs three hours to mow and trim Mrs. Tayler's lawn. One day she asked her friend Peter to work with her. When Peter worked with her, the job took only one hour. How long would it take Peter, in hours, to complete the job himself?

32

Twenty members of a math club are planning a trip to an amusing park that has an admission price of $10 per person. The club members going on the trip must share the $500 cost of a bus and the admission price for 2 supervisors who will accompany them on the trip. What is the cost, in dollars, for each member?

33

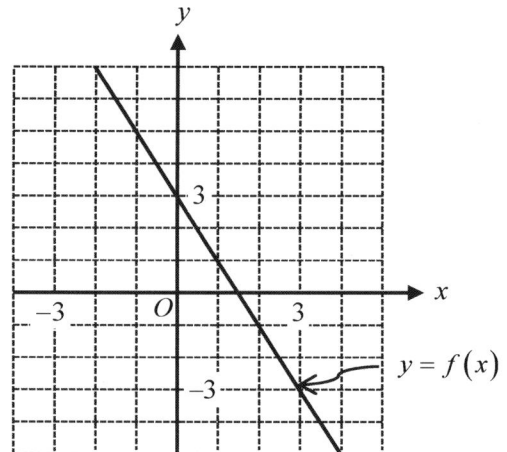

The graph of a linear function f is shown in the xy-plane above. If $f(k) = 1$, what is the value of $f(-2k)$?

34

If the average of $2a$ and b is equal to 50 percent of $4b$, what is the value of $\dfrac{a}{b}$?

CONTINUE

35

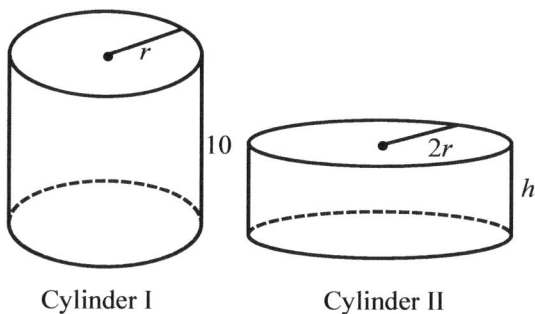

Cylinder I   Cylinder II

Two cylinders shown above have the same volume. If the radius of cylinder II is twice the radius of cylinder I and the height of cylinder I is 10, what is the height h of cylinder II?

36

$$g(x) = \frac{x^2 - 3x + 2}{(x+2)^2 - 8x}$$

For what value of x is the function above undefined?

Questions 37 and 38 refer to the following information.

Suppose Claire deposits a principal amount of P dollars in a bank account that pays compound interest. If the annual interest is r (expressed as a decimal) and the bank makes interest payments n times every year, she would have an amount of money equal to R after t years, given by

$$R(t) = P\left(1 + \frac{r}{n}\right)^{nt}$$

37

If she deposit $2,000 into an account paying 4% annual interest compounded annually, what is the amount of interest after one year? (Disregard the $ sign when gridding your answer.)

38

If she deposits $2,000 into an account paying 4% annual interest compounded quarterly, what is her account balance after one year? (Round your answer to the nearest dollar and disregard the $ sign when gridding your answer.)

STOP

If you finish before time is called, you may check your work on this section only.
Do not turn to any other section in the test.

No Test Material on This Page

SECTION 3

	A B C D		A B C D		A B C D		A B C D		A B C D
1	○○○○	4	○○○○	7	○○○○	10	○○○○	13	○○○○
2	○○○○	5	○○○○	8	○○○○	11	○○○○	14	○○○○
3	○○○○	6	○○○○	9	○○○○	12	○○○○	15	○○○○

16 | | | | |
17 | | | | |
18 | | | | |
19 | | | | |
20 | | | | |

/ ○○ . ○○○○
0 ○○○○ 1 ○○○○ 2 ○○○○ 3 ○○○○ 4 ○○○○ 5 ○○○○ 6 ○○○○ 7 ○○○○ 8 ○○○○ 9 ○○○○

NO CALCULATOR ALLOWED

SECTION 4

	A B C D		A B C D		A B C D		A B C D		A B C D
1	○○○○	7	○○○○	13	○○○○	19	○○○○	25	○○○○
2	○○○○	8	○○○○	14	○○○○	20	○○○○	26	○○○○
3	○○○○	9	○○○○	15	○○○○	21	○○○○	27	○○○○
4	○○○○	10	○○○○	16	○○○○	22	○○○○	28	○○○○
5	○○○○	11	○○○○	17	○○○○	23	○○○○	29	○○○○
6	○○○○	12	○○○○	18	○○○○	24	○○○○	30	○○○○

CALCULATOR ALLOWED

SECTION 4 (Continued)

31

/
.
0
1
2
3
4
5
6
7
8
9

32

/
.
0
1
2
3
4
5
6
7
8
9

33

/
.
0
1
2
3
4
5
6
7
8
9

34

/
.
0
1
2
3
4
5
6
7
8
9

35

/
.
0
1
2
3
4
5
6
7
8
9

36

/
.
0
1
2
3
4
5
6
7
8
9

37

/
.
0
1
2
3
4
5
6
7
8
9

38

/
.
0
1
2
3
4
5
6
7
8
9

CALCULATOR ALLOWED

Math Conversion Table

Raw Score	Scaled Score	Raw Score	Scaled Score
58	800	27	500
57	800	26	490
56	800	25	480
55	800	24	470
54	790	23	460
53	780	22	460
52	770	21	450
51	750	20	440
50	740	19	430
49	730	18	430
48	720	17	420
47	710	16	420
46	700	15	410
45	690	14	400
44	670	13	390
43	680	12	380
42	670	11	370
41	660	10	360
40	650	9	450
39	640	8	340
38	630	7	330
37	620	6	310
36	610	5	290
35	600	4	280
34	590	3	270
33	580	2	260
32	560	1	240
31	550	0	200
30	540		
29	530		
28	520		

Answer Explanations

Test 1   Answers and Explanations

SECTION 3	1	2	3	4	5	6	7	8	9	10
	D	D	D	B	B	D	A	A	C	D
	11	12	13	14	15	16	17	18	19	20
	C	C	A	D	A	$\frac{5}{13}$	36	3.5	12	7

SECTION 4	1	2	3	4	5	6	7	8	9	10
	C	C	B	A	D	A	D	B	B	C
	11	12	13	14	15	16	17	18	19	20
	A	C	A	A	C	B	A	B	A	D
	21	22	23	24	25	26	27	28	29	30
	A	B	B	A	B	B	C	B	A	C
	31	32	33	34	35	36	37	38		
	1.5	36	7	3/2	2.5	2	80	2081		

SECTION 3

1.   D

$$\frac{2x-3}{2} = 5-1 \;\rightarrow\; 2x-3=8 \;\rightarrow\; 2x=11$$

2.   D

$5+3i-8+2i = a+bi \;\rightarrow\; -3+5i = a+bi$. Therefore, $a=-3$ and $b=5$.

3   D

Original price: x, $\;k = \frac{1}{2}x+20 \;\rightarrow\; k-20 = \frac{1}{2}x \;\rightarrow\; x = 2(k-20) \;\rightarrow\; x = 2k-40$

4.   B

$$m+p = \frac{k}{5} \;\rightarrow\; m = \frac{k}{5}-p \;\rightarrow\; m = \frac{k-5p}{5}$$

5.   B

When $P=700$, $700 = 1200-20s \;\rightarrow\; 20s=500 \;\rightarrow\; s=25$.

When $P=900$, $900 = 1200-20s \;\rightarrow\; 20s=300 \;\rightarrow\; s=15$. $25-15=10$.

There is $10 decrease in selling price.

6. D

$$\left(x^2+y^2\right)^2-\left(x^2-y^2\right)^2=x^4+2x^2y^2+y^4-\left(x^4-2x^2y^2+y^4\right)=4x^2y^2$$

7. A

Proportion. $\dfrac{\$}{\text{week}}=\dfrac{k}{1}=\dfrac{P}{x}$ $\rightarrow$ $x=\dfrac{P}{k}$ weeks

8. A

$\dfrac{2a}{b}=5$ $\rightarrow$ $\dfrac{a}{b}=\dfrac{5}{2}$ $\rightarrow$ $\dfrac{b}{a}=\dfrac{2}{5}$. Therefore, $\dfrac{5b}{a}=5\left(\dfrac{b}{a}\right)=5\left(\dfrac{2}{5}\right)=2.$

9. C

In order to have no solution: $\dfrac{2}{a}=\dfrac{b}{4}\neq\dfrac{10}{15}$. Therefore, $\dfrac{2}{a}=\dfrac{b}{4}$ $\rightarrow$ $ab=8$. Because $a=2b$, $(2b)b=8$

$2b^2=8$ $\rightarrow$ $b^2=4$ $\rightarrow$ $b=\pm2$. Now $a=2b=\pm4$. Possible value of a is 4 or –4.

We know that $\dfrac{2}{4}\neq\dfrac{10}{15}$ at $a=4.$

10. D

Since axis of symmetry is $x=0$, the graph is symmetry in y-axis. Therefore $f(5)=f(-5)$.

11. C

Inverse variation: (# of people) $\times$ (# of hours) = Constant. Therefore, $20\times4=10\times p$ $\rightarrow$ $p=8$ people.

12. C

Choice C is the vertex form of the equation.

In order to have vertex form, $f(x)=(x^2-6x+9)+(7-9)$ $\rightarrow$ $f(x)=(x-3)^2-2.$

13. A

Identical equation has infinitely many solutions. Because $(3x+4)(ax-b)=3ax^2+(4a-3b)x-4b,$

$24x^2-kx+16=3ax^2+(4a-3b)x-4b$. Therefore, $3a=24$ $\rightarrow$ $a=8$ and $-4b=16$ $\rightarrow$ $b=-4$.

Now $-k=4a-3b$ $\rightarrow$ $-k=4(8)-3(-4)=48$ $\rightarrow$ $k=-44.$

14. D

Because $x+3y=5$ $\rightarrow$ $y=-\dfrac{1}{3}x+\dfrac{5}{3}$, slope is $-\dfrac{1}{3}$. Slope of the perpendicular line must have slope of 3,

which is negative reciprocal of the other.

15. A

Since $\dfrac{27^a}{3^b} = \dfrac{3^{3a}}{3^b} = 3^{3a-b}$ and $81 = 3^4$, $3^{3a-b} = 3^4$ $\rightarrow$ $3a - b = 4$.

When you solve system of equation by addition,

$$a + b = 8$$
$$\underline{3a - b = 4}$$
$$4a \quad\ = 12 \quad \rightarrow \quad a = 3.$$

16. $\dfrac{5}{13}$ or $.384$ (or $.385$)

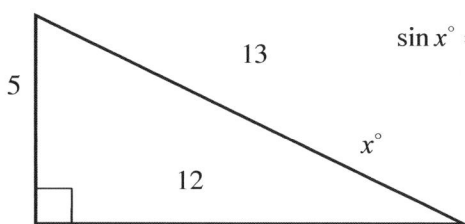

$$\sin x° = \frac{\text{opposite}}{\text{hyp}} = \frac{5}{12}$$

17. 36

Proportion: $\dfrac{4}{10} = \dfrac{k}{10 + 80}$ $\rightarrow$ $10k = 360$ $\rightarrow$ $k = 36$. Or, $\left(\dfrac{4}{k} = \dfrac{10}{90}\right)$

18. 3.5

$a(x + 1) + b(x - 1) = (a + b)x + a - 6$. Then $(a + b)x + a - 6 = 7x$.

Since the equation is true for all real x, both side expressions must be same.

Therefore, $a + b = 7$ and $a - b = 0$. (system of equations)

$$a + b = 7$$
$$\underline{a - b = 0}$$
$$2a \quad\ = 7 \quad \rightarrow \quad a = \frac{7}{2}, \text{ or } 3.5$$

19. 12

From the equation $p = \dfrac{4}{3}k + 81$, $\dfrac{4}{3}$ is slope.

By definition of slope: $\dfrac{\Delta p}{\Delta k} = \dfrac{4}{3}$ $\rightarrow$ $\dfrac{16}{\Delta k} = \dfrac{4}{3}$ $\rightarrow$ $4\Delta k = 48$ $\rightarrow$ $\Delta k = 12$

Or

When $k = 0$, $p = 81$. After increased by 16, $p = 97$.

When $p = 97$, $\rightarrow$ $97 = \dfrac{4}{3}k + 81$, or $\dfrac{4}{3}k = 16$ $\rightarrow k = \dfrac{3}{4}(16) = 12$.

20. 7

Substitution: $\left.\begin{matrix} x^2 + y^2 = 56 \\ y = \sqrt{x} \end{matrix}\right\}$ $\rightarrow$ $x^2 + x = 56$ $\rightarrow$ $x^2 + x - 56 = 0$ $\rightarrow$ $(x+8)(x-7) = 0$

Therefore, $x = -8$ or $x = 7$. But $\sqrt{-8}$ is undefined. Only $x = 7$ is the solution.

You can take a look the graphs.

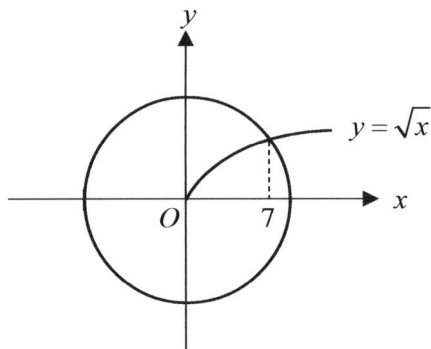

SECTION 4

1. C

When define the linear equation $y = mx + b$, slope $m = 4$. Substitute any point in the table.

You can choose $(1, 6)$. $y = 4x + b$ $\rightarrow$ $6 = 4(1) + b$ $\rightarrow$ $b = 2$. Therefore the equation is $y = 4x + 2$.

When $x = 10$, $f(10) = 4(10) + 2 = 42$.

2. C

$3(a + 2b - c) = 12$ $\rightarrow$ $a + 2b - c = 4$ $\rightarrow$ $a + 2b = c + 4$

3. B

Define: $\angle 2 = x$ and $\angle 1 = 2x$. $\angle 1 + \angle 2 = 180$ $\rightarrow$ $x + 2x = 180$ $\rightarrow$ $x = 60$

Therefore $\angle 1 = 2x = 2(60) = 120$.

4. A

$8^n \times 4^2 = 2^{10}$ $\rightarrow$ $2^{3n} \times 2^4 = 2^{10}$ $\rightarrow$ $2^{3n+4} = 2^{10}$. From the equation $3n + 4 = 10$ $\rightarrow$ $n = 2$.

5. D

Since $|n+4| \geq 0$, $|n+4| + 1 \geq 1$. It cannot be less than 1.

6. A

From the equation $y = mx + 6$, you can see y-intercept is $(0, 6)$. Therefore, slope $m = \dfrac{6-2}{0-3} = -\dfrac{4}{3}$.

Answer Explanations

7.   D

# of boys $= 2 \times ($# of girls$) \rightarrow b = 2g$. But $g \geq 7$ and $b \leq 15$.

You can use a table as follows.

        # of girls $=$ 7   8   9   10$\cdots$

        # of boys $=$ 15 ~~16~~ ~~18~~ ~~20~~$\cdots$

From the table, # of boys cannot be more than 15. Therefore, # of students is $7 + 15 = 22$.

8.   B

$$\text{Average} = \frac{\text{Total score}}{20} = \frac{30(2) + 70(4) + 100(6) + 80(5) + 40(3)}{20} = \frac{1460}{20} = 73$$

9.   B

Since the peak is on axis of symmetry, $t = \dfrac{-b}{2a} = \dfrac{-128}{2(-16)} = 4$.

10.   C

When the ball hit the ground, the height is 0. $-16t^2 + 128t + 320 = 0 \rightarrow t^2 - 8t - 20 = 0$

$(t - 10)(t + 2) = 0 \rightarrow t = -2$ or 10. Therefore, $t = 10$.

11.   A

If width $= x$, then length $= x + 2$. $x + (x + 2) = 27 \rightarrow 2x + 2 = 27 \rightarrow x = 12.5$ and $x + 2 = 14.5$.

Therefore, the area of the rectangle is $12.5 \times 14.5 = 181.25$.

12.   C

You can check by substituting the coordinates into the inequalities.

Choice (C) $(0, -3)$, they are not true.

13.   A

Slopes between any two points are constant. First you need to find slope or the value of k.

$\text{slope} = \dfrac{17 - 14}{(k + 2) - k} = \dfrac{3}{2}$. Therefore, $\dfrac{m - 2}{16 - 0} = \dfrac{3}{2} \rightarrow 2m - 4 = 48 \rightarrow 2m = 52 \rightarrow m = 26$.

                                    Or

The linear equation is $y = \dfrac{3}{2}x + 2$. By substituting $(16, m)$, $m = \dfrac{3}{2}(16) + 2 = 26$.

14.   A

If original price $= k$, then $\dfrac{1}{3}\left(\dfrac{1}{4}k\right) = p \rightarrow \dfrac{1}{12}k = p \rightarrow k = 12p$.

15.   C

The equation of radioactive decay is $p = p_\circ \left(\dfrac{1}{2}\right)^{t/n}$, where n is a half-life period.

16.   B

$$y = 800\left(\frac{1}{2}\right)^{200/50} = 800\left(\frac{1}{2}\right)^4 = 800\left(\frac{1}{16}\right) = 50$$

17.   A

Area: $\frac{1}{2}(k+1)k = 3 \ \rightarrow \ k^2 + k = 6 \ \rightarrow \ k^2 + k - 6 = 0 \ \rightarrow \ (k+3)(k-2) = 0$

$k = -3$ or 2. But $k > 0 \ \rightarrow \ k = 2$.

18.   B

Since $r^2 = (7-4)^2 + (4-0)^2 = 25$, the equation is $(x-4)^2 + y^2 = 25$.

19.   A

When $x = -3$, $y = 3$. When $x = 6$, $y = 5$.

Average rate of change (slope between two points) is $\dfrac{5-3}{6-(-3)} = \dfrac{2}{9}$.

20.   D

Average speed $= \dfrac{\text{total distance}}{\text{total time}}$, $t_1 = \dfrac{60}{50} = 1.2$ on the highway, and $t_2 = \dfrac{16}{20} = 0.8$ on local roads.

Therefore, Average speed $= \dfrac{60+16}{1.2+0.8} = \dfrac{76}{2} = 38\,\text{mph}$.

21.   A

Child ticket $= \$k$ and adult ticket $= \$(k+5)$.

$6(k+5) + 10k = 142 \ \rightarrow \ 16k = 112 \ \rightarrow \ k = 7$ and $k+5 = 12$

For one adult and one child, $7 + 12 = \$19$.

22.   B

Average speed for 10 taxi drivers $\cong$ Slope of the line of best fit.

Slope $= \dfrac{240}{4}$ or $\dfrac{480}{8} = 60$. Therefore, 59 is the closest to 60.

23.   B

$p(-5) = 0$ means "$p(x)$ has a factor of $(x+5)$." $\rightarrow$ Factor Theorem.

One possible example: $p(x) = (x+5)(\ ?\ )$

24.   A

Solution set of system of inequalities is as follows.

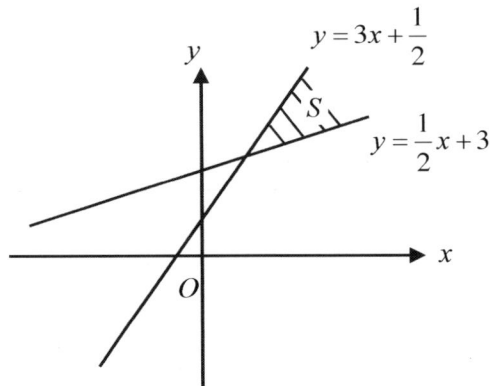

Answer Explanations

25. **B**

For the interval $-2.8 < x < 1$, $f(x) < |f(x)|$.

26. **B**

$a^2 - b^2 = 24 \rightarrow (a+b)(a-b) = 24$. Since a and b are positive integers, only two arrangements are possible.

$a+b = 12$ and $a-b = 2$, or $a+b = 6$ and $a-b = 4$. Solve each system of equations.

27. **C**

Since axis of symmetry is $x = 3$, the y-coordinate is $y = \frac{1}{3}(3)(3-6) = -3$.

Therefore, the coordinates of vertex P is $(3, -3)$.

28. **B**

Special Triangle $30° - 60° - 90°$: $AB = 5$, $AC = 5\sqrt{3}$, and $BC = 10 \rightarrow$ Perimeter $= 15 + 5\sqrt{3} \simeq 23.7$.

29. **A**

$\left.\begin{array}{l} xy = 20 \\ yz = 10 \\ xz = 18 \end{array}\right\}$ multiply bothside $\rightarrow (xyz)^2 = 3600 \rightarrow xyz = 60$. Volume of the solid is $V = xyz = 60$.

30. **C**

Since the area of a whole circle is 4π, radius is 2. Length of the arc is $4\pi \times \frac{3}{4} = 3\pi$.

Therefore, perimeter is $P = 3\pi + 2 + 2 \simeq 13.4$.

31. $\frac{3}{2}$ or 1.5

The rates are equal. If Peter takes x hours to complete the job, $\frac{1}{3} + \frac{1}{x} = \frac{1}{1} \rightarrow \frac{1}{x} = \frac{2}{3} \rightarrow x = \frac{3}{2} = 1.5$.

32. 36

$\$10 + \dfrac{500 + 20}{20} = \$36$

33. 7

Since $f(1) = 1$, $k = 1$. $f(-2k) = f(-2) = 7$.

34. $\frac{3}{2}$ or 1.5

$\dfrac{2a+b}{2} = 0.5(4b) \rightarrow 2a + b = 4b \rightarrow 2a = 3b$. If $a = 3$, then $b = 2$. Therefore, $\dfrac{a}{b} = \dfrac{3}{2}$ or 1.5.

35. 2.5

$$\pi r^2 (10) = \pi (2r)^2 h \quad \rightarrow \quad 10 = 4h \quad \rightarrow \quad h = 2.5$$

36. 2

$$(x+2)^2 - 8x = 0 \quad \rightarrow \quad x^2 + 4x + 4 - 8x = 0 \quad \rightarrow \quad x^2 - 4x + 4 = 0 \quad \rightarrow \quad (x-2)^2 = 0$$

$x = 2$ is the answer.

37. 80

$$\text{Interest} = 2000 \times 0.4 = 80$$

38. 2081

Since $n = 4$, balance $= 2000 \left(1 + \dfrac{0.04}{4}\right)^{4(1)} \simeq 2081$.

PRACTICE TEST 2

Dr. John Chung's SAT Math

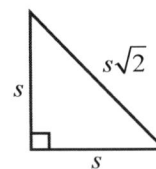

Math Test - No Calculator

25 MINUTES, 20 QUESTIONS

Turn to Section 3 of your answer sheet to answer the questions in this section.

DIRECTIONS

For questions 1–15, solve each problem, choose the best answer from the choices provided, and fill in the corresponding circle on your answer sheet. **For questions 16–20,** solve the problem and enter your answer in the grid on your answer sheet. Please refer to the directions before question 16 on how to enter your answers in the grid. You may use any available space in your test booklet for scratch work.

NOTE

1. The use of a calculator **is not permitted**.

2. All variables and expressions used represent real numbers unless otherwise indicated.

3. Figures provided in this test are drawn to scale unless otherwise indicated.

4. All figures lie in a plane unless otherwise indicated.

5. Unless otherwise indicated, the domain of a given function f is the set of all real numbers x for which $f(x)$ is a real number.

REFERENC

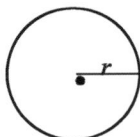
$A = \pi r^2$
$C = 2\pi r$

$A = \ell w$

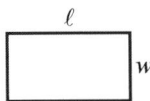
$A = \frac{1}{2}bh$

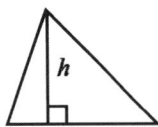
$c^2 = a^2 + b^2$

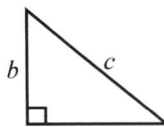

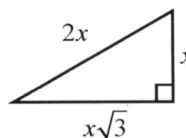
Special Right Triangles

$V = \ell wh$

$V = \pi r^2 h$

$V = \frac{4}{3}\pi r^3$

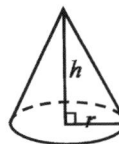
$V = \frac{1}{3}\pi r^2 h$

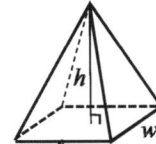
$V = \frac{1}{3}\ell wh$

The number of degrees of arc in a circle is 360.
The number of radians of arc in a circle is 2π.
The number of the measures in degrees of the angles of a triangle is 180.

CONTINUE

1

If $x - 2y = 10$, $y = z + 1$, and $z = 2$, what is the value of x?

A) 12

B) 14

C) 16

D) 18

2

$$2x + 6y = 5$$
$$ax + by = 7$$

If the system of equations above has only one solution, which of the following could be the values of a and b?

A) $a = 1$ and $b = 3$

B) $a = 2$ and $b = 6$

C) $a = 3$ and $b = 8$

D) $a = 4$ and $b = 12$

3

A smart phone company plans to produce and sell p smart phones. The cost of producing p phones is given by $265,000 + 150p$ in dollars. The company receives $400 on the sale of each phone, so the revenue for selling p phones is given by $400p$. For what value of p is the revenue equal to the cost?

A) 500

B) 840

C) 1060

D) 1200

4

$$\left(a + \frac{1}{a}\right)^2 - 2$$

Which of the following is equivalent to the expression above?

A) $a^2 + \dfrac{1}{a^2}$

B) $a^2 + \dfrac{1}{a^2} - 2$

C) $a^2 - 2a + \dfrac{1}{a^2}$

D) $a^2 + 2 + \dfrac{1}{a^2}$

5

$$\sqrt{a^2 - a + 4} = 2$$

If a is a positive number in the equation above, what is the value of a?

A) 10

B) 8

C) 4

D) 1

CONTINUE

6

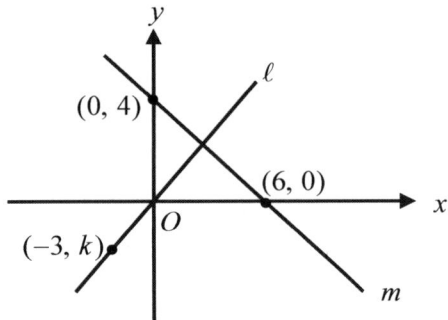

In the xy-plane above, line ℓ is perpendicular to line m. What is the value of k?

A) -1

B) -2

C) -3

D) -4.5

7

If $4a = 2b = c$, what is the average (arithmetic mean) of a, b, and c in terms of a?

A) $\dfrac{4a}{3}$

B) $2a$

C) $\dfrac{7a}{3}$

D) $3a$

8

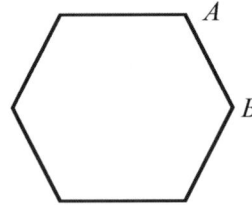

The figure above shows a regular hexagon. If the length of $\overline{AB}$ is 4, what is the area of the hexagon?

A) 24

B) $24\sqrt{3}$

C) 32

D) $32\sqrt{3}$

9

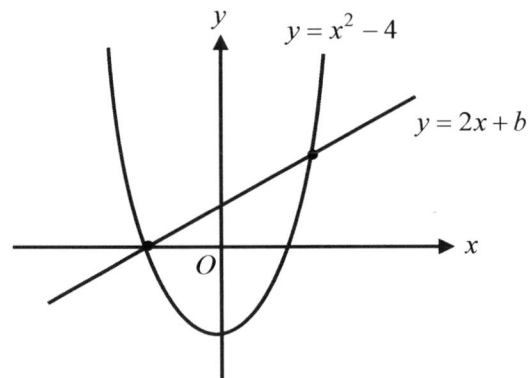

In the xy-plane above, two graphs intersect at two points. What is the value of b?

A) 1

B) 2

C) 3

D) 4

CONTINUE

10

$$\frac{1}{2}\left(\frac{1}{x-1}-\frac{1}{x+1}\right)$$

Which of the following is equivalent to the expression above?

A) $\frac{1}{2}\left(\frac{1}{x^2-1}\right)$

B) $\frac{1}{x^2-1}$

C) $\frac{-2}{x^2-1}$

D) $\frac{-2x}{x^2-1}$

11

The surface area S of a cylinder with radius r and height h is $S = 2\pi r^2 + 2\pi rh$. If the surface area of the cylinder is 20π and the height is 3, what is the value of r?

A) 1

B) 2

C) 4

D) 5

12

$$R = \frac{(m_1 + m_2)}{m_1}$$

The ratio for the kinetic energy between two objects of mass m_1 and m_2 before and after the collision is given above. Which of the following is equivalent to the expression for m_1?

A) $\frac{m_2}{R}$

B) $\frac{R-1}{m_2}$

C) $\frac{m_2}{R-1}$

D) $\frac{m_2 - R}{R}$

13

$$f(x) = x^2 + ax - 10$$

If $f(2) = 0$ in the quadratic function above, which of the following must be true?

A) $f(-5) = 0$

B) $f(-2) = 0$

C) $f(-1) = 0$

D) $f(0) = 0$

CONTINUE

14

$$2a + (4a + 2)i = b - 10i$$

If $i = \sqrt{-1}$ in the equation above, where a and b are constants, what is the value of b?

A)   6

B)   4

C) −3

D) −6

15

Esposito tried to compute the average of his 10 math scores. He mistakenly divided the correct total S of his scores by 8. The result was 5 more than what it should have been. Which of the following would determine the value of S?

A) $10S = 7S + 5$

B) $\dfrac{S}{10} = \dfrac{S}{8} + 5$

C) $\dfrac{S}{8} - \dfrac{S}{10} = 5$

D) $\dfrac{S+5}{10} = \dfrac{S}{8}$

CONTINUE

DIRECTIONS

For questions 16–20, solve the problem and enter your answer in the grid, as described below, on the answer sheet.

1. Although not required, it is suggested that you write your answer in the boxes at the top of the columns to help you fill in the circles accurately. You will receive credit only if the circles are filled in correctly.
2. Mark no more than one circle in any column.
3. No question has a negative answer.
4. Some problems may have more than one answer.

5. **Mixed numbers** such as $3\frac{1}{2}$ must be gridded as 3.5 or 7/2. (If $\boxed{3\,1\,/\,2}$ is entered into the grid, it will be interpreted as $\frac{31}{2}$, not $3\frac{1}{2}$.)

6. **Decimal answers:** If you obtain a decimal answer with more digits than the grid can accommodate, it may be either rounded or truncated, but it must fill the entire grid.

Answer: $\frac{7}{12}$

Write answer in boxes. → Fraction line

Grid in result. →

Answer: 2.5

← Decimal point

Acceptable ways to grid $\frac{2}{3}$ are:

Answer: 201
Either position is correct.

Note: You may start your answers in any column, space permitting. Columns you don't need to use should be left blank.

CONTINUE

16

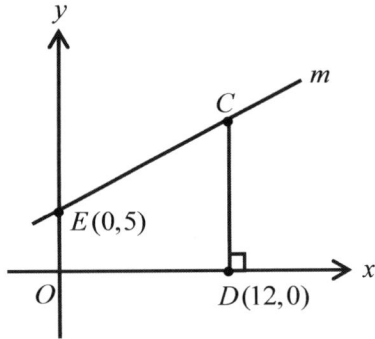

If the slope of line m in the xy-plane above is $\dfrac{1}{3}$, what is the area of quadrilateral $OECD$?

17

Claire and Peter both want to buy new smart phones. Claire has already saved 100 dollars and plans to save 5 dollars per week until she can buy the phone. Peter has 25 dollars and plans to save 8 dollars per week. In how many weeks will Claire and Peter have saved the same amount of money?

18

$$(a-8)x^2 + (b-5)x + c + 2 = 0$$

In the equation above, a, b, and c are constants. If the equation is true for all values of x, what is the value of $a + b + c$?

19

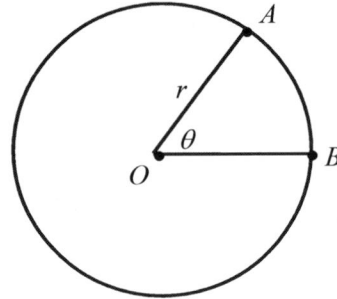

In the figure above, O is the center of the circle with radius r, and the measure of θ is $\dfrac{\pi}{5}$ radians. If the length of minor arc AB is 3π, what is the value of r?

20

In a certain class of 70 students, $\dfrac{4}{7}$ of the students are boys, and the ratio of students 10 years or older to students less than 10 years is 2:3. If $\dfrac{2}{3}$ of the girls are less than 10 years old, how many boys are 10 years old or older?

STOP

If you finish before time is called, you may check your work on this section only.
Do not turn to any other section in the test.

No Test Material on This Page

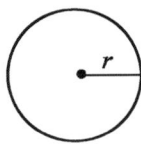
Math Test - Calculator

55 MINUTES, 38 QUESTIONS

Turn to Section 4 of your answer sheet to answer the questions in this section.

DIRECTIONS

For questions 1-30, solve each problem, choose the best answer from the choices provided, and fill in the corresponding circle on your answer sheet. **For questions 31-38,** solve the problem and enter your answer in the grid on your answer sheet. Please refer to the directions before question 31 on how to enter your answers in the grid. You may use any available space in your test booklet for scratch work.

NOTE

1. The use of a calculator **is permitted**.

2. All variables and expressions used represent real numbers unless otherwise indicated.

3. Figures provided in this test are drawn to scale unless otherwise indicated.

4. All figures lie in a plane unless otherwise indicated.

5. Unless otherwise indicated, the domain of a given function f is the set of all real numbers x for which $f(x)$ is a real number.

REFERENCE

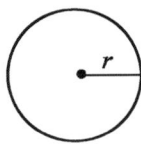 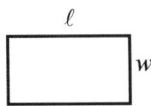 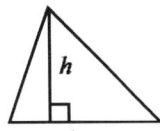 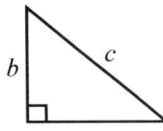 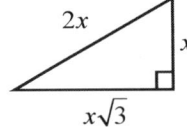 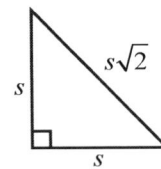

$A = \pi r^2$
$C = 2\pi r$

$A = \ell w$

$A = \frac{1}{2}bh$

$c^2 = a^2 + b^2$

Special Right Triangles

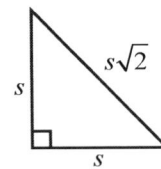   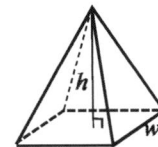

$V = \ell wh$

$V = \pi r^2 h$

$V = \frac{4}{3}\pi r^3$

$V = \frac{1}{3}\pi r^2 h$

$V = \frac{1}{3}\ell wh$

The number of degrees of arc in a circle is 360.

The number of radians of arc in a circle is 2π.

The number of the measures in degrees of the angles of a triangle is 180.

CONTINUE

1

During its Labor Day sale, a store advertises that $40 will be deducted from every purchase over $200. In addition, after the deduction is taken, the store offers an early-bird discount of 40% to any person who makes a purchase before 9 a.m. If Claire makes a purchase of k dollars, $k > 200$, at 8 a.m., which of the following expressions represents the cost of her purchase?

A) $0.4k - 16$

B) $0.4k - 24$

C) $0.6k - 24$

D) $0.6k - 40$

2

On a map, 3 centimeters represents k kilometers. How many kilometers are represented by p centimeters?

A) $3pk$

B) $\dfrac{k}{3p}$

C) $\dfrac{3k}{p}$

D) $\dfrac{pk}{3}$

3

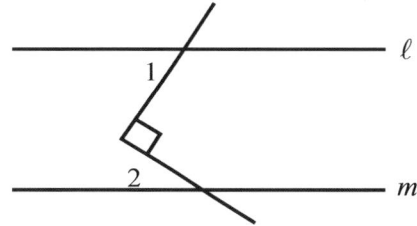

In the figure above, lines ℓ and m are parallel. If the measure of $\angle 1$ is $20°$ more than the measure of $\angle 2$, what is the measure of $\angle 1$?

A) $35°$

B) $45°$

C) $55°$

D) $75°$

4

$$T = 150 + 20w$$

Cassy plans to buy a new computer, and plans to save $20 each week for the next w weeks. The total amount of money she saved is represented by the equation above, where T is the total amount. Which of the following is the best interpretation of the number 150 in the equation?

A) The new computer costs $150.

B) She saved $150 each week.

C) She wants to buy a computer when she saves $150.

D) She has already saved $150 toward the cost of a new computer.

CONTINUE

Questions 5 and 6 refer to the following information.

Dog age (D)	0	2	4	6	8	$\cdots$	15
Human age (H)	a	10	20	30	40	$\cdots$	b

The chart above shows equivalent ages for dogs and humans. Human age is directly proportional to dog age.

5

What is the value of $a + b$?

A) 60

B) 75

C) 80

D) 85

6

Which of the following graphs best represents the relationship between dog and human ages?

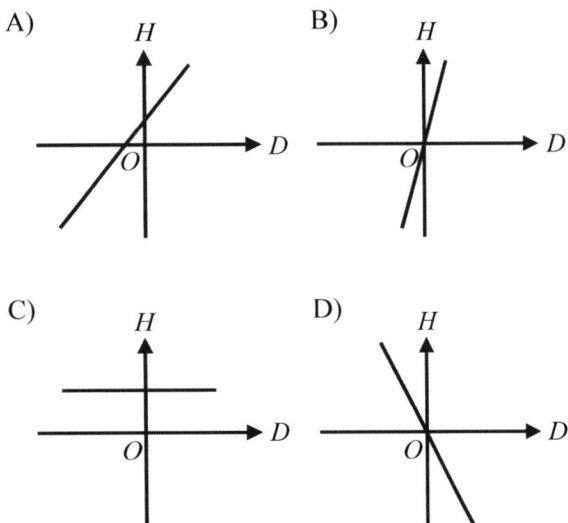

A)

B)

C)

D)

7

In the fraction $\dfrac{a-5}{2b}$, a is 5 less than two times b. If the fraction is equal to $\dfrac{1}{2}$, what is the value of a?

A) 15

B) 20

C) 25

D) 30

8

Tyler spent 60 dollars at an amusement park for admission and rides. If he paid $10 for admission, and rides cost $3 each, what is the maximum number of rides that he went on?

A) 16

B) 17

C) 18

D) 20

CONTINUE

9

For a school summer concert, one type of ticket costs $5 and another costs $10. The supervisor of the concert can sell at most 500 tickets, but the gross receipts must total at least $3,000 in order for the concert to be held. Which of the following systems of inequalities could represent this relationship?

A) $\begin{cases} 5x + 10y \geq 3000 \\ x + y \leq 500 \\ x \geq 0 \\ y \geq 0 \end{cases}$

B) $\begin{cases} \dfrac{5}{x} + \dfrac{10}{y} \leq 500 \\ x + y \leq 3000 \\ x \geq 0 \\ y \geq 0 \end{cases}$

C) $\begin{cases} 5x + 10y \leq 3000 \\ x + y \leq 500 \\ x \geq 0 \\ y \geq 0 \end{cases}$

D) $\begin{cases} 5x + 10y > 3000 \\ x + y < 500 \\ x > 0 \\ y > 0 \end{cases}$

10

If a linear function f satisfies $f(3) = 10$ and $f(7) = 18$, what is the value of $f(5)$?

A) 12

B) 14

C) 15

D) 16

Questions 11 and 12 refer to the following information.

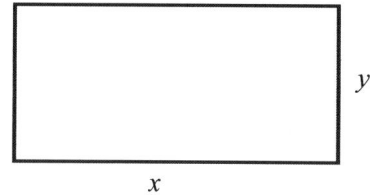

A rancher has 100 feet of fencing to enclose rectangular region as shown above. The length and width are represented by x and y respectively.

11

Which of the following expressions represents the area of the rectangular region as a function of x?

A) $100x - x^2$

B) $50x - x^2$

C) $50x + x^2$

D) $50x^2$

12

If the value of y is 25, what is the area of the rectangular region in square feet?

A) 325

B) 625

C) 1250

D) 2500

13

Factory Workers over 60		
Year	Percent of Men	Percent of Women
1990	19.6	13.5
2000	23.6	10.8

The table above shows the percent of men and women 60 years and older who were working in a certain factory in the U.S. in the given years. If the rate of increase or decrease every year is constant, which of the following represents the percent of men over 60 who were working in the factory in the year 2015?

A) 26.6

B) 27.2

C) 29.6

D) 30.5

14

Mary is making a rectangle whose perimeter is less than 100 inches. If the dimensions of the rectangle are integers, what is the largest possible area for the rectangle in square inches?

A) 600

B) 625

C) 650

D) 800

15

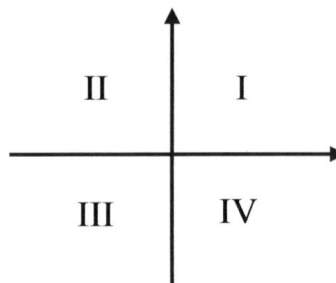

If $z = 3 + 2i$ is in the first quadrant of the complex number plane above, then which quadrant contains z^2 ?

A) Quadrant I

B) Quadrant II

C) Quadrant III

D) Quadrant IV

16

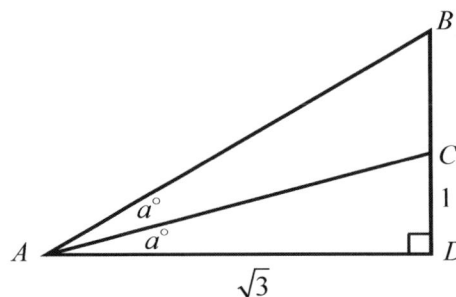

In the figure above, $AD = \sqrt{3}$ and $CD = 1$. What is the length of $\overline{AB}$?

A) 2

B) 3

C) $2\sqrt{3}$

D) $3\sqrt{3}$

CONTINUE

Questions 17 and 18 refer to the following information.

The number of bacteria in a controlled laboratory environment is defined by the function

$f(x) = 1000 \times b^x$, where x is the time in hours.

The graph of f is shown in the xy-plane below.

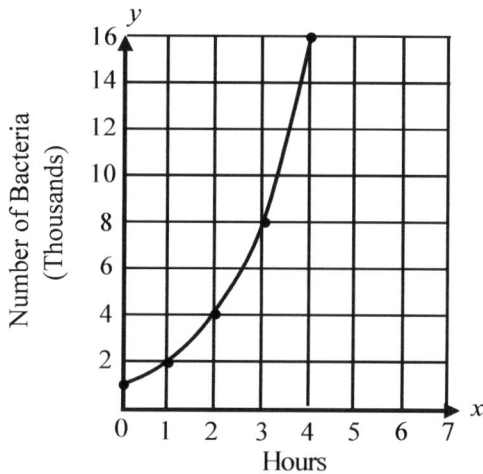

17

What is the value of b?

A) 1

B) 2

C) 3

D) 4

18

What is the number of bacteria in 5 hours?

A) 27,000

B) 32,000

C) 40,000

D) 64,000

19

A school nurse chose 50 girls from the seventh grade and measured their weight, in pounds, shown in the table below.

Measure of Weight	Frequency
70	5
75	8
78	20
80	10
85	5
90	2

If there is a total of 500 girls in the seventh grade, what could be the possible number of girls in the median measure of weight for the entire grade?

A) 80

B) 100

C) 150

D) 200

20

$$x^2 - 2x + y^2 = 10$$

The equation of a circle in the xy-plane is shown above. What is the center of the circle?

A) $(1, -1)$

B) $(1, 1)$

C) $(1, 0)$

D) $(2, 0)$

CONTINUE

21

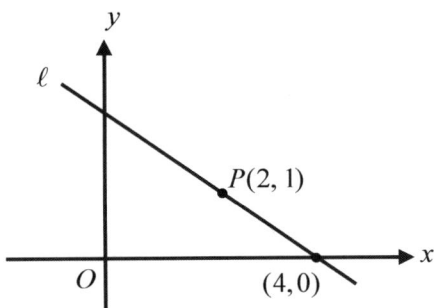

The graph of line ℓ is shown in the xy-plane above.

Line m (not shown) has the equation $y = ax + b$, where a and b are constants. If line m is perpendicular to line ℓ and passes through point P, what is the value of b?

A) 0

B) −1

C) −2

D) −3

22

$$2^{3k-3} = 64$$

In the equation above, what is the value of 2^k ?

A) 4

B) 8

C) 16

D) 32

23

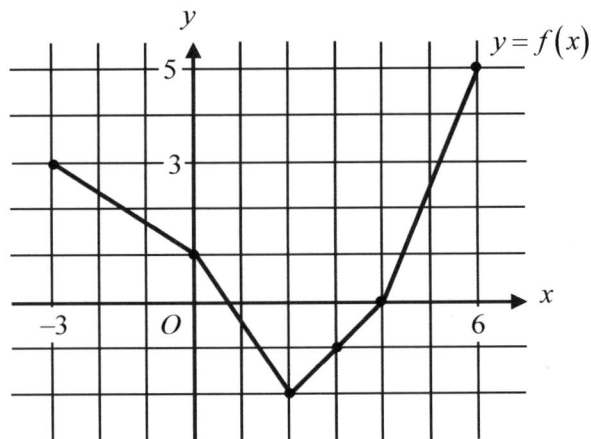

The complete graph of the function f is shown in the xy-plane above. Which of following is true?

A) $f(0) > |f(0)|$

B) $f(2.2) < |f(2.2)|$

C) $f(3) > |f(3)|$

D) $f(-2) < |f(-2)|$

24

$$y = -\frac{1}{10}x^2 + k$$
$$y = 5$$

In the system of equations above, k is a constant. For which of the following values of k does the system of equations have no real solution?

A) 10

B) 8.5

C) 5

D) −0.05

CONTINUE

25

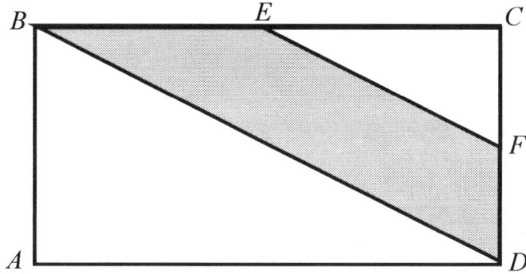

In the figure above, E and F are the mid points of two sides of a rectangle. If the area of $\triangle CEF$ is 10, what is the area of the shaded region?

A) 15

B) 20

C) 25

D) 30

26

$$x - y = 5$$
$$xy = 10$$

In the equations above, what is the value of $x^2 + y^2$?

A) 15

B) 25

C) 36

D) 45

27

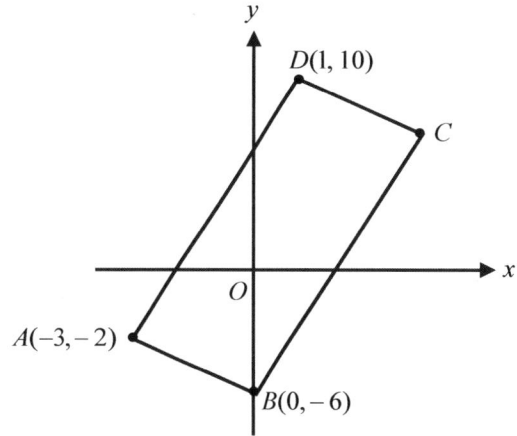

In the xy-plane above, the figure shows the coordinates of points A, B, C, and D of a parallelogram. Which of the following are the coordinates of point C?

A) $(3, 5)$

B) $(4, 6)$

C) $(5, 5)$

D) $(6, 5)$

28

$$a^2 + b^2 \leq 25$$
$$b \geq 3$$

In the inequalities above, what is the greatest possible value of a?

A) -4

B) -3

C) $\ \ 3$

D) $\ \ 4$

CONTINUE

29

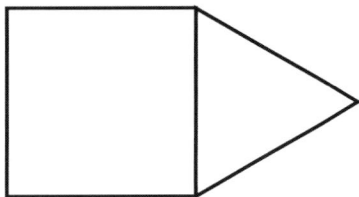

The figure above shows a square and an equilateral triangle. If the area of the triangle is $25\sqrt{3}$ square inches, what is the area, in square inches, of the square?

A) $50\sqrt{3}$

B) 100

C) $100\sqrt{3}$

D) 125

30

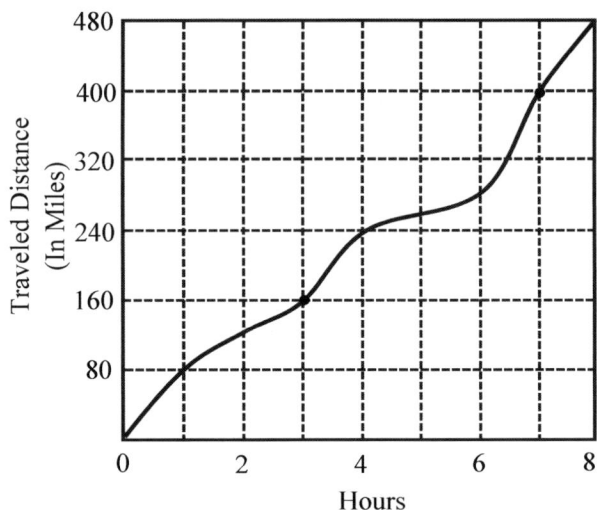

The graph above compares the distance with the number of hours that a car traveled. Which of the following is the average speed, in miles per hour, of the car during the time between 3 and 7 hours?

A) 50

B) 55

C) 60

D) It cannot be determined from the given information.

CONTINUE

DIRECTIONS

For questions 31-38, solve the problem and enter your answer in the grid, as described below, on the answer sheet.

Answer: $\frac{7}{12}$

Write answer in boxes.

Fraction line

Grid in result.

Answer: 2.5

Decimal point

1. Although not required, it is suggested that you write your answer in the boxes at the top of the columns to help you fill in the circles accurately. You will receive credit only if the circles are filled in correctly.

2. Mark no more than one circle in any column.

3. No question has a negative answer.

4. Some problems may have more than one answer.

5. **Mixed numbers** such as $3\frac{1}{2}$ must be gridded as 3.5 or 7/2. (If $\boxed{3\,1\,/\,2}$ is entered into the grid, it will be interpreted as $\frac{31}{2}$, not $3\frac{1}{2}$.)

6. **Decimal answers:** If you obtain a decimal answer with more digits than the grid can accommodate, it may be either rounded or truncated, but it must fill the entire grid.

Acceptable ways to grid $\frac{2}{3}$ are:

Answer: 201
Either position is correct.

Note: You may start your answers in any column, space permitting. Columns you don't need to use should be left blank.

CONTINUE

31

Jackie goes on a 30-mile bike ride every Sunday. He rides the distance in 3 hours. At this rate, how many miles can he ride in 5 hour and 30 minutes?

32

The average of a set of 8 consecutive odd integers is 18. What is the greatest of these 8 integers?

33

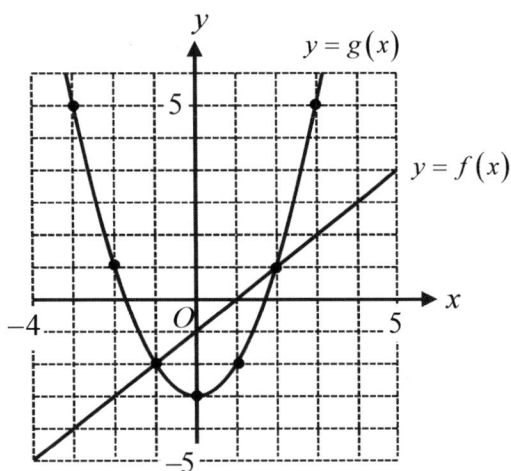

The graphs of a linear function f and a quadratic function g are shown in the xy-plane above. If $f(g(k)) = -3$, what is the value of $|k|$?

34

In the xy-plane, line $x = 2$ is the axis of symmetry of the graph of $f(x) = 5x^2 - kx + 2$. What is the value of k?

35

Twenty grams of solution P is 10% alcohol and 30 grams of solution Q is 20% alcohol by mass. If these two solutions are mixed together, what is the percent of alcohol in the mixture? (Disregard the % sign when gridding your answer.)

CONTINUE

36

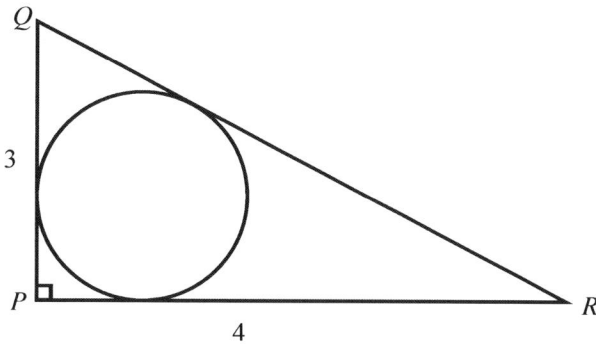

In the figure above, a circle is inscribed in $\triangle PQR$.
If $PQ = 3$ and $PR = 4$, what is the radius of the circle?

Questions 37 and 38 refer to the following information.

The total cost of an internet phone-call is the sum of

(1) a basic fixed charge for using the internet and
(2) an additional charge for each minute that is used.

The total cost of a 20 minute-call is $24 and the total cost of a 35 minute-call is $31.50.
(Disregard the $ sign when gridding your answer.)

37

What is the basic fixed charge, in dollars, for using the internet?

38

What is the total cost, in dollars, of a 40 minute-call?

STOP

**If you finish before time is called, you may check your work on this section only.
Do not turn to any other section in the test.**

Math Conversion Table

Raw Score	Scaled Score	Raw Score	Scaled Score
58	800	27	500
57	800	26	490
56	800	25	480
55	800	24	470
54	790	23	460
53	780	22	460
52	770	21	450
51	750	20	440
50	740	19	430
49	730	18	430
48	720	17	420
47	710	16	420
46	700	15	410
45	690	14	400
44	670	13	390
43	680	12	380
42	670	11	370
41	660	10	360
40	650	9	450
39	640	8	340
38	630	7	330
37	620	6	310
36	610	5	290
35	600	4	280
34	590	3	270
33	580	2	260
32	560	1	240
31	550	0	200
30	540		
29	530		
28	520		

Answer Explanations

Test 2   Answers and Explanations

SECTION 3	1	2	3	4	5	6	7	8	9	10
	C	C	C	A	D	D	C	B	D	B
	11	12	13	14	15	16	17	18	19	20
	B	C	A	D	C	84	25	11	15	18

SECTION 4	1	2	3	4	5	6	7	8	9	10
	C	D	C	D	B	B	A	A	A	B
	11	12	13	14	15	16	17	18	19	20
	B	B	C	A	A	C	B	B	D	C
	21	22	23	24	25	26	27	28	29	30
	D	B	B	D	D	D	B	D	B	C
	31	32	33	34	35	36	37	38		
	55	25	1	20	16	1	14	34		

SECTION 3

1. **C**

    $z = 2 \;\rightarrow\; y = z + 1 = 3 \;\rightarrow\; x - 2(3) = 10 \;\rightarrow\; x = 16$

2. **C**

    To have only one solution, the slopes should be different. $\dfrac{2}{a} \neq \dfrac{6}{b} \;\rightarrow\;$ C) $\dfrac{2}{3} \neq \dfrac{6}{8}$

3. **C**

    $265000 + 150p = 400p \;\rightarrow\; 265000 = 250p \;\rightarrow\; p = 1060$

4. **A**

    $\left(a + \dfrac{1}{a}\right)^2 - 2 = a^2 + 2 + \dfrac{1}{a^2} - 2 = a^2 + \dfrac{1}{a^2}$

5. **D**

    $\sqrt{a^2 - a + 4} = 2 \;\rightarrow\; a^2 - a + 4 = 4 \;\rightarrow\; a(a - 1) = 0 \;\rightarrow\; a = 0 \text{ or } 1 \; (a > 0)$

Answer Explanations

6. **D**

   Slope of line $m = -\dfrac{2}{3}$ → slope of line $m = \dfrac{3}{2}$ → equation of line ℓ is $y = \dfrac{3}{2}x$.

   For $(-3, k)$, → $k = \dfrac{3}{2}(-3)$ → $k = -\dfrac{9}{2}$

7. **C**

   Since $b = 2a$ and $c = 4a$, → $\dfrac{a+b+c}{3} = \dfrac{a+2a+4a}{3} = \dfrac{7a}{3}$.

8. **B**

   The hexagon has 6 equilateral triangles with side of 4. The area of a equilateral triangle with side s is $\dfrac{s^2\sqrt{3}}{4}$.

   Therefore, the area oh the hexagon is $\dfrac{16\sqrt{3}}{4} \times 6 = 24\sqrt{3}$.

9. **D**

   $x^2 - 4 = 0$ → $(x+2)(x-2) = 0$ → $x = 2, -2$

   The graph of $y = 2x + b$ passes through point $(-2, 0)$. → $0 = 2(-2) + b$ → $b = 4$

10. **B**

    $\dfrac{1}{2}\left(\dfrac{1}{x-1} - \dfrac{1}{x+1}\right) = \dfrac{1}{2}\left(\dfrac{x+1-(x-1)}{x^2-1}\right) = \dfrac{1}{2}\left(\dfrac{2}{x^2-1}\right) = \dfrac{1}{x^2-1}$

11. **B**

    $S = 2\pi r^2 + 2\pi rh$ → $20\pi = 2\pi r^2 + 2\pi r(3)$ → $r^2 + 3r - 10 = 0$ → $(r+5)(r-2) = 0$ → $r = 2$

12. **C**

    $R = \dfrac{(m_1 + m_2)}{m_1}$ → $Rm_1 = m_1 + m_2$ → $Rm_1 - m_1 = m_2$ → $m_1(R-1) = m_2$ → $m_1 = \dfrac{m_2}{R-1}$

13. **A**

    $f(x) = x^2 + ax - 10$ → $f(2) = 4 + 2a - 10 = 0$ → $a = 3$ → Therefore, $f(x) = x^2 + 3x - 10$

    $x^2 + 3x - 10 = 0$ → $(x+5)(x-2) = 0$ → $x = -5, 2$ → $f(-5) = 0$

14. **D**

    $2a = b$ and $4a + 2 = -10$ → $a = -3$ → $b = -6$

15. **C**

    $\dfrac{S}{8} = \dfrac{S}{10} + 5$

16. **84**

The equation of line m $\rightarrow$ $y = \dfrac{1}{3}x + 5$ $\rightarrow$ $f(12) = \dfrac{1}{3}(12) + 5 = 9$ $\rightarrow$ $CD = 9$

$\text{Area} = \dfrac{(5+9) \times 12}{2} = 84$

17. **25**

$100 + 5x = 25 + 8x$ $\rightarrow$ $3x = 75$ $\rightarrow$ $x = 25 \text{ weeks}$

18. **11**

$a = 8$, $b = 5$, and $c = -2$ $\rightarrow$ Therefore, $a + b + c = 11$.

19. **15**

$r\theta = \ell$ $\rightarrow$ $r\left(\dfrac{\pi}{5}\right) = 3\pi$ $\rightarrow$ $r = 15$

20. **18**

	Boys	Girls	Total
10 ↑	18	10	28
10 ↓	22	20	42
Total	40	30	70

SECTION 4

1. **C**

$0.6(k - 40) = 0.6k - 24$

2. **D**

$\dfrac{3}{k} = \dfrac{p}{x}$ $\rightarrow$ $x = \dfrac{pk}{3}$

3. **C**

$\angle 1 + \angle 2 = 90$ and $\angle 1 = \angle 2 + 20$ $\rightarrow$ $\angle 2 + 20 + \angle 2 = 90$ $\rightarrow$ $\angle 2 = 35$ $\rightarrow$ $\angle 1 = 35 + 20 = 55$

4. **D**

5. **B**

Slope is $\dfrac{20 - 10}{4 - 2} = 5$ $\rightarrow$ The equation is $y = 5x + a$. $\rightarrow$ $a = 0$ $\rightarrow$ when $x = 15$, $b = 5(15) = 75$.

Therefore, $a + b = 0 + 75 = 75$.

Answer Explanations

6. **B**

   The graph of $y = 5x$ is B.

7. **A**

   Since $a = 2b - 5$, $\dfrac{a-5}{2b} = \dfrac{2b-5-5}{2b} = \dfrac{2b-10}{2b} = \dfrac{1}{2}$. $\rightarrow$ $b = 10$ $\rightarrow$ $a = 2(10) - 5 = 15$

8. **A**

   $\dfrac{60-10}{3} \approx 16.6$ $\rightarrow$ 16 rides

9. **A**

10. **B**

    Slope $= \dfrac{18-10}{7-3} = 2$ $\rightarrow$ $\dfrac{f(5)-f(3)}{5-3} = 2$ $\rightarrow$ $\dfrac{f(5)-10}{2} = 2$ $\rightarrow$ $f(5) = 14$

11. **B**

    Since $x + y = 50$, $y = 50 - x$. $\rightarrow$ Area $= xy = x(50-x) = 50x - x^2$

12. **B**

    If $y = 25$, $x = 25$. Area $= 25 \times 25 = 625$

13. **C**

    Since 4 increases over 10 years, $\dfrac{4}{10} = \dfrac{x}{15}$ $\rightarrow$ $x = 6$ from 2000 $\rightarrow$ $23.6 + 6 = 29.6$

14. **A**

    Since $a + b = 99$, the largest possible area is $24 \times 25 = 600$.

15. **A**

    $z = 3 + 2i$ $\rightarrow$ $z^2 = (3+2i)^2 = 5 + 12i$ $\rightarrow$ Quadrant I

16. **C**

    $a = 30°$ and $\angle A = 60°$ $\rightarrow$ $AB = 2\sqrt{3}$

17. **B**

    At $x = 3$, $8000 = 1000b^3$. $\rightarrow$ $b = 2$

18. **B**

    $f(5) = 1,000(2)^5 = 32,000$

19. **D**

There are 20 girls in the median measure. Therefore, total number of girls is $\dfrac{20}{50} = \dfrac{x}{500} \rightarrow x = 200$.

20. C

$$x^2 - 2x + y^2 = 10 \quad \rightarrow \quad (x-1)^2 + y^2 = 11 \quad \rightarrow \quad \text{Center is at } (1, 0)$$

21. D

Slope of line ℓ is

$$-\dfrac{1}{2}. \quad \rightarrow \quad a = 2 \quad \rightarrow \quad y = 2x + b \quad \rightarrow \text{Putting } (2,1) \text{ in the equation} \quad \rightarrow \quad 1 = 2(2) + b \quad \rightarrow \quad b = -3$$

22. B

$$2^{3k-3} = 64 \quad \rightarrow \quad 2^{3k-3} = 2^6 \quad \rightarrow \quad 3k - 3 = 6 \quad \rightarrow \quad 3k = 9 \quad \rightarrow \quad k = 3 \quad \rightarrow \quad \text{Therefore, } 2^3 = 8.$$

23. B

24. D

The quadratic graph must be below the graph of $y = 5$. Therefore, D is correct.

25. D

Ratio of areas of $\triangle BDC : \triangle EFC = 4 : 1$. The area of the shaded region is three times the area of $\triangle EFC$. Therefore, the area is $3 \times 10 = 30$.

26. D

$$x - y = 5 \quad \rightarrow \quad (x-y)^2 = 25 \quad \rightarrow \quad x^2 + y^2 - 2xy = 25 \quad \rightarrow \quad x^2 + y^2 = 25 + 2xy = 25 + 20 = 45$$

27. B

If $C(a,b) \rightarrow$ Mid point $\left(\dfrac{a + (-3)}{2}, \dfrac{b + (-2)}{2} \right) = \left(\dfrac{1+0}{2}, \dfrac{10 + (-6)}{2} \right)$.

$a + (-3) = 1 \rightarrow a = 4$ and $b + (-2) = 10 + (-6) \rightarrow b = 6 \rightarrow C(4, 6)$

28. D

If $b = 3$, a is the greatest. $a^2 + 9 = 25 \rightarrow a^2 = 16 \rightarrow a = 4$

Or

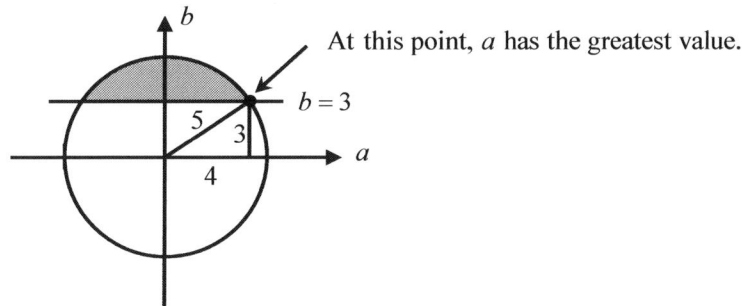

At this point, a has the greatest value.

$b = 3$

29. B

Since $\dfrac{s^2 \sqrt{3}}{4} = 25\sqrt{3}$, $s = 10$. $\rightarrow$ Therefore, the area of the aquare is 100.

30. C

$$\frac{400-160}{7-3}=60$$

31. 55

$$\frac{30}{3}=\frac{x}{5.5} \quad \rightarrow \quad x=55$$

32. 25

Median = average $\rightarrow$ The fifth number is 19. $\rightarrow$ Therefore, $a_8=19+2\times3=25$.

33. 1

$$f\big(g(k)\big)=-3 \quad \rightarrow \quad g(k)=-2 \quad \rightarrow \quad k=1 \text{ or } -1 \quad \rightarrow \quad |k|=1$$

34. 20

Axis of symmetry $=\dfrac{-(-k)}{2(5)}=\dfrac{k}{10}=2 \quad \rightarrow \quad k=20$

35. 16

The amount of alcohol before and after must be same.

$$\frac{10}{100}\times20+\frac{20}{100}\times30=\frac{x}{100}\times50 \quad \rightarrow \quad 200+600=50x \quad \rightarrow \quad x=16 \quad \rightarrow \quad 16\%$$

36. 1

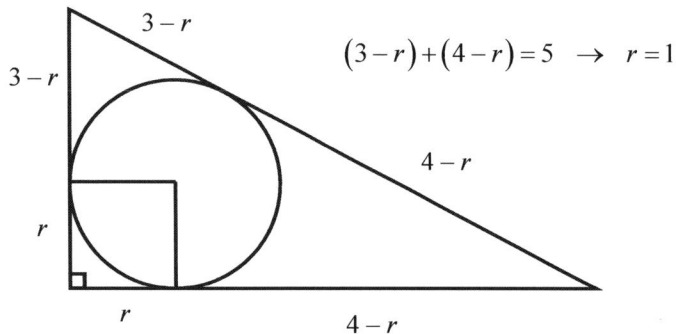

$$(3-r)+(4-r)=5 \quad \rightarrow \quad r=1$$

37. 14

Define: fixed charge $=a$ and charge per minute $=b$

$C=a+20b=24$ and $C=a+35b=31.5 \quad \rightarrow \quad a=14$ and $b=0.5$

38. 34

$$C=14+40(0.5)=\$34$$

PRACTICE TEST 3

Dr. John Chung's SAT Math

Math Test - No Calculator

25 MINUTES, 20 QUESTIONS

Turn to Section 3 of your answer sheet to answer the questions in this section.

DIRECTION

For questions 1–15, solve each problem, choose the best answer from the choices provided, and fill in the corresponding circle on your answer sheet. **For questions 16–20,** solve the problem and enter your answer in the grid on your answer sheet. Please refer to the directions before question 16 on how to enter your answers in the grid. You may use any available space in your test booklet for scratch work.

NOTE

1. The use of a calculator **is not permitted**.

2. All variables and expressions used represent real numbers unless otherwise indicated.

3. Figures provided in this test are drawn to scale unless otherwise indicated.

4. All figures lie in a plane unless otherwise indicated.

5. Unless otherwise indicated, the domain of a given function f is the set of all real numbers x for which $f(x)$ is a real number.

REFERENC

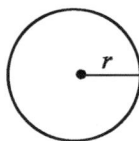 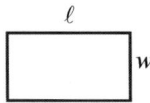 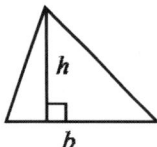 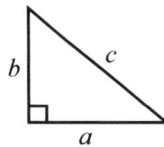 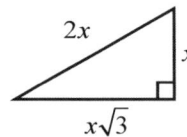 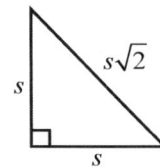

$A = \pi r^2$
$C = 2\pi r$

$A = \ell w$

$A = \dfrac{1}{2}bh$

$c^2 = a^2 + b^2$

Special Right Triangles

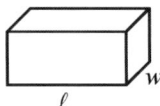  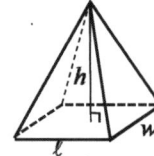

$V = \ell w h$

$V = \pi r^2 h$

$V = \dfrac{4}{3}\pi r^3$

$V = \dfrac{1}{3}\pi r^2 h$

$V = \dfrac{1}{3}\ell w h$

The number of degrees of arc in a circle is 360.
The number of radians of arc in a circle is 2π.
The number of the measures in degrees of the angles of a triangle is 180.

CONTINUE ➡

1

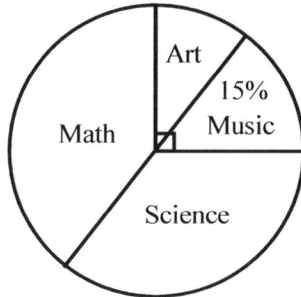

A total of 40 students in Mr. Lee's class voted for their favorite subject. The results are shown in the pie chart above. How many students voted for math?

A) 12

B) 14

C) 16

D) 18

2

If $3r + 5 = 10$, what is the value of $6r + 5$?

A) 10

B) 15

C) 20

D) 21

3

If $a^{-2} = \dfrac{1}{5}$, what is the value of $5a^2$?

A) 1

B) 5

C) 10

D) 25

4

When a certain number p is divided by 10, the quotient is k and the remainder is r. Which of the following expressions represents r?

A) $r = p - 10k$

B) $r = 10p - k$

C) $r = 10(k - p)$

D) $r = 10k - p$

CONTINUE

5

If $\dfrac{5}{12} = \dfrac{1}{a} + \dfrac{1}{b}$ and $ab = 24,$ what is the value of $a + b?$

A) 25

B) 13

C) 11

D) 10

6

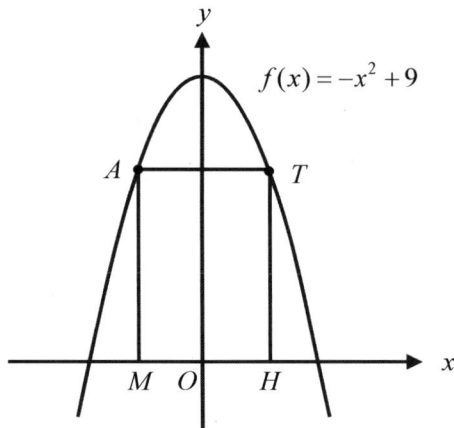

$f(x) = -x^2 + 9$

The graph of function f is shown in the xy-plane above. If length of $\overline{MA}$ of the rectangle $MATH$ is 5, what is the length of $\overline{AT}$?

A) 2

B) 2.5

C) 3

D) 4

7

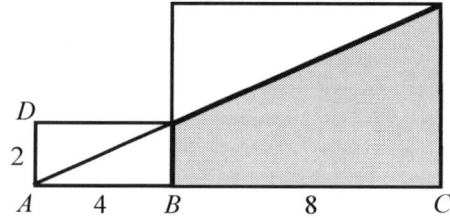

Two rectangles are shown in the figure above. If $AB = 4$, $AD = 2$, and $BC = 8$, what is the area of the shaded region?

A) 32

B) 36

C) 48

D) 64

8

$$ax - by = 9$$
$$3x + y = 3$$

If the system of linear equations above has infinitely many solutions, what is the value of $a + b?$

A) −3

B) 6

C) 9

D) 12

CONTINUE

x	$g(x)$
-3	6
-2	0
0	-6
2	-2
3	0
4	6

The function g is defined by a polynomial. Some selected values of x and $g(x)$ are shown in the table above. Which of the following is true?

    I. $(x-3)$ is a factor of $g(x)$.

    II. $(x-2)$ is a factor of $g(x)$.

    III. $(x+2)$ is a factor of $g(x)$.

A) I and II only

B) I and III only

C) II and III only

D) I, II, and III

10

If y is inversely proportional to x^2, and $y=10$ when $x=2$, what is the value of y when $x=10$?

A) $\dfrac{2}{5}$

B) 2

C) 50

D) 250

$$y = k(x-4)(x+2)$$

The graph of the quadratic equation above, where k is a constant, has a vertex at point (a, b) in the xy-plane. Which of the following is equal to a?

A) -1

B) 0

C) 1

D) 2

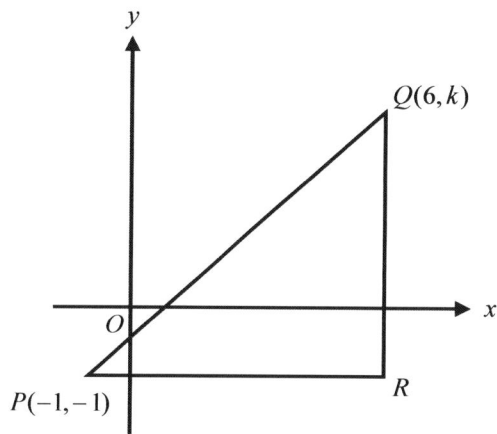

The figure PQR in the xy-plane is an isosceles right triangle. Which of the following is equal to k?

A) 6

B) 7

C) 8

D) 9

13

$$\frac{2i}{1-i} = a + bi$$

If $i = \sqrt{-1}$ in the equation above, where a and b are constants, what is the value of a?

A) -1

B) 1

C) 2

D) 3

14

$$\frac{1}{x} = \frac{x}{2x+1}$$

What are the solutions to the equation above?

A) $x = -1 \pm \sqrt{2}$

B) $x = 1 \pm \sqrt{2}$

C) $x = 1 \pm \sqrt{3}$

D) $x = \dfrac{1 \pm \sqrt{2}}{2}$

15

$$P = \frac{9}{2}K + 40$$

The equation above shows how the value of P relates to the value of K. Based on the equation, which of the following must be true?

I.   When the value of K increases by 1, the value of P increases by 40.

II.   When the value of K increases by 2, the value of P increases by 9.

III. When the value of K increases by 4, the value of P increases by 18.

A) I and II only

B) I and III only

C) II and III only

D) I, II, and III

CONTINUE

DIRECTIONS

For questions 16–20, solve the problem and enter your answer in the grid, as described below, on the answer sheet.

1. Although not required, it is suggested that you write your answer in the boxes at the top of the columns to help you fill in the circles accurately. You will receive credit only if the circles are filled in correctly.
2. Mark no more than one circle in any column.
3. No question has a negative answer.
4. Some problems may have more than one answer.
5. **Mixed numbers** such as $3\frac{1}{2}$ must be gridded as 3.5 or 7/2. (If $\boxed{3\ 1\ /\ 2}$ is entered into the grid, it will be interpreted as $\frac{31}{2}$, not $3\frac{1}{2}$.)
6. **Decimal answers:** If you obtain a decimal answer with more digits than the grid can accommodate, it may be either rounded or truncated, but it must fill the entire grid.

Answer: $\frac{7}{12}$ Answer: 2.5

Write answer in boxes.

Grid in result.

Fraction line

Decimal point

Acceptable ways to grid $\frac{2}{3}$ are:

Answer: 201
Either position is correct.

Note: You may start your answers in any column, space permitting. Columns you don't need to use should be left blank.

CONTINUE

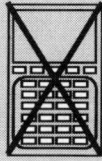

16

$$x^2 - ax = -10$$

The quadratic equation above has two real solutions. If one of the solutions is 5 and a is a constant, what is the other solution?

17

$$\frac{15}{x-1} - 7 = 3 - \frac{5}{x-1}$$

If $x > 1$, what is the solution to the equation above?

18

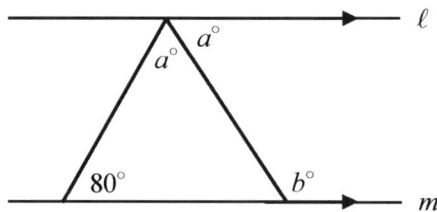

In the figure above, line ℓ is parallel to line m. What is the value of b?

19

At a certain party, an executive committee provided one soda for 8 people, one large bag of chips for 4 people, and one cheese cake for 6 people. If the total number of sodas, large bag of chips, and cheese cakes was 78, how many people were at the party?

20

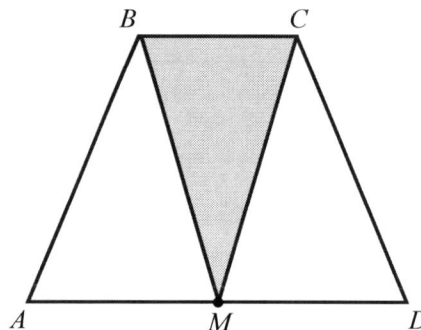

The figure above shows trapezoid $ABCD$. If M is the midpoint of $\overline{AD}$ and $AD = 3 \cdot BC$, what fraction of the area of the trapezoid is shaded?

STOP

If you finish before time is called, you may check your work on this section only.
Do not turn to any other section in the test.

No Test Material on This Page

Math Test - Calculator

55 MINUTES, 38 QUESTIONS

Turn to Section 4 of your answer sheet to answer the questions in this section.

DIRECTIONS

For questions 1-30, solve each problem, choose the best answer from the choices provided, and fill in the corresponding circle on your answer sheet. **For questions 31-38**, solve the problem and enter your answer in the grid on your answer sheet. Please refer to the directions before question 31 on how to enter your answers in the grid. You may use any available space in your test booklet for scratch work.

NOTE

1. The use of a calculator **is permitted**.

2. All variables and expressions used represent real numbers unless otherwise indicated.

3. Figures provided in this test are drawn to scale unless otherwise indicated.

4. All figures lie in a plane unless otherwise indicated.

5. Unless otherwise indicated, the domain of a given function f is the set of all real numbers x for which $f(x)$ is a real number.

REFERENCE

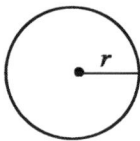 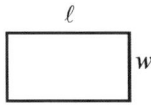 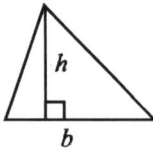 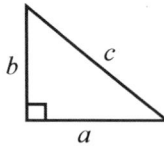 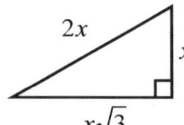 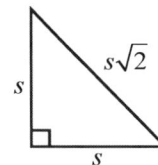

$A = \pi r^2$
$C = 2\pi r$ $A = \ell w$ $A = \frac{1}{2}bh$ $c^2 = a^2 + b^2$ Special Right Triangles

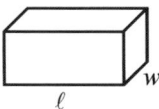 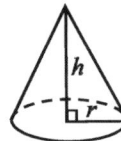 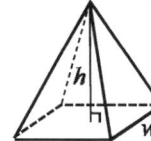

$V = \ell wh$ $V = \pi r^2 h$ $V = \frac{4}{3}\pi r^3$ $V = \frac{1}{3}\pi r^2 h$ $V = \frac{1}{3}\ell wh$

The number of degrees of arc in a circle is 360.
The number of radians of arc in a circle is 2π.
The number of the measures in degrees of the angles of a triangle is 180.

CONTINUE

1

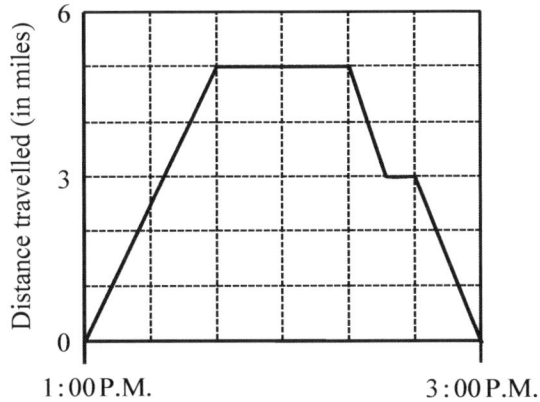

Bernard began to ride a bicycle to the town library, and then rode to the book store to buy a novel. After 10 minutes, he began to ride home again. If the graph above shows his trip, how long did he stay in the library?

A) 10 minutes

B) 20 minutes

C) 30 minutes

D) 40 minutes

2

If $\dfrac{2}{k} = 9$ and $9k + h = 20$, what is the value of h?

A) 9.5

B) 12

C) 15.5

D) 18

3

n	−1	0	1	2	a
$f(n)$	0	3	6	9	b

The table above shows some values of the linear function f. Which of the following defines b?

A) $b = a + 3$

B) $b = a + 5$

C) $b = 2a + 4$

D) $b = 3a + 3$

4

	Subject		Total
Gender	Art	Music	
Males	30		65
Females		20	
Total			100

The incomplete table above shows the results of a survey about subject preference given to 100 students. What is the probability of art students being females?

A) $\dfrac{7}{25}$

B) $\dfrac{1}{3}$

C) $\dfrac{1}{4}$

D) $\dfrac{2}{5}$

CONTINUE

Questions 5 and 6 refer to the following information.

TEST RESULTS

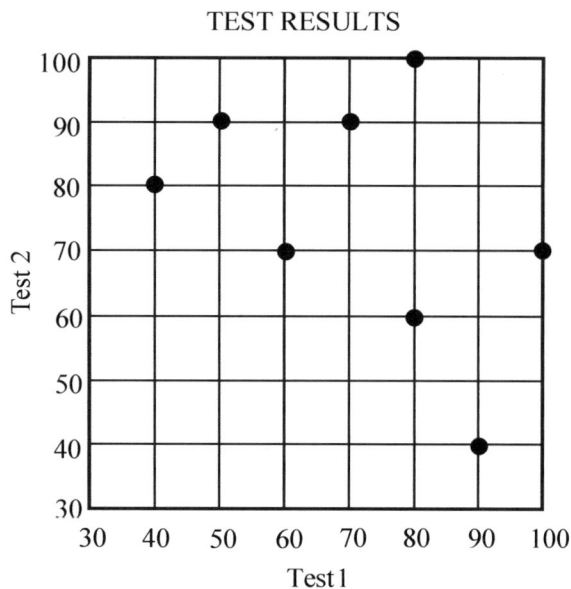

The scatterplot above relates two sets of data on a graph and shows the results of a class of students' last two algebra tests. Both the vertical and horizontal axes show the scores.

5

What is the average (arithmetic mean) score for Test 1?

A) 68.35
B) 70.50
C) 71.25
D) 74.75

6

Which of the following is the greatest change in scores between test 1 and test 2?

A) 60
B) 50
C) 40
D) 30

7

$$L = 0.2(t - 2010) + 10$$

The lifespan of a certain bird has been tracked from the year 2010, and the average lifespan is modeled by the equation above. In 2010 the lifespan of the bird was 10 years. What is the meaning of the number 0.2 in the equation?

A) The lifespan in the year 2010

B) The life span increase each year from 2010

C) The lifespan increase every 10 year

D) The life span decrease each year from 2010

8

$$x^2 - 2x + y^2 + 2y - 3 = 0$$

The equation of a circle in the xy-plane is shown above. What is the diameter of the circle?

A) $\sqrt{5}$

B) $2\sqrt{5}$

C) 5

D) 10

CONTINUE

9

$$x - 4y = -3$$
$$4x - y = 12$$

In the system of equations above, what is the value of $x + y$?

A) 5

B) 6

C) 8

D) 9

10

$$\left(a^k\right)^{\frac{2}{3}} = \frac{1}{a^2}$$

In the equation above, if $a > 0$, what is the value of k?

A) −3

B) −1

C) 1

D) 3

11

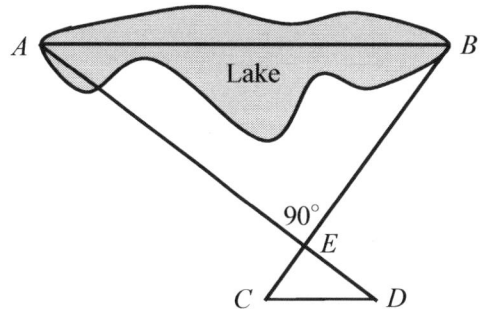

Jackson wants to measure the length AB of a lake. In the figure above, $\overline{AB}$ is parallel to $\overline{CD}$, $DE = 6$ feet, $CD = 10$ feet, and $BE = 300$ feet. What is the length of the lake?

A) 250 feet

B) 275 feet

C) 375 feet

D) 500 feet

12

$$2x^2y - 3xy^2 - xy(3x + 5y - 2)$$

Which of the following is equivalent to the expression above?

A) $xy(x - 8y - 2)$

B) $xy(x + 8y - 2)$

C) $-xy(x - 8y + 2)$

D) $-xy(x + 8y - 2)$

CONTINUE

13

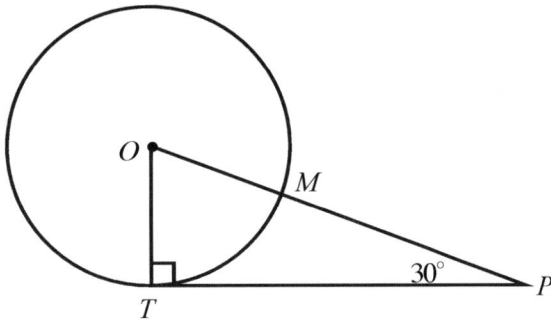

In the figure above, point O is the center of the circle. If the length of $\overline{TP}$ is $10\sqrt{3}$, what is the length of minor arc $\overarc{TM}$?

A) $\dfrac{5\pi}{3}$

B) $\dfrac{7\pi}{3}$

C) $\dfrac{8\pi}{3}$

D) $\dfrac{10\pi}{3}$

14

A certain number is proportional to another number in the ratio $3:7$. If 12 is subtracted from the sum of the numbers, the result is 38. What is the average (arithmetic mean) of the numbers?

A) 10

B) 12

C) 25

D) 40

15

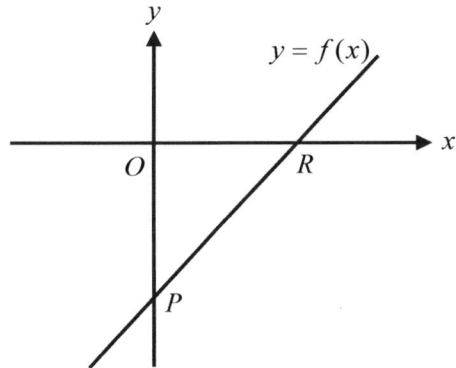

The function f, defined by $f(x) = mx - m$, is graphed in the xy-plane above. Which of the following expressions represents the area of triangle OPR?

A) $\dfrac{m}{2}$

B) m

C) $\dfrac{m^2}{2}$

D) m^2

16

If pipe S can fill a certain water tank in 3 hours and pipe U can empty it in 4 hours, how long, in hours, would it take to fill the empty tank when both pipes are open?

A) 6

B) 8

C) 10

D) 12

CONTINUE

17

$$\frac{1}{R} + \frac{1}{S} = \frac{1}{T}$$

When electrical circuits are connected in parallel, the reciprocal of the total resistance is found by adding the reciprocals of each resistance as shown above. Which of the following gives S in terms of R and T?

A) $S = \dfrac{R - T}{RT}$

B) $S = \dfrac{T - R}{RT}$

C) $S = \dfrac{RT}{R - T}$

D) $S = \dfrac{RT}{T - R}$

18

$$h(t) = 36t - 6t^2$$

The function h above shows the height, in feet, of an object thrown upward after t seconds. How long, in seconds, does the object stay in the air higher than 48 feet?

A) 2

B) 3

C) 4

D) 5

Questions 19 and 20 refer to the following information.

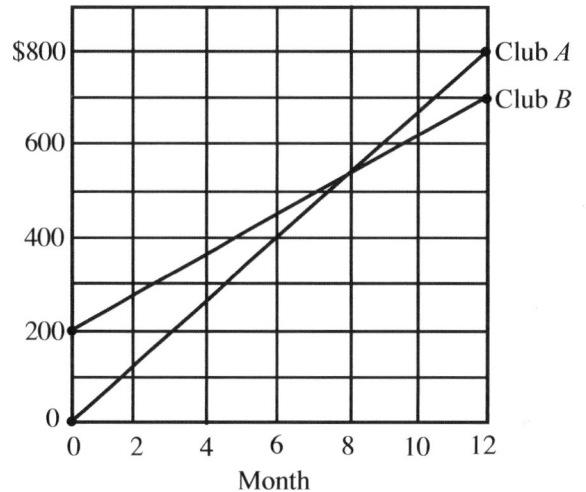

Two health clubs offer different membership plans. The graph above shows the yearly cost, including a membership fee plus a monthly charge, for each club.

19

Which of the following is closest to the monthly charge, in dollars, for club B?

A) 42

B) 67

C) 70

D) 72

20

Which of the following best approximates the total cost, in dollars, for club B when both plans are the same?

A) 510

B) 525

C) 533

D) 550

CONTINUE

21

$$y = a(x-2)^2 + b$$
$$y = 5$$

In the system of equations above, for which of the following values of a and b does the system have no solution?

A) $a = 1$ and $b = -4$

B) $a = 2$ and $b = 5$

C) $a = -1$ and $b = 6$

D) $a = -2$ and $b = 4$

22

$$D(t) = 30 - at^2$$

An apple falls from the branch of a tree to the ground 30 feet below. The distance, D, the apple is from the ground is represented by the equation above, where a is a constant and t is time in seconds.
If $D(0.1) - D(0.2) = 6$, what is the value of a?

A) 160

B) 180

C) 200

D) 240

23

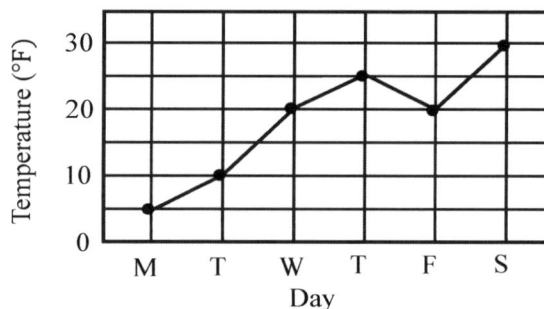

The graph above shows the daily high temperatures in Albany, New York, for 6 days in January. Which of the following describes the data?

    I.   mean = median

    II.   mean = mode

    III.   median = mode

A) I and II only

B) II and III only

C) III only

D) I, II, and III

24

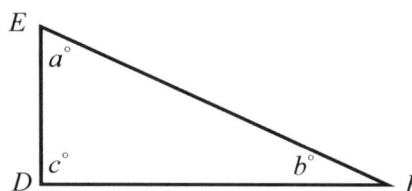

Note: Figure not drawn to scale.

In the figure above, if $\sin(a^\circ) = \cos(b^\circ)$, which of the following must be true?

A) $a = b$

B) $a > b$

C) $a = 60$

D) $c = 90$

CONTINUE

25

In an art class, $\frac{2}{3}$ of the students are girls and $\frac{2}{5}$ of girls are seniors. If $\frac{1}{3}$ of senior girls have passed the final art test, which of the following could be the number of students in this class?

A) 20

B) 30

C) 45

D) 60

26

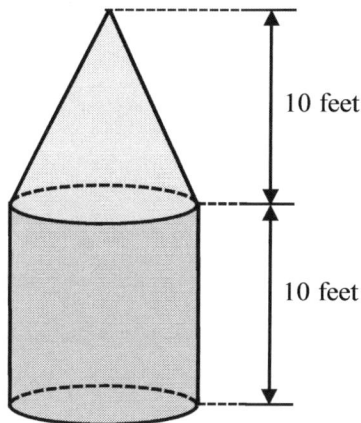

The figure above shows a silo built from a right circular cone and a right circular cylinder. If the volume of the cylinder is 1911 cubic feet, what is the volume of the silo, in cubic feet?

A) 2125

B) 2548

C) 2684

D) 3017

27

$$k = x^2 - 5x$$

In the equation above, for how many integers x is the number k negative?

A) 2

B) 3

C) 4

D) 5

28

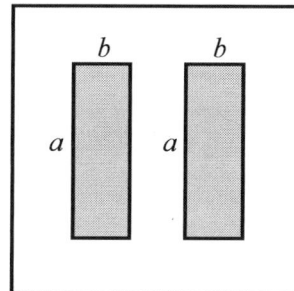

In figure above, two identical rectangles lie inside a square and the dimensions of the rectangle are a and b respectively. If the distance from the rectangles to the square and each other are 4 inches, and $a:b = 5:2$. What is the area of the square in square inches?

A) 625

B) 676

C) 729

D) 784

CONTINUE

29

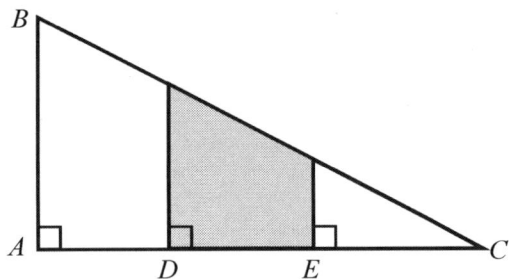

In the figure above, $AD = DE = EC$. If the area of triangle ABC is 81, what is the area of the shaded region?

A) 24

B) 27

C) 30

D) 40.5

30

$$a - b + 3i\sqrt{5} = \sqrt{5} + (a+b)i$$

In the equation above, a and b are constants. If $i = \sqrt{-1}$, what is the value of $a^2 - b^2$?

A) $8\sqrt{3}$

B) 12

C) 15

D) $12\sqrt{3}$

CONTINUE

DIRECTIONS

For questions 31-38, solve the problem and enter your answer in the grid, as described below, on the answer sheet.

1. Although not required, it is suggested that you write your answer in the boxes at the top of the columns to help you fill in the circles accurately. You will receive credit only if the circles are filled in correctly.

2. Mark no more than one circle in any column.

3. No question has a negative answer.

4. Some problems may have more than one answer.

5. **Mixed numbers** such as $3\frac{1}{2}$ must be gridded as 3.5 or 7/2. (If $\boxed{3\,1\,/\,2}$ is entered into the grid, it will be interpreted as $\frac{31}{2}$, not $3\frac{1}{2}$.)

6. **Decimal answers:** If you obtain a decimal answer with more digits than the grid can accommodate, it may be either rounded or truncated, but it must fill the entire grid.

Answer: $\frac{7}{12}$

Write answer in boxes. → Fraction line

Grid in result.

Answer: 2.5

Decimal point

Acceptable ways to grid $\frac{2}{3}$ are:

Answer: 201
Either position is correct.

Note: You may start your answers in any column, space permitting. Columns you don't need to use should be left blank.

CONTINUE

x	$f(x)$
1	7
3	13
5	19
a	b

x	$g(x)$
0	12
1	14
2	16
a	b

The tables above show some values of the linear functions f and g. What is the value of $a+b$?

Mr. Benjamin has brought grammar work books to distribute to the students in his reading class. If he gives each student 5 books, he will have 10 books left over, and if he gives each student 7 books, he will need an additional 20 books. How many students are in the class?

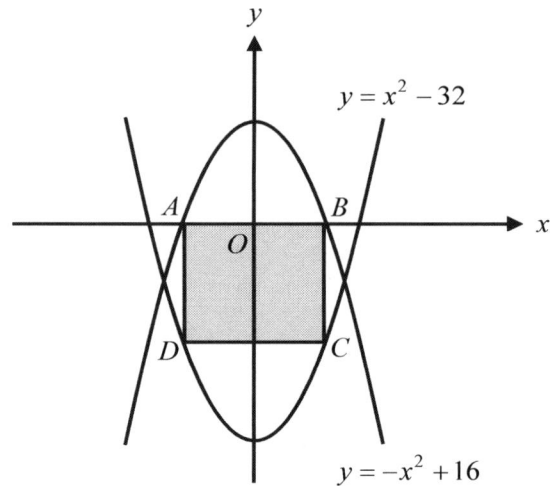

In the xy-plane above, what is the area of rectangle $ABCD$?

$$f(x) = x^2 + ax + b$$
$$g(x) = f(x-3)$$

In the functions above, a and b are constants. If $g(3) = 5$ and $g(4) = 10$, what is the value of a?

CONTINUE

35

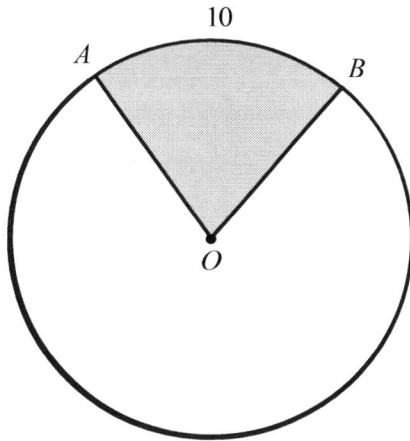

10

A   B

O

In the figure above, central angle AOB has a

measure of $\dfrac{\pi}{3}$ radians. If the length of minor arc

$\overset{\frown}{AB}$ is 10, what is the area of the shaded sector?

(Round your answer to the nearest tenth.)

36

$$P(x) = x^2 + 4x - k$$

In the quadratic function above, if $P(0) = 5$, what is
the minimum value of P?

Questions 5 and 6 refer to the following information.

$$R = 100x$$
$$C = 85x + 2000$$

A smartphone production company expressed a
relationship between revenue (R) and cost (C) for selling
x units of a product as shown above.

37

For what value of x will the product start to return a
profit?

38

For what value of x, will the company achieve a profit
of $100,000?

STOP

**If you finish before time is called, you may check your work on this section only.
Do not turn to any other section in the test.**

Math Conversion Table

Raw Score	Scaled Score	Raw Score	Scaled Score
58	800	27	500
57	800	26	490
56	800	25	480
55	800	24	470
54	790	23	460
53	780	22	460
52	770	21	450
51	750	20	440
50	740	19	430
49	730	18	430
48	720	17	420
47	710	16	420
46	700	15	410
45	690	14	400
44	670	13	390
43	680	12	380
42	670	11	370
41	660	10	360
40	650	9	450
39	640	8	340
38	630	7	330
37	620	6	310
36	610	5	290
35	600	4	280
34	590	3	270
33	580	2	260
32	560	1	240
31	550	0	200
30	540		
29	530		
28	520		

Answer Explanations

Test 3 Answers and Explanations

	1	2	3	4	5	6	7	8	9	10
SECTION 3	C	B	D	A	D	D	A	B	B	A
	11	12	13	14	15	16	17	18	19	20
	C	A	A	B	C	2	3	130	144	1/4

	1	2	3	4	5	6	7	8	9	10
SECTION 4	D	D	D	B	C	B	B	B	A	A
	11	12	13	14	15	16	17	18	19	20
	C	D	D	C	A	D	C	A	A	C
	21	22	23	24	25	26	27	28	29	30
	D	C	C	D	C	B	C	D	B	C
	31	32	33	34	35	36	37	38		
	36	15	128	4	47.7	1	134	6800		

SECTION 3

1. **C**

   40% of $40 = 0.4 \times 40 = 16$

2. **B**

   $3r = 5 \rightarrow 6r = 10 \rightarrow 6r + 5 = 15$

3. **D**

   $\dfrac{1}{a^2} = \dfrac{1}{5} \rightarrow a^2 = 5 \rightarrow 5a^2 = 25$

4. **A**

   $p = 10k + r \rightarrow r = p - 10k$

5. **D**

   $\dfrac{5}{12} = \dfrac{1}{a} + \dfrac{1}{b} \rightarrow \dfrac{5}{12} = \dfrac{a+b}{ab} \rightarrow \dfrac{5}{12} = \dfrac{a+b}{24} \rightarrow a+b = 10$

6. **D**

   $-x^2 + 9 = 5 \rightarrow x^2 = 4 \rightarrow x = \pm 2 \rightarrow$ Therefore, $AT = 4$.

Answer Explanations

7. A

$$\text{Area} = \frac{(2+6)8}{2} = 32$$

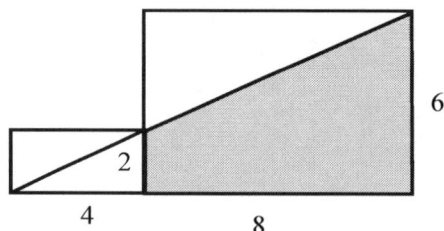

8. B

$$\frac{a}{3} = \frac{-b}{1} = \frac{9}{3} = \frac{3}{1} \quad \rightarrow \quad a = 9 \text{ and } b = -3 \quad \rightarrow \quad a+b = 6$$

9. B

$g(3) = 0$ and $g(-2) = 0 \quad \rightarrow \quad (x-3)$ and $(x+2)$ are factore of $g(x)$.

10. A

$$x^2 y = k \quad \rightarrow \quad 4 \times 10 = 100y \quad \rightarrow \quad y = \frac{40}{100} = \frac{2}{5}$$

11. C

$$y = k(x-4)(x+2) \quad \rightarrow \quad \text{Axis of symmetry} = \frac{4+(-2)}{2} = 1 = a$$

12. A

$$PR = QR = 7 \quad \rightarrow \quad k = -1 + 7 = 6$$

13. A

$$\frac{2i}{1-i} = a + bi \quad \rightarrow \quad \frac{2i(1+i)}{(1-i)(1+i)} = \frac{-2+2i}{2} = -1 + i \quad \rightarrow \quad a = -1$$

14. B

$$\frac{1}{x} = \frac{x}{2x+1} \quad \rightarrow \quad x^2 - 2x - 1 = 0 \quad \rightarrow \quad x = \frac{2 \pm \sqrt{8}}{2} = 1 \pm \sqrt{2}$$

15. C

16. 2

$$x^2 - ax = -10 \quad \rightarrow \quad x^2 - ax + 10 = 0 \quad \rightarrow \quad \text{Product of the roots is 10.} \quad \rightarrow \quad 5 \times (2) = 10$$

The other root is 2.

Or

Five is one of the roots.

$$5^2 - 5a = -10 \quad \rightarrow \quad a = 7 \quad \rightarrow \quad x^2 - 7x + 10 = 0 \quad \rightarrow \quad (x-5)(x-2) = 0 \quad \rightarrow \quad \text{The other root is 2.}$$

17. 3

$$\frac{15}{x-1} - 7 = 3 - \frac{5}{x-1} \quad \rightarrow \quad 15 - 7(x-1) = 3(x-1) - 5 \quad \rightarrow \quad x = 3$$

Answer Explanations

18. 130

$2a = 180 - 80 = 100 \rightarrow a = 50 \rightarrow b = 130$

19. 144

The number of people $= n \rightarrow \dfrac{n}{8} + \dfrac{n}{4} + \dfrac{n}{6} = 78 \rightarrow \dfrac{13n}{24} = 78 \rightarrow n = 144$

20. $\dfrac{1}{4}$

Ratio of the bases of the three triangles is $2:3:3. \rightarrow$ Ratio of their areas is also $2:3:3$.

Therefore, $\dfrac{2}{2+3+3} = \dfrac{1}{4}$.

SECTION 4

1. D

Each scale has 20 minutes. $20 \times 2 = 40$ minutes

2. D

Since $9k = 2$, $9k + h = 20 \rightarrow 2 + h = 20 \rightarrow h = 18$.

3. D

4. B

$\dfrac{15}{45} = \dfrac{1}{3}$

Gender	Subject		Total
	Art	Music	
Males	30		65
Females	15	20	35
Total	45		100

5. C

$\dfrac{40 + 50 + 60 + 70 + 80 + 80 + 90 + 100}{8} = 71.25$

6. B

The greatest change in scores is $90 - 40 = 50$.

7. B

8. B

$x^2 - 2x + y^2 + 2y - 3 = 0 \rightarrow (x-1)^2 + (y+1)^2 = 5 \rightarrow$ Radius is $\sqrt{5}. \rightarrow$ Diameter is $2\sqrt{5}$.

Answer Explanations

9. A

$$4x - y = 12$$
$$-\underline{\begin{array}{l} x - 4y = -3 \end{array}}$$
$$3x + 3y = 15 \qquad \rightarrow \quad 3(x+y) = 15 \quad \rightarrow \quad x + y = 5$$

10. A

$$\left(a^k\right)^{\frac{2}{3}} = \frac{1}{a^2} \quad \rightarrow \quad a^{\frac{2}{3}k} = a^{-2} \quad \rightarrow \quad \frac{2}{3}k = -2 \quad \rightarrow \quad k = -3$$

11. C

Since $CE = 8$, $\dfrac{8}{300} = \dfrac{10}{AB}$. $\quad \rightarrow \quad AB = \dfrac{3000}{8} = 375$

12. D

$$2x^2 y - 3xy^2 - xy(3x + 5y - 2) = 2x^2 y - 3xy^2 - 3x^2 y - 5xy^2 + 2xy = -x^2 y - 8xy^2 + 2xy$$
$$= -xy(x + 8y - 2)$$

13. D

Since OT (radius) $= 10$ and $\angle TOM = 60°$, $\overset{\frown}{TM} = 20\pi \times \dfrac{1}{6} = \dfrac{10\pi}{3}$.

14. C

Two numbers are $3k$ and $7k$. $\rightarrow$ $10k - 12 = 38$ $\rightarrow$ $k = 5$ $\rightarrow$ Therefore, the average is $\dfrac{3k + 7k}{2} = 25$.

15. A

Since y-intercept is $-m$ and x-intercept is 1, $OP = m$ and $OR = 1$.

Therefore, the area of $\triangle OPR$ is $\dfrac{m \times 1}{2} = \dfrac{m}{2}$.

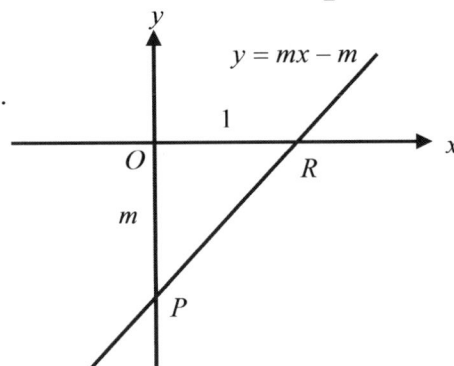

16. D

Combined rate $= \dfrac{1}{3} - \dfrac{1}{4} = \dfrac{1}{12}$ per hour

Therefore, $1 \div \dfrac{1}{12} = 12$ hours.

17. C

$$\dfrac{1}{R} + \dfrac{1}{S} = \dfrac{1}{T} \quad \rightarrow \quad \dfrac{1}{S} = \dfrac{1}{T} - \dfrac{1}{R} = \dfrac{R - T}{TR} \quad \rightarrow \quad S = \dfrac{TR}{R - T}$$

18. A

$$36t - 6t^2 \geq 48 \quad \rightarrow \quad t^2 - 6t + 8 \leq 0 \quad \rightarrow \quad (t - 2)(t - 4) \leq 0 \quad \rightarrow \quad 2 \leq t \leq 4 \quad \rightarrow \quad 4 - 2 = 2 \text{ seconds}$$

19. A

$$\frac{700-200}{12} \approx 42$$

20. C

Club A: $c = \frac{800}{12}x$ Club B: $c = \frac{500}{12}x + 200$ $\rightarrow$ $\frac{500}{12}x + 200 = \frac{800}{12}x$ $\rightarrow$ $200 = 25x$ $\rightarrow$ $x = 8$

Therefore, the total cost each club is $\frac{500}{12} \times 8 + 200 \approx 533$.

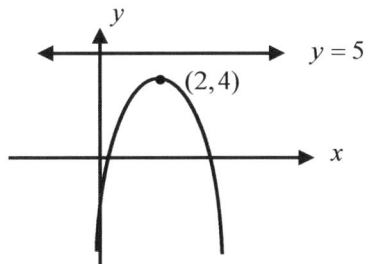

21. D

Graph opens downward and its maximum is less than 5.

22. C

$$D(0.1) - D(0.2) = 6 \;\rightarrow\; (30 - 0.01a) - (30 - 0.04a) = 6 \;\rightarrow\; 0.03a = 6 \;\rightarrow\; a = \frac{6}{0.03} = 200$$

23. C

5 10 20 20 25 30 $\rightarrow$ mean $= 18.3$, median $= 20$, and mode $= 20$

24. D

Since $a + b = 90$, $\triangle DEF$ is a right triangle.

25. C

$n =$ number of students: $\rightarrow$ $\frac{1}{3} \times \frac{2}{5} \times \frac{2}{3}(n) = \frac{4}{45}n =$ integer $\rightarrow$ n must be multiples of 45.

26. B

volume of the cylinder : $\pi r^2(10) = 1911$ $\rightarrow$ volume of the circular cone: $\frac{\pi r^2(10)}{3} = 637$

Volume of the silo is $1911 + 637 = 2548$.

27. C

$$x^2 - 5x < 0 \;\rightarrow\; x(x-5) < 0 \;\rightarrow\; 0 < x < 5 \;\rightarrow\; \text{integers of } x : 1, 2, 3, 4$$

28. D

$2b + 12 = a + 8$ $\rightarrow$ $a = 2b + 4$

$2a = 5b$ $\rightarrow$ $a = 2.5b$ Putting in the equation

$2.5b = 2b + 4$ $\rightarrow$ $0.5b = 4$ $\rightarrow$ $b = 8$

Therefore, $x = 2(8) + 12 = 28$.

Area of the square is $28^2 = 784$.

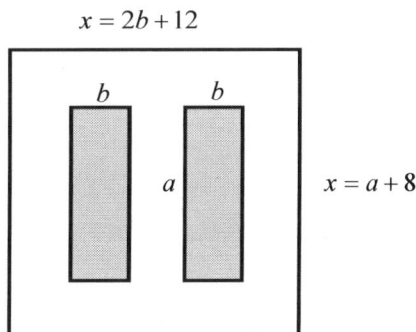

29. B

Ratio of corresponding sides of the similar triangles is $3:2:1$. Ratio of their areas is $9:4:1$.

The areas in terms of k are $9k$, $4k$, and k. Since $9k = 81 \rightarrow k = 9$, the area of the shaded region is $3k = 27$.

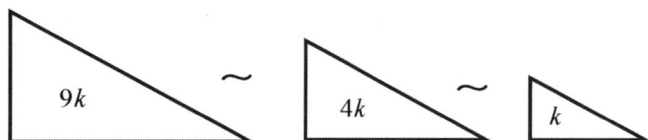

30. C

$a - b = \sqrt{5}$ and $3\sqrt{5} = a + b \rightarrow a^2 - b^2 = (a+b)(a-b) = (\sqrt{5})(3\sqrt{5}) = 15$

31. 36

(a, b) is a point of intersection of two graphs.

$f(x) = 3x + 4$ and $g(x) = 2x + 12 \rightarrow 3x + 4 = 2x + 12 \rightarrow x = 8$ and $y = 28$

Therefore, $a + b = 8 + 28 = 36$.

32. 15

$n =$ number of students $\rightarrow$ The number of books: $5n + 10 = 7n - 20 \rightarrow n = 15$

33. 128

Find the zeros of $y = -x^2 + 16 \rightarrow x^2 = 16 \rightarrow x = \pm 4$

When $x = 4$, $y = 4^2 - 32 = 16$. Therefore, $AB = 8$ and $BC = 16$. Area of $\square ABCD$ is $8 \times 16 = 128$.

34. 4

$g(3) = f(0) = b = 5$ and $g(4) = f(1) = 1 + a + b = 10 \rightarrow 1 + a + 5 = 10 \rightarrow a = 4$

35. 47.7

Since $r\theta = 10 \rightarrow r = \dfrac{10}{\dfrac{\pi}{3}} = \dfrac{30}{\pi} \rightarrow$ Area of the sector $= \dfrac{1}{2}r^2\theta = \dfrac{1}{2}\left(\dfrac{30}{\pi}\right)^2\left(\dfrac{\pi}{3}\right) \approx 47.7$

Or

$\dfrac{\pi}{3} = 60° \rightarrow 2\pi r \times \dfrac{1}{6} = 10 \rightarrow r = \dfrac{30}{\pi} \rightarrow$ Area of the shaded region $= \pi r^2 \times \dfrac{1}{6} = \pi\left(\dfrac{30}{\pi}\right)^2 \times \dfrac{1}{6} = \dfrac{900}{6\pi} \approx 47.7$

36. 1

$P(x) = x^2 + 4x - k \rightarrow P(0) = -k = 5 \rightarrow k = -5 \rightarrow P(x) = x^2 + 4x + 5$

$P(x) = (x+2)^2 + 1 \rightarrow P$ has a minimum 1 at $x = -2$.

37. 134

Profit $= R - C \geq 0 \rightarrow 100x \geq 85x + 2000 \rightarrow 15x \geq 2000 \rightarrow x \geq 133.333 \rightarrow x = 134$

38. 6800

$100x - 85x - 2,000 \geq 100,000 \rightarrow 15x \geq 102,000 \rightarrow x \geq 6800 \rightarrow x = 6800$

PRACTICE TEST 4

Dr. John Chung's SAT Math

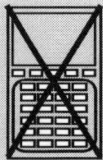

Math Test - No Calculator

25 MINUTES, 20 QUESTIONS

Turn to Section 3 of your answer sheet to answer the questions in this section.

DIRECTIONS

For questions 1–15, solve each problem, choose the best answer from the choices provided, and fill in the corresponding circle on your answer sheet. **For questions 16–20,** solve the problem and enter your answer in the grid on your answer sheet. Please refer to the directions before question 16 on how to enter your answers in the grid. You may use any available space in your test booklet for scratch work.

NOTE

1. The use of a calculator **is not permitted**.

2. All variables and expressions used represent real numbers unless otherwise indicated.

3. Figures provided in this test are drawn to scale unless otherwise indicated.

4. All figures lie in a plane unless otherwise indicated.

5. Unless otherwise indicated, the domain of a given function f is the set of all real numbers x for which $f(x)$ is a real number.

REFERENC

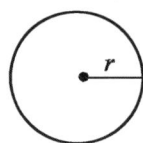 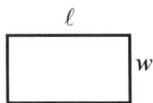 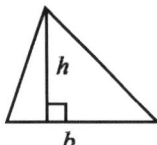 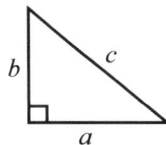 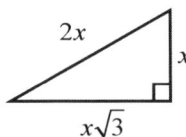 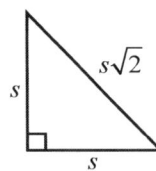

$A = \pi r^2$ $\qquad A = \ell w \qquad A = \dfrac{1}{2}bh \qquad c^2 = a^2 + b^2 \qquad$ Special Right Triangles
$C = 2\pi r$

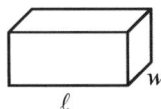 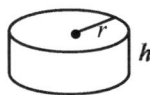  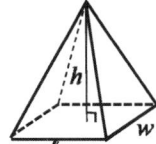

$V = \ell wh \qquad V = \pi r^2 h \qquad V = \dfrac{4}{3}\pi r^3 \qquad V = \dfrac{1}{3}\pi r^2 h \qquad V = \dfrac{1}{3}\ell wh$

The number of degrees of arc in a circle is 360.
The number of radians of arc in a circle is 2π.
The number of the measures in degrees of the angles of a triangle is 180.

CONTINUE

1

Which of the following expressions cannot be equal to 0 for some value of x?

A) $x^2 - 2$

B) $x^2 + 1$

C) $1 - x^2$

D) $2 - x^2$

2

$$f(x) = mx + b$$

In the function above, m and b are constants.

If $\dfrac{f(5) - f(2)}{3} = 2$, what is the values of m?

A) 2

B) 3

C) 4

D) 5

3

The line passing through the points $(a, 3)$ and $(b, -2)$ is parallel to the graph of $y = \dfrac{1}{2}x - 10$.
What is the value of $a - b$?

A) 5

B) 7

C) 8

D) 10

4

$$y = mx - \frac{2}{5}$$
$$2x + 3y = 4$$

In the system of equations, a is a constant. If the system has no solution, what is the value of m?

A) $-\dfrac{2}{3}$

B) $-\dfrac{3}{2}$

C) $\dfrac{2}{3}$

D) $\dfrac{3}{2}$

CONTINUE

5

If $f(x) = (x-1)^2 - (x-1) - 1$, which of the following expressions is equal to $f(1-x)$?

A) $f(1-x) = (x+1)^2 - (x+1) - 1$

B) $f(1-x) = (1-x)^2 - (1-x) - 1$

C) $f(1-x) = x^2 - x - 1$

D) $f(1-x) = x^2 + x - 1$

6

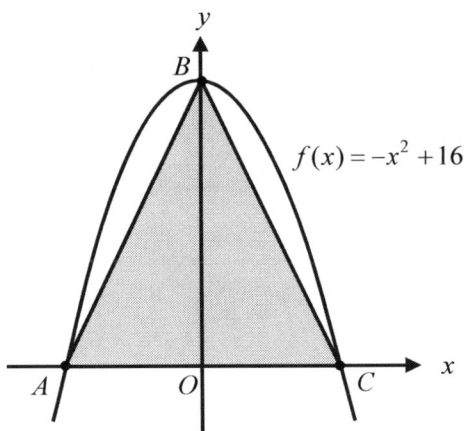

The graph of a function $f(x)$ is shown in the xy-plane above. What is the area of triangle ABC?

A) 16

B) 32

C) 64

D) 128

7

If $\dfrac{a-2b}{b} = \dfrac{2}{3}$, which of the following is equal to $\dfrac{a}{b}$?

A) $\dfrac{2}{3}$

B) $\dfrac{4}{3}$

C) $\dfrac{5}{3}$

D) $\dfrac{8}{3}$

8

$$y = m\sqrt{x}$$
$$y = mx - k$$

In the system of equations, m and k are constants. If $(4,16)$ is a solution to the system of equations above, what is the value of k?

A) 4

B) 8

C) 12

D) 16

CONTINUE

9

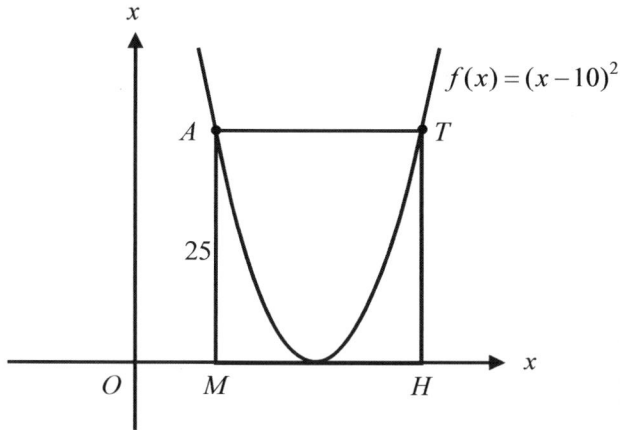

The graphs of f and rectangle $MATH$ are shown in the xy-plane above. If $MA = 25$, what is the length of $\overline{AT}$?

A) 5

B) 10

C) 15

D) 20

10

Which of the following equations has no solution?

A) $10x - 5x = 3$

B) $7x = 9x - 2x + 10$

C) $10x - 6 = 8x + 2x - 6$

D) $10x - 6 = 5x + 4x - 1$

11

$$f(x) = k(x+4)(x-10)$$

In the quadratic function f above, k is a constant. The graph of the function in the xy-plane is a parabola with vertex (a,b). If $b = -7$, which of the following is equal to k?

A) $\dfrac{1}{49}$

B) $\dfrac{1}{7}$

C) 7

D) 49

12

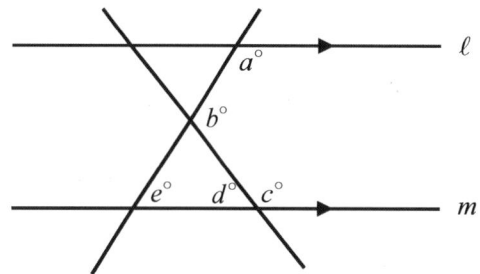

Note: Figure not drawn to scale.

In the figure above, if line ℓ and m are parallel, which of the following must be true?

A) $a + b + c = 180$

B) $d + e = b$

C) $b + e = c$

D) $a + b = 180$

CONTINUE

13

$$\frac{5x^2 + kx + 1}{x - 1} = ax + 1 + \frac{2}{x - 1}$$

The equation above is true for all values of x except 1, where k and a are constants. What is the value of k?

A) -4

B) -2

C) 2

D) 4

14

What are the solutions to $4(x - 2)^2 - 1 = 5$?

A) $x = 2 \pm \dfrac{\sqrt{6}}{4}$

B) $x = 2 \pm \dfrac{\sqrt{6}}{2}$

C) $x = \sqrt{2} \pm \dfrac{\sqrt{6}}{2}$

D) $x = 2 \pm \sqrt{6}$

15

Grade	For	Against	Total
Junior	60		
Senior			
Total	130		300

A supervisor surveyed students in his school to see if they were for or against building a fast-food restaurant in the school. The incomplete table above shows the results of his survey. If 40% of juniors are against it, how many seniors are in the school?

A) 100

B) 120

C) 170

D) 200

CONTINUE

DIRECTIONS

For questions 16–20, solve the problem and enter your answer in the grid, as described below, on the answer sheet.

1. Although not required, it is suggested that you write your answer in the boxes at the top of the columns to help you fill in the circles accurately. You will receive credit only if the circles are filled in correctly.

2. Mark no more than one circle in any column.

3. No question has a negative answer.

4. Some problems may have more than one answer.

5. **Mixed numbers** such as $3\frac{1}{2}$ must be gridded as 3.5 or 7/2. (If $\boxed{3\,1\,/\,2}$ is entered into the grid, it will be interpreted as $\frac{31}{2}$, not $3\frac{1}{2}$.)

6. **Decimal answers:** If you obtain a decimal answer with more digits than the grid can accommodate, it may be either rounded or truncated, but it must fill the entire grid.

Answer: $\frac{7}{12}$

Write answer in boxes.

Fraction line

Grid in result.

Answer: 2.5

Decimal point

Acceptable ways to grid $\frac{2}{3}$ are:

Answer: 201
Either position is correct.

Note: You may start your answers in any column, space permitting. Columns you don't need to use should be left blank.

CONTINUE

16

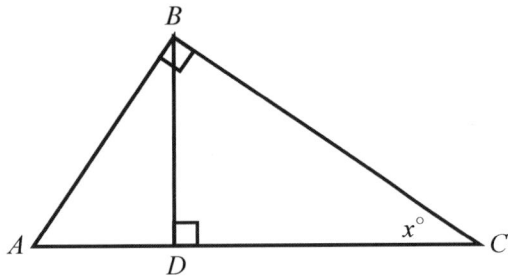

In the triangle above, the value of $\cos x°$ is 0.8. If the length of $\overline{AC}$ is 20, what is the length of $\overline{BD}$?

17

$$2x^3 - 10x^2 + 5x - 25 = 0$$

For what real value of x is the equation above true?

18

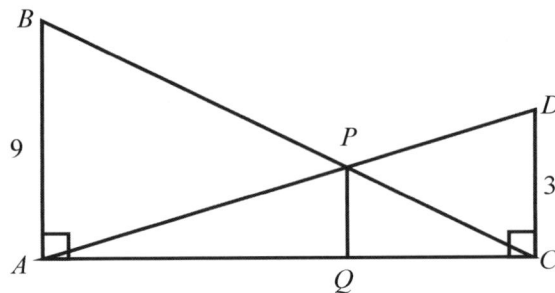

Note: Figure not drawn to scale.

In the figure above, $AB = 9$, $CD = 3$, and $AC = 12$. What is the length of $\overline{PQ}$?

CONTINUE

Question 19 and 20 refer to the following information.

A T-Mobile telephone company offers domestic texting plans as follows.

Plan A	Plan B
$0.25 per domestic text with no plan	Any 200 domestic texts for $40 per month with an aditional cost of $0.15 per text over 200.

19

For what number of texts do the two plans cost the same per month?

20

If Angela uses 400 texts per month, how much money, in dollars, will she save per month by using the less expensive plan? (Disregard the $ sign when gridding your answer.)

STOP

If you finish before time is called, you may check your work on this section only.
Do not turn to any other section in the test.

Math Test - Calculator

55 MINUTES, 38 QUESTIONS

Turn to Section 4 of your answer sheet to answer the questions in this section.

DIRECTIONS

For questions 1-30, solve each problem, choose the best answer from the choices provided, and fill in the corresponding circle on your answer sheet. **For questions 31-38,** solve the problem and enter your answer in the grid on your answer sheet. Please refer to the directions before question 31 on how to enter your answers in the grid. You may use any available space in your test booklet for scratch work.

NOTE

1. The use of a calculator **is permitted**.

2. All variables and expressions used represent real numbers unless otherwise indicated.

3. Figures provided in this test are drawn to scale unless otherwise indicated.

4. All figures lie in a plane unless otherwise indicated.

5. Unless otherwise indicated, the domain of a given function f is the set of all real numbers x for which $f(x)$ is a real number.

REFERENCE

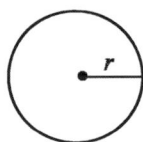 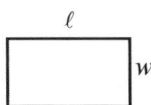 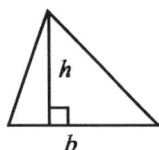 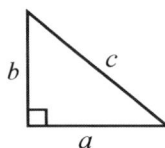 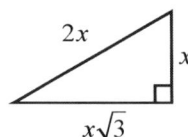 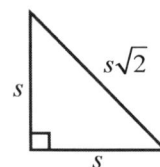

$A = \pi r^2$
$C = 2\pi r$
　　$A = \ell w$　　$A = \dfrac{1}{2}bh$　　$c^2 = a^2 + b^2$　　Special Right Triangles

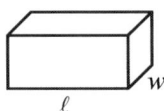 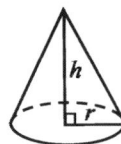 

$V = \ell w h$　　$V = \pi r^2 h$　　$V = \dfrac{4}{3}\pi r^3$　　$V = \dfrac{1}{3}\pi r^2 h$　　$V = \dfrac{1}{3}\ell w h$

The number of degrees of arc in a circle is 360.
The number of radians of arc in a circle is 2π.
The number of the measures in degrees of the angles of a triangle is 180.

CONTINUE

1

At a local video store, Angel rented two movies and three games for a total of $20. The next day, she rented three movies and 2 games for a total of $15. How much money, in dollars, is needed to rent a combination of one movie and one game?

A) 7

B) 10

C) 12

D) 16

2

CLASSES STUDENTS ARE TAKING

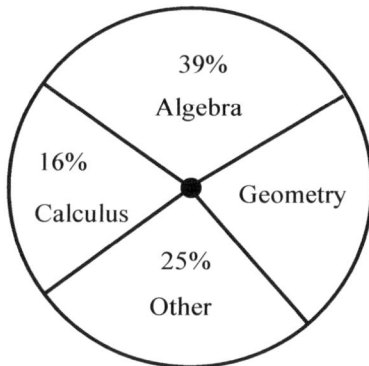

The circle graph above shows the percent of which 200 students are taking each subject. How many more students are taking Algebra than Geometry?

A) 30

B) 34

C) 36

D) 38

3

x	−1	0	1	2
$f(x)$	a	5	k	b

The table above shows some values of a linear function f. If $b - a = 9$, what is the value of k?

A) 7

B) 8

C) 9

D) 12

4

Gender	For	Against	Total
Boys	35		
Girls		23	55
Total			100

Ted surveyed a random sample of 100 students in his high school to see if they were for or against purchasing an additional grand piano for the school music concert. The incomplete table above shows the results of his survey. Based on this information, about how many of the 800 students in the school would be expected to be against the purchasing the piano?

A) 200

B) 264

C) 320

D) 350

CONTINUE

Questions 5 and 6 refer to the following information.

Average Height for Boys

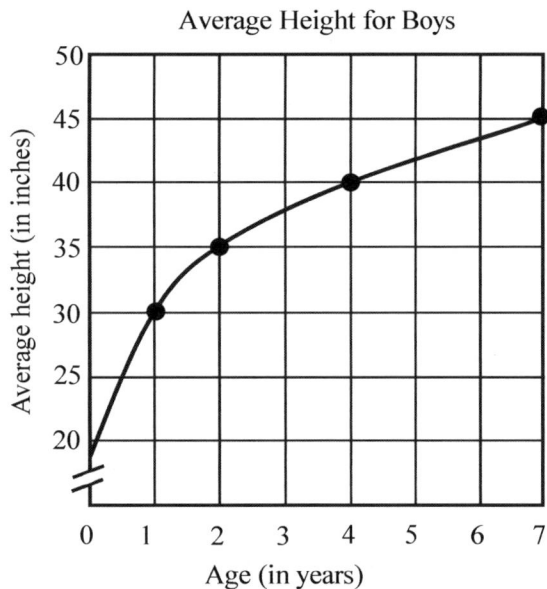

The graph above shows the average height for boys ages 0 to 7 in a certain state of the last year.

5

What is the annual average growth, in inches, between ages 2 and 7?

A) 2

B) 5

C) 10

D) 12

6

By what percent does the average height increase from age 1 to age 7?

A) 60

B) 50

C) 40

D) 30

7

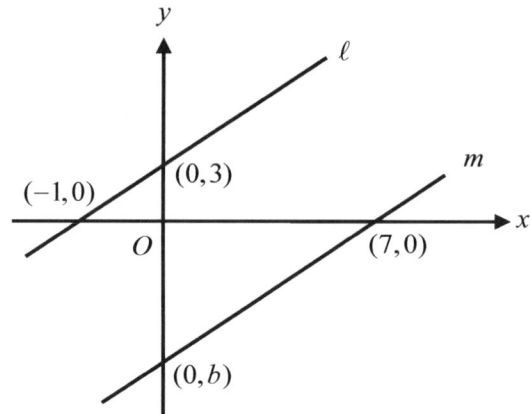

Note: Figure not drawn to scale.

In the xy-plane above, line ℓ is parallel to line m. What is the value of b?

A) $-\dfrac{7}{3}$

B) -14

C) -21

D) -28

8

$$x^2 - 8x + y^2 = 0$$

The equation of a circle in the xy-plane is shown above. What is the area of the circle?

A) 2π

B) 4π

C) 8π

D) 16π

CONTINUE

9

If $f(x-5) = 3x - 10$ for all values of x, what is the value of $f(-2)$?

A) -16

B) -10

C) -5

D) -1

10

Lee's family starts a trip with a supply of 20 pounds of coffee. When they arrive at their destination, 8 days later, they have found only 4 pounds left. They consume coffee at a constant rate per day. If T is amount of coffee remaining as a function of days d, which of the following represents the function $T(d)$?

A) $T(d) = 8d$

B) $T(d) = 2d + 20$

C) $T(d) = 20 - 8d$

D) $T(d) = 20 - 2d$

11

$$8x + y = 300$$

The elevator in a trade center is moving down from a height of 300 feet. The equation above can be used to model the height of the elevator, y, above the lobby, where x is the time in seconds. If the ordered pair (x, y) satisfies the equation, what does $(37.5, 0)$ mean?

A) The elevator stops at a height of 37.5 feet.

B) The elevator is moving down at a constant speed of 37.5 feet per second.

C) The elevator moves 37.5 feet from the lobby.

D) The elevator takes 37.5 seconds to move down to the lobby.

12

Thompson invested \$10,000 in stocks for two years. During the first year he suffered a 30 percent loss, but during the second year the remaining investment showed a 30 percent gain. Over the two-year period, how did Thompson's investment change?

A) His investment did not change.

B) His investment increased by 10 percent.

C) His investment decreased by 10 percent.

D) His investment decreased by 9 percent.

CONTINUE

13

The graph of $ax + by = 5$ in the xy-plane contains points from each of Quadrants I, II, and III, but no points from Quadrant IV. Which of the following must be true?

A) $a > 0$ and $b > 0$

B) $a > 0$ and $b < 0$

C) $a < 0$ and $b > 0$

D) $a < 0$ and $b < 0$

14

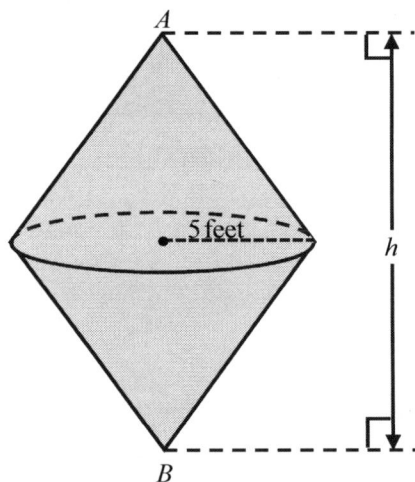

A water tank is built from two right circular cones with a radius 5 feet. If the volume of the tank is 200π cubic feet, what is the length h, in feet, from the bottom to the top of the tank?

A) 6

B) 12

C) 18

D) 24

15

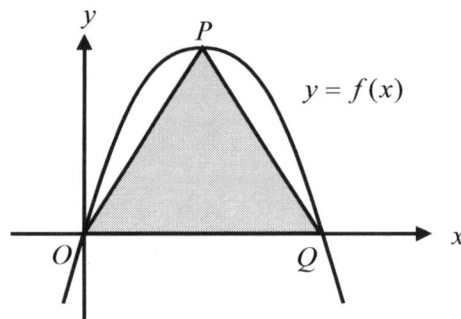

Note: Figure not drawn to scale.

The function f, defined by $f(x) = -x^2 + 6x$ is graphed in the xy-plane above. An isosceles triangle OPQ with $OP = PQ$ is built on the x-axis. What is the area of the triangle?

A) 13.5

B) 27

C) 40.5

D) 54

16

Cathy can do a job in 8 hours while Danny can do the same job in 6 hours. If Cathy and Danny work together for three hours, what fraction of the job is left to be finished?

A) $\dfrac{1}{12}$

B) $\dfrac{1}{8}$

C) $\dfrac{1}{6}$

D) $\dfrac{1}{4}$

CONTINUE

17

In a plane, the distance between points X and Y is 10, the distance between points X and P is 3, and the distance between points Y and Q is 4. Which of the following CANNOT be the length of $\overline{PQ}$?

A) 2

B) 3

C) 15

D) 17

18

The town library is planning to order student desks for the next school year. The costs to purchase student desks are as follows.

Two desks for $50, four desks for $80, six desks for $110, eight desks for $140, and so on.

If the town library wants to purchase 200 student desks, what would be the total cost in dollars?

A) 5000

B) 3020

C) 2860

D) 2500

Questions 19 and 20 refer to the following information.

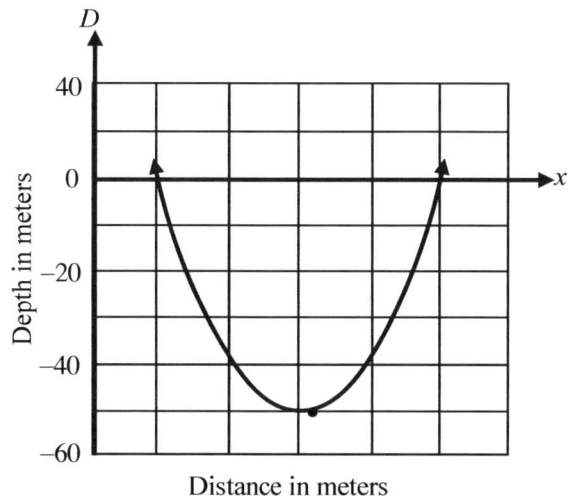

Distance in meters

The cross-section view of a river in Los Angeles is modeled by the graph above. The equation represented by the graph is defined by $D(x) = k(x-10)(x-50)$, where

19

Based on the graph above, how wide is the river in meters?

A) 20

B) 25

C) 40

D) 60

20

Based on the equation above, what is the value of k ?

A) 8

B) 4

C) $\dfrac{1}{4}$

D) $\dfrac{1}{8}$

CONTINUE

21

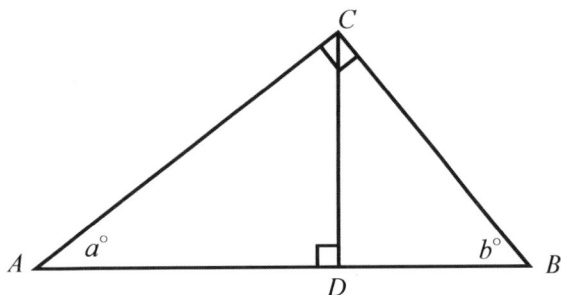

In the right triangle ABC above, the length of $\overline{BC}$ is 20. If the value of $\sin(a°)$ is 0.35, what is the length of $\overline{BD}$?

A) 5

B) 7

C) 8

D) 10

22

Which of the following polynomials has a factor of $x-1$?

A) $p(x) = x^3 + x^2 - 2x + 1$

B) $q(x) = 2x^3 - x^2 + x - 1$

C) $r(x) = 3x^3 - x - 2$

D) $s(x) = -3x^3 + 3x + 1$

23

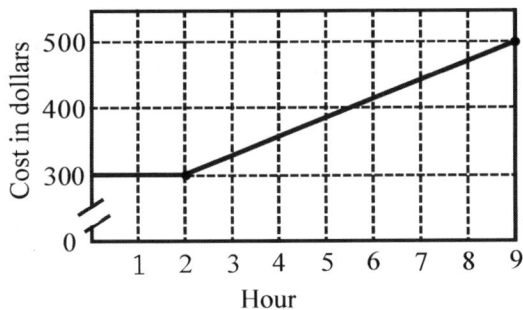

A decorating consultant charges consultation costs based on the graph above. If the consultant works for x hours ($x > 2$), for the consultation, which of the following represents the total cost?

A) $C(x) = 300$

B) $C(x) = 300 + 100x$

C) $C(x) = 300 + \dfrac{200}{7}x$

D) $C(x) = 300 + \dfrac{200}{7}(x - 2)$

24

$$|x| = k$$

If the equation above has a real solution set, which of the following must be true?

  I. $k \geq 0$
  II. $k < 0$
  III. $x > 0$

A) I only

B) II only

C) I and II only

D) II and III only

25

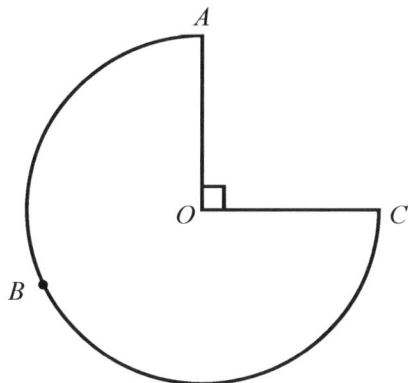

In the sector above, segment AO is a radius. If the length of arc $\overset{\frown}{ABC}$ is 12, what is the area of the sector?

A) 48π

B) 16π

C) $\dfrac{64}{\pi}$

D) $\dfrac{48}{\pi}$

26

$$(k-1)x + 3k = ax + 24$$

If the equation above is true for all real values of x, where k and a are constants, what is the value of a?

A) 1

B) 5

C) 7

D) 8

27

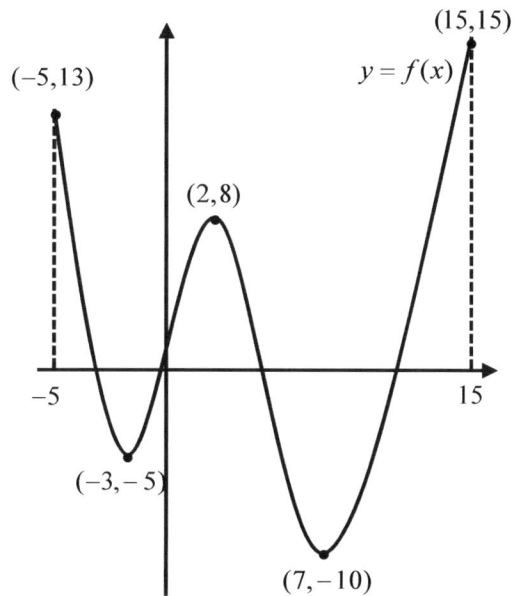

$$y = f(x)$$
$$y = k$$

The function f is graphed in the xy-plane above. If the system of equations above has exactly three real solutions for $-5 \le x \le 15$, which of the following could be the value of k?

A) 10

B) 5

C) −5

D) −8

CONTINUE

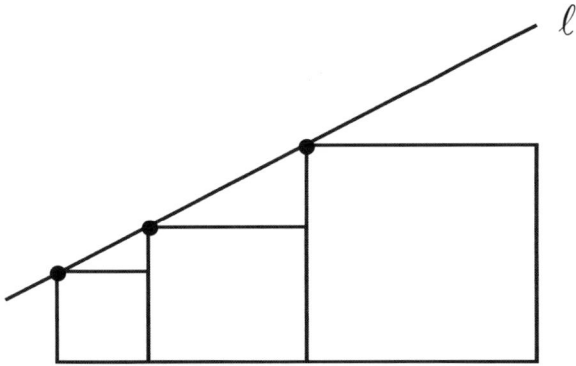

The figure above shows three squares with areas of 16, 64, and k respectively. If line ℓ passes through the vertex of each square, what is the value of k?

A) 81

B) 144

C) 196

D) 256

29

If the average of a and $2b$ is 26, the average of b and $2c$ is 41, and the average of c and $2a$ is 23, what is the average of a, b, and c?

A) 12

B) 16

C) 20

D) 24

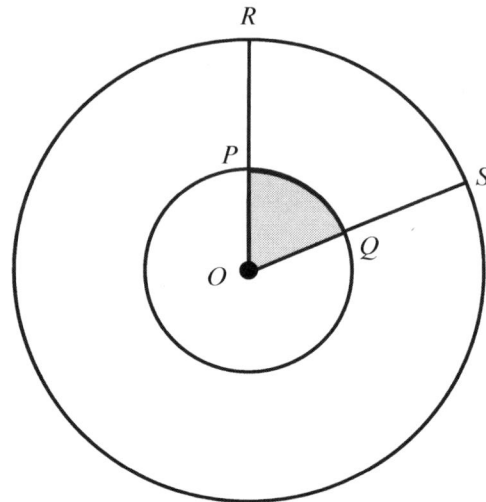

In the figure above, two circles have a common center O, and two rays from the center intercept the circles at points P, Q, R, and S. The measure of angle POQ is $\dfrac{2\pi}{5}$ and the area of the shaded region of sector OPQ is 20π. If $OP:PR = 2:3$, what is the length of minor arc $\overset{\frown}{RS}$?

A) 5π

B) 10π

C) 15π

D) 20π

CONTINUE

Answer: $\frac{7}{12}$

Answer: 2.5

Write answer in boxes. →

← Fraction line

← Decimal point

Grid in result. →

DIRECTIONS

For questions 31-38, solve the problem and enter your answer in the grid, as described below, on the answer sheet.

1. Although not required, it is suggested that you write your answer in the boxes at the top of the columns to help you fill in the circles accurately. You will receive credit only if the circles are filled in correctly.
2. Mark no more than one circle in any column.
3. No question has a negative answer.
4. Some problems may have more than one answer.
5. **Mixed numbers** such as $3\frac{1}{2}$ must be gridded as 3.5 or 7/2. (If $\boxed{3\,|\,1\,|\,/\,|\,2}$ is entered into the grid, it will be interpreted as $\frac{31}{2}$, not $3\frac{1}{2}$.)
6. **Decimal answers:** If you obtain a decimal answer with more digits than the grid can accommodate, it may be either rounded or truncated, but it must fill the entire grid.

Acceptable ways to grid $\frac{2}{3}$ are:

Answer: 201

Either position is correct.

Note: You may start your answers in any column, space permitting. Columns you don't need to use should be left blank.

CONTINUE ►

31

$$g(x) = 2f(x) - 1$$

In the equation above, if $g(1) = 3$, what is the value of $f(1)$?

32

For all values of a and b, let $a \nabla b$ be defined by $a \nabla b = ab - a + 1$. If $k \nabla (k - 2) = 2 \nabla 3$, what is the positive value of k?

33

If $4m + 5n$ is equal to 250 percent of $4n$, what is the value of $\dfrac{m+n}{m-n}$?

34

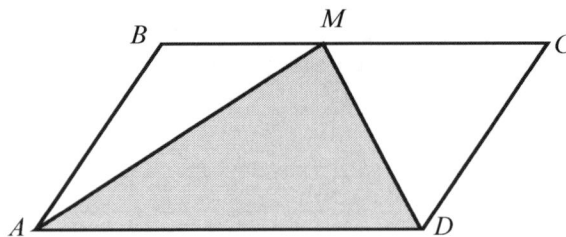

In the parallelogram above, $BM : MC = 2 : 3$. If the area of triangle ABM is 20, what is the area of the shaded region?

35

$$P(x) = 23,500 - 250x$$

The population of a certain town has been declining since the year 2,000. Scientists chose a linear decay model for the decline and arrived at the function above, where x is the number of years since 2,000. In how many years, will the population be decreased by 2,000?

CONTINUE

36

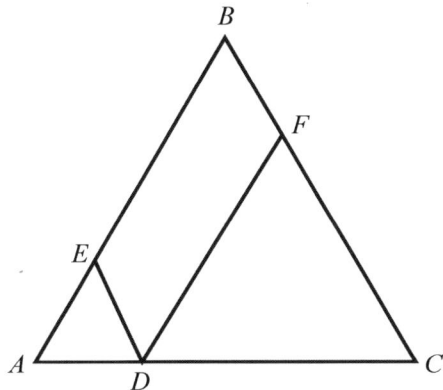

The length of a side of equilateral triangle ABC above is 10. In the figure, $\overline{ED} \parallel \overline{BC}$ and $\overline{DF} \parallel \overline{AB}$. If the ratio of DE to DF is 1:3, what is the perimeter of triangle CDF?

37

$$3x^2 - 8x + 4 = 0$$

If a and b are two solutions of the equation above, what is the value of $\dfrac{1}{a} + \dfrac{1}{b}$?

38

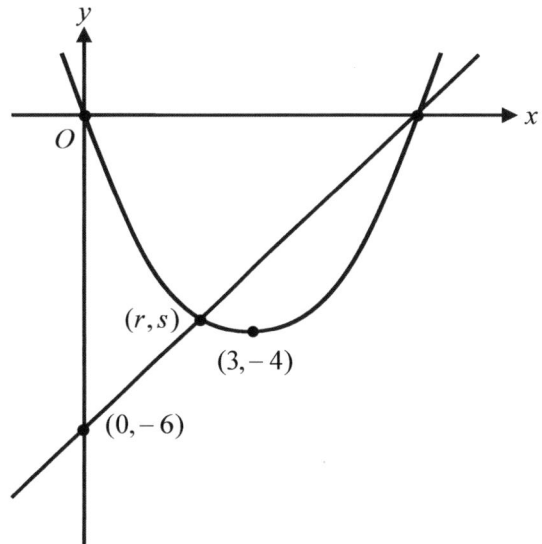

The xy-plane above shows two points of intersection of the graphs of a linear function and a quadratic function. The vertex of the graph of the quadratic function is at $(3, -4)$ and (r, s) is one of points of intersection of the graphs. What is the value of r?

STOP

If you finish before time is called, you may check your work on this section only.
Do not turn to any other section in the test.

Math Conversion Table

Raw Score	Scaled Score	Raw Score	Scaled Score
58	800	27	500
57	800	26	490
56	800	25	480
55	800	24	470
54	790	23	460
53	780	22	460
52	770	21	450
51	750	20	440
50	740	19	430
49	730	18	430
48	720	17	420
47	710	16	420
46	700	15	410
45	690	14	400
44	670	13	390
43	680	12	380
42	670	11	370
41	660	10	360
40	650	9	450
39	640	8	340
38	630	7	330
37	620	6	310
36	610	5	290
35	600	4	280
34	590	3	270
33	580	2	260
32	560	1	240
31	550	0	200
30	540		
29	530		
28	520		

Test 4   Answers and Explanations

SECTION 3	1	2	3	4	5	6	7	8	9	10
	B	A	D	A	D	C	D	D	B	B
	11	12	13	14	15	16	17	18	19	20
	B	B	A	B	D	9.6	5	2.25	160	30

SECTION 4	1	2	3	4	5	6	7	8	9	10
	A	D	B	B	A	B	C	D	D	D
	11	12	13	14	15	16	17	18	19	20
	D	D	C	D	B	B	A	B	C	D
	21	22	23	24	25	26	27	28	29	30
	B	C	D	A	D	C	C	D	C	B
	31	32	33	34	35	36	37	38		
	2	4	9	50	8	22.5	2	9/4		

SECTION 3

1. **B**

   $x^2 + 1 \geq 1$ for all real numbers x.

2. **A**

   $$m = \frac{y_2 - y_1}{x_2 - x_1} = \frac{f(x_2) - f(x_1)}{x_2 - x_1} \quad \rightarrow \quad \frac{f(5) - f(2)}{3} = \frac{f(5) - f(2)}{5 - 2} = 2$$

3. **D**

   Same slope; $\dfrac{3 - (-2)}{a - b} = \dfrac{1}{2} \quad \rightarrow \quad \dfrac{5}{a - b} = \dfrac{1}{2} \quad \rightarrow \quad a - b = 10$

4. **A**

   In order to have no solution, two lines have same slopes but different y-intercepts.

   $3y = -2x + 4 \quad \rightarrow \quad y = -\dfrac{2}{3}x + \dfrac{4}{3} \quad \rightarrow \quad m = -\dfrac{2}{3}$

5. **D**

   Replace x with $(1 - x)$: $\quad f(1 - x) = (1 - x + 1)^2 - (1 - x - 1) - 1 = x^2 + x - 1$

6. C

Find the zeros; $-x^2 + 16 = 0 \rightarrow x^2 - 16 = 0 \rightarrow (x+4)(x-4) = 0 \rightarrow x = 4, -4 \rightarrow AB = 8$

$OB = f(0) = 16$. The area of $\triangle ABC$ is $\dfrac{8 \times 16}{2} = 64$.

7. D

$$\frac{a-2b}{b} = \frac{a}{b} - 2 \rightarrow \frac{a}{b} - 2 = \frac{2}{3} \rightarrow \frac{a}{b} = 2 + \frac{2}{3} = \frac{8}{3}$$

8. D

Since $(4,16)$ is the solution; $16 = m\sqrt{4}$ and $16 = 4m - k$. $16 = 2m \rightarrow m = 8$. Put this number in the equation. $16 = 4(8) - k \rightarrow k = 16$

9. B

Find the x-coordinates of points A and T. $(x-10)^2 = 25 \rightarrow x - 10 = 5, -5 \rightarrow x = 15$ or 5

Therefore, $MH = 15 - 5 = 10$.

10. B

$7x = 9x - 2x + 10 \rightarrow 7x = 7x + 10 \rightarrow 0 = 10(?)$ For any values of x, it CANNOT be equal.

11. B

The axis of symmetry is $x = a$, which is the midpoint of the zeros. $a = \dfrac{-4 + 10}{2} = 3$

$b = f(a) = f(3) = k(3+4)(3-10) = -7 \rightarrow -49k = -7 \rightarrow k = \dfrac{1}{7}$

12. B

Exterior angle theorem.

13. A

$$\frac{5x^2 + kx + 1}{x-1} = ax + 1 + \frac{2}{x-1} \rightarrow \frac{5x^2 + kx + 1}{x-1} = \frac{(ax+1)(x-1) + 2}{x-1} \rightarrow$$

$$\frac{5x^2 + kx + 1}{x-1} = \frac{ax^2 + (1-a)x + 1}{x-1} \rightarrow a = 5 \text{ and } 1 - a = k \rightarrow \text{Therefore, } k = -4.$$

14. B

$$4(x-2)^2 - 1 = 5 \rightarrow 4(x-2)^2 = 6 \rightarrow (x-2)^2 = \frac{6}{4} \rightarrow x - 2 = \pm\frac{\sqrt{6}}{2} \rightarrow x = 2 \pm \frac{\sqrt{6}}{2}$$

15 D

Grade	For	Against	Total
Junior	60	40	100
Senior	70	130	200
Total	130	170	300

16. **9.6**

$\dfrac{BC}{20} = \cos x = 0.8 \rightarrow BC = 16$ By Pythagorean Theorem; $AB = 12$

The area of the triangle; $\dfrac{12 \times 16}{2} = \dfrac{20 \times BD}{2} \rightarrow BD = \dfrac{12 \times 16}{20} = 9.6$

17. **5**

$2x^3 - 10x^2 + 5x - 25 = 0 \rightarrow 2x^2(x-5) + 5(x-5) = 0 \rightarrow (x-5)(2x^2+5) = 0 \rightarrow x = 5$

18. **2.25 or $\dfrac{9}{4}$**

$\triangle ABP$ and $\triangle CDP$ are similar. The ratio of the corresponding sides are $9:3$. The ratio of h_1 to h_2 is also $9:3$. Therefore, $h_1 = 9$ and $h_2 = 3$.

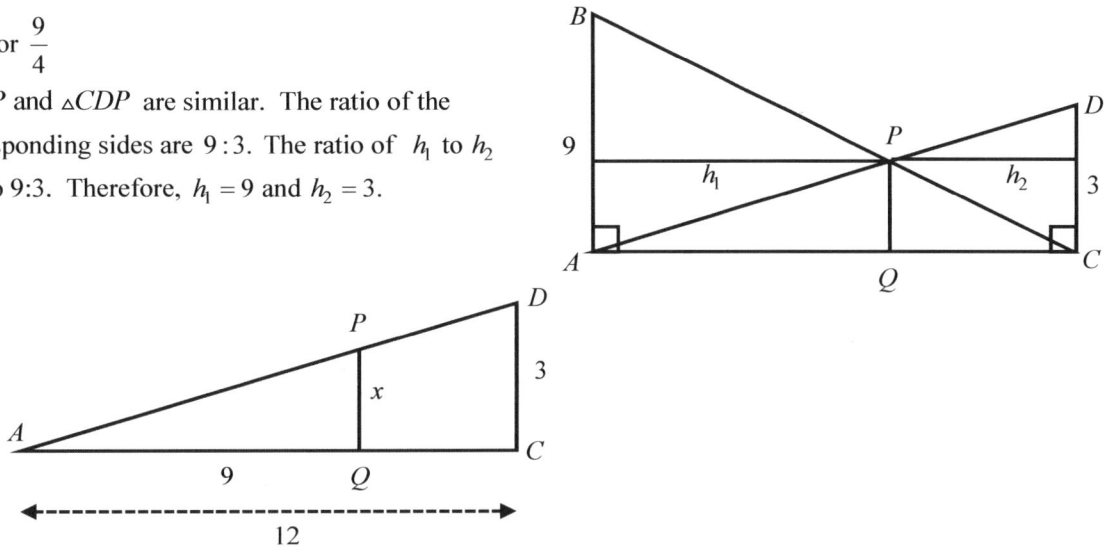

$\triangle APQ$ and $\triangle ADC$ are similar. $\dfrac{PQ}{3} = \dfrac{9}{12} \rightarrow PQ = \dfrac{3 \times 9}{12} = \dfrac{9}{4} = 2.25$

19. **160**

Plan A: $C = 0.25x$ Plan B: $\begin{cases} C = 40 & \text{if } 0 \le x \le 200 \\ C = 40 + 0.15(x - 200) & \text{if } x > 200 \end{cases}$ $x =$ number of texts

First you need to check the value of x in the interval $0 \le x \le 200$.

$0.25x = 40 \rightarrow x = \dfrac{40}{0.25} = 160 \,(\text{OK})$

The graphs of the two plans will be as follows.

20. **30**

For plan A: $C = 0.25 \times 400 = \$100$

For plan B: $C = 40 + 0.15(400 - 200) = \$70$

$100 - 70 = 30$

She can save $30.

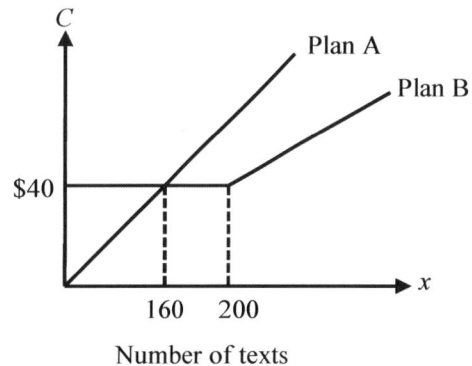

Answer Explanations

SECTION 4

1. A

$$\begin{cases} 2m+3g=20 \\ 3m+2g=15 \end{cases} \rightarrow 5m+5g=35 \rightarrow m+g=7$$

2. D

% of Geometry $= 100-(39+16+25)=20$ Algebra has 19% more than Geometry.

Therefore, the number of students $= 200 \times 0.19 = 38$.

3. B

Slope $= \dfrac{b-a}{2-(-1)} = \dfrac{9}{3} = 3$, slope between $(1,k)$ and $(0,5)$ is also 3. Therefore, $\dfrac{k-5}{1-0} = 3 \rightarrow k = 8$.

4. B

$$P(\text{against}) = \frac{33}{100}$$

Expected number $= 800 \times \dfrac{33}{100} = 264$

Gender	For	Against	Total
Boys	35	10	45
Girls	32	23	55
Total	67	33	100

5. A

$(2, 35) \rightarrow (7, 45)$

Average growth is $\dfrac{45-35}{7-2} = 2$ inches per year

6. B

$$\frac{45-30}{30} \times 100 = 50\%$$

7. C

The slopes are equal. $\dfrac{0-b}{7-0} = \dfrac{3-0}{0-(-1)} \rightarrow -\dfrac{b}{7} = \dfrac{3}{1} \rightarrow b = -21$

8. D

$x^2 - 8x + y^2 = 0 \rightarrow (x-4)^2 + y^2 = 16 \rightarrow$ Therefore, the area of the circle is $\pi r^2 = 16\pi$

9. D

When $x = 3$, $f(3-5) = f(-2) = 3(3) - 10 = -1$

10. D

Rate $= \dfrac{4-20}{8} = -2 \rightarrow$ Therefore, $T = 20 - 2d$.

11. D

12.   D

$$A = 10,000(1-0.3)(1+0.3) = 10,000(1-0.09) \rightarrow 9\% \text{ decreased.}$$

13.   C

$$y = -\frac{a}{b}x + \frac{5}{b} \rightarrow \text{Positive slope and positive } y\text{-intercept} \rightarrow b > 0 \text{ and } a < 0$$

14.   D

$$\frac{\pi r^2 h}{3} = 200\pi \rightarrow \frac{\pi(25)h}{3} = 200\pi \rightarrow h = 24$$

15.   B

$$-x^2 + 6x = 0 \rightarrow x(x-6) = 0 \rightarrow OQ = 6 \text{ and the height of } \triangle OPQ = 9 \ (\because f(3) = -9 + 18 = 9)$$

$$\text{Area} = \frac{6 \times 9}{2} = 27$$

16.   B

$$\frac{1}{8} + \frac{1}{6} = \frac{7}{24} \rightarrow \frac{7}{24} \times 3 = \frac{7}{8} \rightarrow 1 - \frac{7}{8} = \frac{1}{8}$$

17.   A

$$3 \le PQ \le 17$$

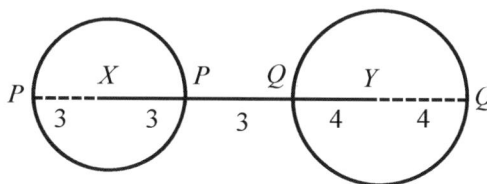

18.   B

$$\text{Slope} = \frac{30}{2} = 15 \rightarrow y = 15x + 20 \rightarrow \text{For 200 desks } y = 15(200) + 20 = \$3020$$

19.   C

$$D(x) = k(x-10)(x-50) = 0 \rightarrow x = 10, 50 \rightarrow 50 - 10 = 40 \text{ meters}$$

20.   D

$$\text{Putting } (30, -50) \text{ in the equation} \rightarrow -50 = k(30-10)(30-50) \rightarrow k = \frac{-50}{-400} = \frac{1}{8}$$

21.   B

Since $\sin a = \cos b$, $BD = BC \times \cos b = 20 \times 0.35 = 7$

22.   C

$$r(x) = 3x^3 - x - 2 \rightarrow r(1) = 3 - 1 - 2 = 0$$

23.   D

24.   A

25. D

Since circumference is 16, $2\pi r = 16 \ \rightarrow \ r = \dfrac{8}{\pi} \ \rightarrow$ Area of the sector $= \pi r^2 \times \dfrac{3}{4} = \pi\left(\dfrac{64}{\pi^2}\right)\times\left(\dfrac{3}{4}\right) = \dfrac{48}{\pi}$

26. C

$k - 1 = a$ and $3k = 24 \ \rightarrow \ k = 8$ and $a = 7$

27. C

$y = 8$ or $y = -5$ have three solutions with $y = f(x)$.

28. D

Slopes between two points are constant. $\dfrac{8-4}{4} = \dfrac{\sqrt{k}-8}{8} \ \rightarrow \ \sqrt{k}-8 = 8 \ \rightarrow \ k = 256$

29. C

$a + 2b = 52$, $b + 2c = 82$, and $c + 2a = 46 \ \rightarrow$ addition $\rightarrow \ 3(a+b+c) = 180 \ \rightarrow \ a+b+c = 60$

Therefore, the average is $\dfrac{a+b+c}{3} = \dfrac{60}{3} = 20$.

30. B

Since the area of a sector is $\dfrac{1}{2}r^2\theta$, where θ is in radian. $\ \rightarrow \ \dfrac{1}{2}r^2\left(\dfrac{2\pi}{5}\right) = 20\pi$

$r^2 = 100 \ \rightarrow \ r = 10 \ \rightarrow \ OR = 10\times\dfrac{5}{2} = 25 \ \rightarrow \ \overarc{RS} = 25\times\dfrac{2\pi}{5} = 10\pi$

31. 2

$g(1) = 2f(1) - 1 = 3 \ \rightarrow \ f(1) = 2$

32. 4

$k\nabla(k-2) = 2\nabla3 \ \rightarrow \ k(k-2)-k+1 = (2)(3)-2+1 \ \rightarrow \ k^2-3k+1 = 5 \ \rightarrow \ k^2-3k-4 = 0$

$\rightarrow (k-4)(k+1) = 0 \ \rightarrow \ k = 4$ or $-1 \ \rightarrow$ Therefore, $k = 4$.

33. 9

$4m + 5n = 2.5(4n) \ \rightarrow \ 4m = 5n \ \rightarrow$ when $m = 5$, $n = 4$

Therefore, $\dfrac{m+n}{m-n} = \dfrac{5+4}{5-4} = 9$

34. 50

Since the area of $\triangle ABM$ is 20, the area of $\triangle MDB$ is $20 \times \dfrac{3}{2} = 30$.

Therefore, the area of $\triangle AMD$ is 50.

($\because$ The area of the shaded region is equal to the area of the unshaded region.)

Answer Explanations

35. 8

Since $\triangle P = 2000$, $\dfrac{\triangle P}{\triangle t} = 250 \rightarrow \dfrac{2000}{\triangle t} = 250 \rightarrow \dfrac{2000}{250} = 8$

36. 22.5

Since $DE + DF = 10$, $DF = 10 \times \dfrac{3}{4} = 7.5$.

Therefore, the perimeter is $7.5 \times 3 = 22.5$

37. 2

Since sum of the root is $\dfrac{8}{3}$ and product of the roots is $\dfrac{4}{3}$, $\dfrac{1}{a} + \dfrac{1}{b} = \dfrac{a+b}{ab} = \dfrac{8/3}{4/3} = 2$.

Or

$3x^2 - 8x + 4 = 0 \rightarrow (3x-2)(x-2) = 0 \rightarrow x = \dfrac{2}{3}, 2 \rightarrow \dfrac{1}{a} + \dfrac{1}{b} = \dfrac{3}{2} + \dfrac{1}{2} = 2$

38. $\dfrac{9}{4}$

Quadratic function: $y = a(x-3)^2 - 4 \rightarrow$ putting $(0,0)$ in the equation $\rightarrow 0 = 9a - 4 \rightarrow a = \dfrac{4}{9}$

The zeros of the quadratic function are 0 and 6. Therefore, the equation of the line is $y = x - 6$.

$\dfrac{4}{9}(x-3)^2 - 4 = x - 6 \rightarrow 4(x-3)^2 - 36 = 9x - 54 \rightarrow 4x^2 - 33x + 54 = 0 \rightarrow (4x-9)(x-6) = 0$

$x = \dfrac{9}{4}$ or $x = 6 \rightarrow$ Therefore, $r = \dfrac{9}{4}$.

No Test Material on This Page

PRACTICE TEST 5

Dr. John Chung's SAT Math

Math Test - No Calculator

25 MINUTES, 20 QUESTIONS

Turn to Section 3 of your answer sheet to answer the questions in this section.

DIRECTIONS

For questions 1–15, solve each problem, choose the best answer from the choices provided, and fill in the corresponding circle on your answer sheet. **For questions 16–20**, solve the problem and enter your answer in the grid on your answer sheet. Please refer to the directions before question 16 on how to enter your answers in the grid. You may use any available space in your test booklet for scratch work.

NOTE

1. The use of a calculator **is not permitted**.

2. All variables and expressions used represent real numbers unless otherwise indicated.

3. Figures provided in this test are drawn to scale unless otherwise indicated.

4. All figures lie in a plane unless otherwise indicated.

5. Unless otherwise indicated, the domain of a given function f is the set of all real numbers x for which $f(x)$ is a real number.

REFERENC

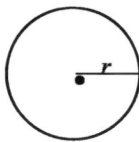 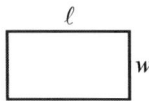 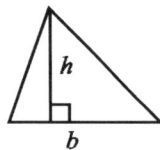 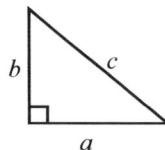 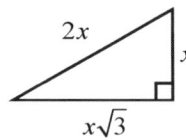 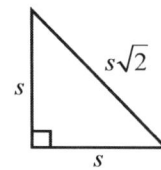

$A = \pi r^2$
$C = 2\pi r$

$A = \ell w$

$A = \dfrac{1}{2}bh$

$c^2 = a^2 + b^2$

Special Right Triangles

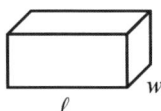 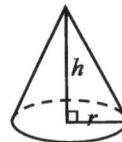 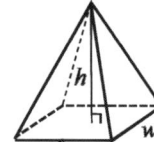

$V = \ell w h$

$V = \pi r^2 h$

$V = \dfrac{4}{3}\pi r^3$

$V = \dfrac{1}{3}\pi r^2 h$

$V = \dfrac{1}{3}\ell w h$

The number of degrees of arc in a circle is 360.
The number of radians of arc in a circle is 2π.
The number of the measures in degrees of the angles of a triangle is 180.

CONTINUE

1

If $4r - 35 = 4s + 13$, what is the value of $r - s$?

A) 9

B) 10

C) 12

D) 16

2

$$x^2 - y^2 = 35$$
$$x + y = 5$$

In the system of equations above, which of the following is the value of x?

A) 5

B) 6

C) 7

D) 8

3

$$\left(x - \frac{1}{x}\right)^2 + 4$$

Which of the following is equivalent to the expression shown above?

A) $x^2 - \dfrac{1}{x^2} + 4$

B) $x^2 + \dfrac{1}{x^2} + 4$

C) $\left(x + \dfrac{1}{x}\right)^2$

D) $\left(x + \dfrac{1}{x}\right)^2 - 4$

4

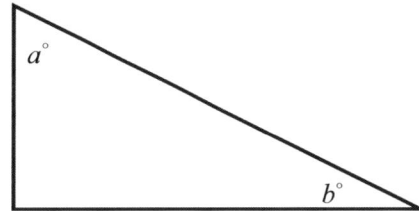

Note: Figure not drawn to scale.

In the triangle above, $a = 3x + 20$ and $b = x - 10$. If $\cos a^\circ = \sin b^\circ$, what is the value of x?

A) 20

B) 25

C) 28

D) 30

CONTINUE

5

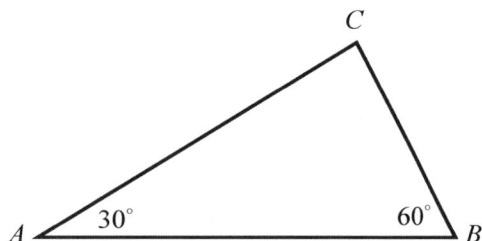

In the triangle above, the length of $\overline{AB}$ is 20. What is the area of triangle ABC?

A) $50\sqrt{3}$

B) $25\sqrt{3}$

C) $\dfrac{25\sqrt{3}}{2}$

D) $\dfrac{25\sqrt{3}}{2}$

6

$$\sqrt{-6} \cdot \sqrt{-24}$$

If $i = \sqrt{-1}$, which of the following is equivalent to the expression shown above?

A) 12

B) −12

C) 12i

D) −12i

Questions 7 and 8 refer to the following information.

$$P(t) = b + at$$

Jessie purchased a micro oven for \$750. After 10 years, the value of the oven will be \$0. The value P of the oven during year t is modeled by the equation above, where a and b are constants.

7

Based on the information above, what is the value of a?

A)   75

B)   50

C) −10

D) −75

8

In how many years will the value of the micro oven be decreased by \$180?

A) 2.4

B) 4

C) 4.5

D) 5

CONTINUE

9

$$P = \frac{A - d}{B + d}$$

A tire repair center uses the formula above to calculate the pressure of tire, where d is the diameter of the tire. Which of the following expresses d in terms of the other variables?

A) $d = \dfrac{P - PB}{A - 1}$

B) $d = \dfrac{A - P}{B - 1}$

C) $d = \dfrac{A - PB}{P + 1}$

D) $d = \dfrac{A - 1}{P - B}$

10

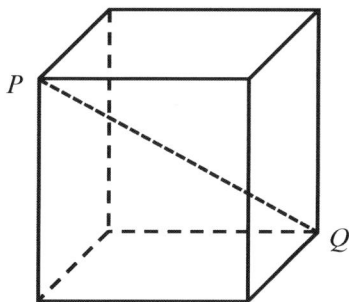

In the cube above, the length of diagonal $\overline{PQ}$ is 12.
What is the surface area of the cube?

A) 27

B) 64

C) 150

D) 288

11

On a car trip Adam drove 50 miles more than half the number of miles Benjamin drove. If together they drove 500 miles, how many miles did Adam drive?

A) 200

B) 250

C) 300

D) 350

12

Plan	Monthly Fee	Cost/Minute
A	$25	$0.20
B	$40	$0.08

A cellular phone company offers two different phone plans shown in the table above. What is the number of minutes when the total cost is the same for both plans?

A) 80

B) 95

C) 100

D) 125

CONTINUE

13

$$\frac{10x + 5}{x - 1}$$

Which of the following is equivalent to the expression above?

A) -15

B) $\dfrac{5}{x-1} - 10$

C) $\dfrac{5}{x-1} + 10$

D) $\dfrac{15}{x-1} + 10$

14

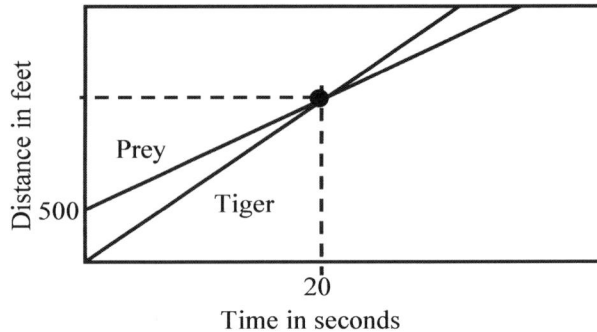

A tiger is 500 feet from its prey. It starts to sprint toward its prey at 88 feet per second. At the same time, the prey starts to sprint in the same direction at p feet per second. The tiger catches its prey in 20 seconds. The graphs shown above represent this relationship. Based on the graphs, what is the value of p?

A) 50

B) 63

C) 70

D) 72

15

$$P(x) = (x - 2)Q(x) + R$$

The equation above shows when $P(x)$ is divided by $(x - 2)$, the remainder is R, where $Q(x)$ is the quotient. If $P(x) = 5x^2 - 3x + 4$, what is the value of R?

A) 4

B) 6

C) 12

D) 18

CONTINUE

DIRECTIONS

For questions 16–20, solve the problem and enter your answer in the grid, as described below, on the answer sheet.

Answer: $\dfrac{7}{12}$ Answer: 2.5

Write answer in boxes. → Fraction line ← Decimal point ←

Grid in result. →

1. Although not required, it is suggested that you write your answer in the boxes at the top of the columns to help you fill in the circles accurately. You will receive credit only if the circles are filled in correctly.
2. Mark no more than one circle in any column.
3. No question has a negative answer.
4. Some problems may have more than one answer.
5. **Mixed numbers** such as $3\dfrac{1}{2}$ must be gridded as 3.5 or 7/2. (If ⌊3 1 / 2⌋ is entered into the grid, it will be interpreted as $\dfrac{31}{2}$, not $3\dfrac{1}{2}$.)
6. **Decimal answers:** If you obtain a decimal answer with more digits than the grid can accommodate, it may be either rounded or truncated, but it must fill the entire grid.

Acceptable ways to grid $\dfrac{2}{3}$ are:

Answer: 201
Either position is correct.

Note: You may start your answers in any column, space permitting. Columns you don't need to use should be left blank.

CONTINUE →

16

$$R^2 - S^2 = 19$$

In the equation above, if R and S are positive integers, what is the value of R ?

17

In reading group A with 90 students, there are 4 boys for every 5 girls. In the other reading group, B, there are 3 boys for every 2 girls. If these two groups are combined, the ratio of boys to girls will be 10:9. How many students are in the reading group B?

18

$$(a-1)x^2 + (b-2)x + c = 0$$

In the equation above, a, b, and c are constants. If the equation is true for all values of x, what is the value of $a + b + c$?

19

$$3x + py = 12$$
$$rx + 5y = 6$$

In the system of equations above, p and r are constants. If the system has infinitely many solutions, what is the value of $\dfrac{p}{r}$?

20

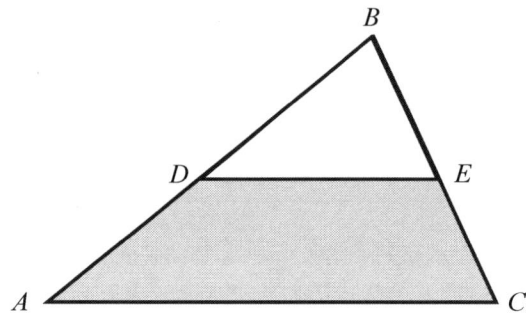

In the figure above, D and E are the midpoints of $\overline{AB}$ and $\overline{BC}$ respectively. If the area of the shaded region is 42, what is the area of triangle ABC ?

STOP

If you finish before time is called, you may check your work on this section only.
Do not turn to any other section in the test.

No Test Material on This Page

Math Test - Calculator

55 MINUTES, 38 QUESTIONS

Turn to Section 4 of your answer sheet to answer the questions in this section.

DIRECTIONS

For questions 1-30, solve each problem, choose the best answer from the choices provided, and fill in the corresponding circle on your answer sheet. **For questions 31-38**, solve the problem and enter your answer in the grid on your answer sheet. Please refer to the directions before question 31 on how to enter your answers in the grid. You may use any available space in your test booklet for scratch work.

NOTE

1. The use of a calculator **is permitted**.

2. All variables and expressions used represent real numbers unless otherwise indicated.

3. Figures provided in this test are drawn to scale unless otherwise indicated.

4. All figures lie in a plane unless otherwise indicated.

5. Unless otherwise indicated, the domain of a given function f is the set of all real numbers x for which $f(x)$ is a real number.

REFERENCE

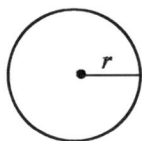 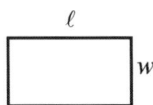 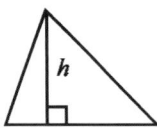 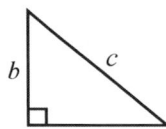 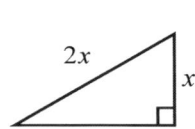 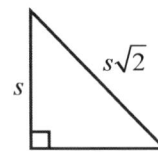

$A = \pi r^2$
$C = 2\pi r$

$A = \ell w$

$A = \frac{1}{2}bh$

$c^2 = a^2 + b^2$

Special Right Triangles

$V = \ell wh$

$V = \pi r^2 h$

$V = \frac{4}{3}\pi r^3$

$V = \frac{1}{3}\pi r^2 h$

$V = \frac{1}{3}\ell wh$

The number of degrees of arc in a circle is 360.
The number of radians of arc in a circle is 2π.
The number of the measures in degrees of the angles of a triangle is 180.

CONTINUE

1

A local telephone company charges $30 for the first 400 texts with additional texts over 400 costing $0.08 per text. If Jessie uses n texts, $n > 400$, which of the following expressions represents her total cost in dollars?

A) $0.08n + 30$

B) $0.08n + 30(400)$

C) $0.08(n - 400) + 30$

D) $0.08(n - 400) + 30(400)$

2

Robert earns P dollars in 4 days. At this rate how many days will it take him to earn S dollars?

A) $4S$

B) $\dfrac{4P}{S}$

C) $\dfrac{S}{4P}$

D) $\dfrac{4S}{P}$

3

If $f(x - 5) = 5x - 14$, which of the following is the value of $f(2)$?

A) -4

B) -3

C) 15

D) 21

4

Gender	Seniors	Juniors	Total
Boys	15		22
Girls		23	
Total	45		

A certain reading group consists of only senior and junior students. The incomplete table above shows the number of students. How many students are in the reading group?

A) 68

B) 75

C) 79

D) 85

CONTINUE

Questions 5 and 6 refer to the following information.

The length of a spring varies directly as the amount of weight attached to it. When a weight of 10 grams is attached, the spring is stretched to 25 centimeters.

5

Which of the following is the equation that relates the weight W and the length L of the spring?

A) $L = 15W$

B) $L = 0.8W$

C) $L = 2.5W$

D) $L = 2.5W + 25$

6

What is the number of grams that stretches a spring 33 centimeters?

A) 12.8

B) 13.2

C) 15

D) 18

7

$$p(x) = 20x - k$$

The profit p, in dollars, from a car wash is given by the function above, where x is the number of cars washed and k is a constant. When 40 cars were washed today, the profit was $500. If the owner wants to make a profit of at least $650, how many more cars should be washed?

A) 7

B) 8

C) 23

D) 25

8

If $4^{a+b} = 8$ and $9^{a-b} = 81$, what is the value of $a^2 - b^2$?

A) 3

B) 8

C) 12

D) 15

CONTINUE

9

If $f(x-2)=3x-5$ for all values of x, which of the following is the expression for $f(x)$?

A) $f(x)=3x-1$

B) $f(x)=3x+1$

C) $f(x)=3x+2$

D) $f(x)=3x+3$

10

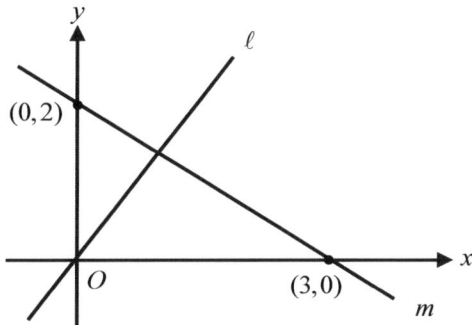

In the xy-plane above, line ℓ is perpendicular to line m. Which of the following points lies on line ℓ?

A) $(1, 2)$

B) $(3, 4)$

C) $(5, 7)$

D) $(6, 9)$

11

$$5a+b+4i=(a-2b)+ki$$

In the equation above, a, b, and k are constants. If $i=\sqrt{-1}$, what is the value of $\dfrac{a}{b}$?

A) $\dfrac{4}{3}$

B) $\dfrac{3}{4}$

C) $-\dfrac{3}{4}$

D) $-\dfrac{4}{3}$

12

$$v(t)=490-9.8t$$

A bullet is shot up into the air from ground level. The equation above shows the velocity, v, of the bullet, in meters per second, after t seconds. According to the model, what is the meaning of the 9.8 in the equation?

A) For every increase of 1 second, the velocity increases by 9.8 meters per second.

B) For every increase of 1 second, the velocity decreases by 9.8 meters per second.

C) For every decrease of 1 second, the velocity decreases by 9.8 meters per second.

D) For every decrease of 9.8 second, the velocity increases by 490 meters per second.

CONTINUE

13

$$ax + by = 5$$

In the equation above, a and b are non-zero constants. If $a + b = 0$, which of the following must be true about the graph in the xy-plane ?

A) The slope of the graph is negative.

B) The slope of the graph is positive.

C) The slope of the graph is zero.

D) The slope of the graph is undefined.

14

Claire first walked one third of the way from home to her friend's house for a birthday party. For the rest of the way to her friend's house, she ran 4 times as fast as she walked. If she took 14 minutes to walk one third of the way, how many minutes did it take her to get from home to her friend's house?

A) 21

B) 24

C) 28

D) 35

15

$$\frac{1}{8}x - \frac{1}{4}y = 1$$
$$\frac{1}{10}x + \frac{1}{5}y = 2$$

In the system of equations above, point (a, b) is the solution of the system. What is the value of $a + b$?

A) 10

B) 13

C) 17

D) 20

16

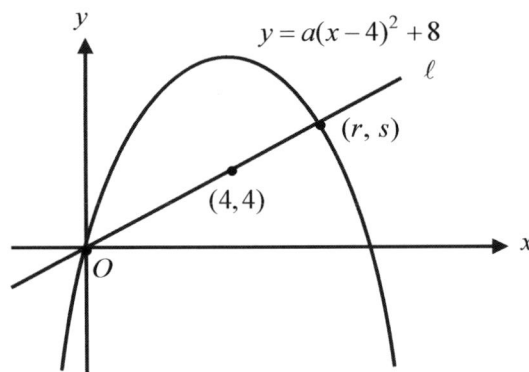

The xy-plane shows graphs of a linear function and a quadratic function, where a is a constant. If (r, s) is the point of intersection, what is the value of r?

A) 6

B) 6.5

C) 7

D) 7.5

CONTINUE

17

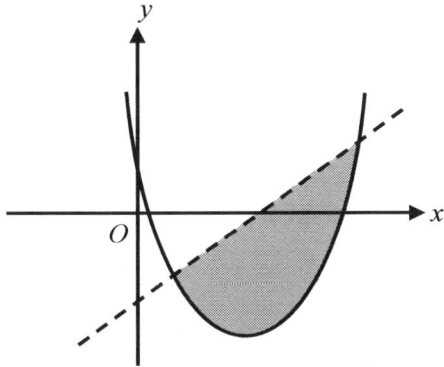

In the xy-plane above, the shaded region represents the solution set of a system of inequalities. Which of the following could be the system of inequalities?

A) $\begin{cases} 4x - 5y - 10 \geq 0 \\ y \geq x^2 - 6x + 5 \end{cases}$

B) $\begin{cases} 4x - 5y - 10 > 0 \\ y \geq x^2 + 6x + 5 \end{cases}$

C) $\begin{cases} 4x - 5y - 10 > 0 \\ y \geq x^2 - 6x + 5 \end{cases}$

D) $\begin{cases} 4x + 5y - 10 > 0 \\ y \geq x^2 - 6x + 5 \end{cases}$

18

$$\frac{x-1}{3} = kx + 2$$

In the equation above, k is a constant. If the equation has no solution, what is the value of k?

A) $\dfrac{1}{3}$

B) $\dfrac{1}{2}$

C) 2

D) 3

19

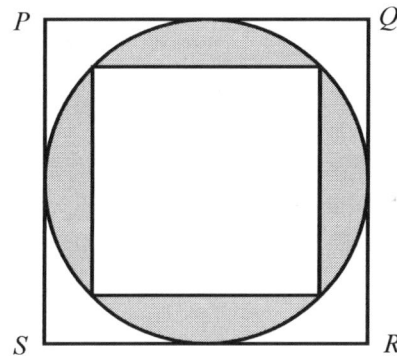

The figure above shows two squares and a circle. If the length of $\overline{PS}$ is a, what is the area of the shaded region?

A) $\dfrac{a^2(\pi - 2)}{4}$

B) $\dfrac{a^2(\pi - 4)}{4}$

C) $\dfrac{\pi a^2}{4} - 4$

D) $\pi a^2 - 8$

CONTINUE

20

$$ax + by - 2 = 0$$

In the function above, a and b are constants. If the graph of the function has a negative slope and a negative y-intercept, which of the following is true?

A) $a = 0$

B) $a > 0$

C) $a < 0$

D) $a \geq 0$

21

$$R = \frac{f(b) - f(a)}{b - a}$$

The average rate of change, R, of function f between a and b is defined by the equation above. If $f(2) = 5$ and $f(5) = -3$, what is the value of R?

A) $-\dfrac{8}{3}$

B) $-\dfrac{3}{8}$

C) $\dfrac{3}{8}$

D) $\dfrac{8}{3}$

22

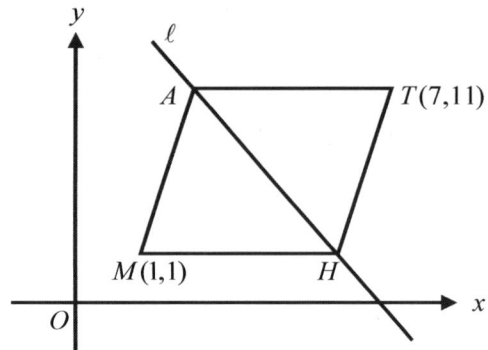

In the xy-plane above, $MATH$ is a rhombus and line ℓ passes through points A and H. Which of the following is the equation of line ℓ?

A) $y = -\dfrac{3}{5}x + 6$

B) $y = -\dfrac{5}{3}x + 6$

C) $y = -\dfrac{3}{5}x + \dfrac{36}{5}$

D) $y = -\dfrac{3}{5}x + \dfrac{42}{5}$

23

$$x^2 + y^2 - 2x - 2y = 7$$
$$y = k$$

In the system of equation above, k is a constant. For which of the following values of k does the system of equations have exactly two real solutions?

A) $k = 6$

B) $k = 5$

C) $k = 4$

D) $k = 3$

CONTINUE

24

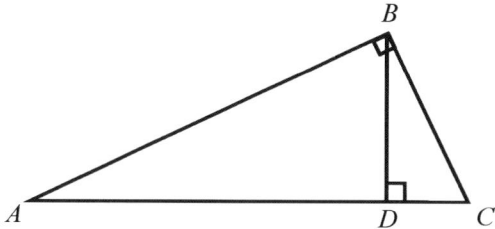

In the right triangle ABC above, $AC = 10$ and the value of $\sin A$ is 0.4. What is the length of $\overline{DC}$?

A) 1.6

B) 2.4

C) 2.5

D) 2.8

25

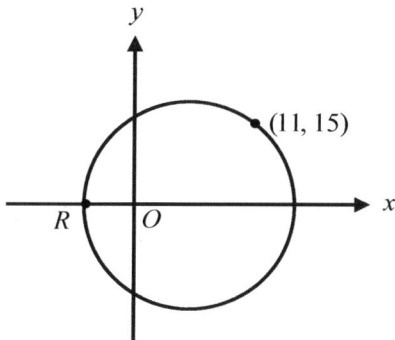

Note: Figure not drawn to scale.

The circle shown in the xy-plane above has a center at $(3, 0)$. Which of the following are the coordinates of point R ?

A) $(-14, 0)$

B) $(-10, 0)$

C) $(-8, 0)$

D) $(-6, 0)$

26

$$(a+b)x^2 + (a-2b)x + k = (k-1)x^2 + 5x + 3$$

In the equation above, a, b, and k are constants. If the equation is true for all real values of x, what is the value of a?

A) 2

B) 3

C) 5

D) 8

27

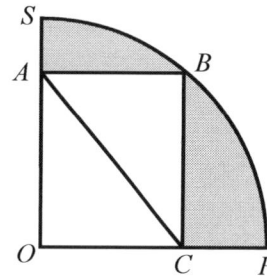

In the figure above, $\overline{OP}$ and $\overline{OS}$ of sector OSP are radii and the length of $\overline{AC}$ of rectangle $ABCO$ is 10. If the measure of angle ACO is $60°$, which of the following is closest to the area of the shaded region?

A) 30

B) 32

C) 35

D) 40

CONTINUE

Questions 28 and 29 refer to the following information.

$$h = v_0 t - \frac{1}{2} g t^2 + 40$$

A rocket is launched from a height of 40 meters with an initial speed of 196 meters per second. The equation above describes the height h and the initial speed v_0 of the rocket, where t is the time elapsed since the rocket is launched and g is the acceleration due to gravity $(9.8 \ m/s^2)$.

28

How long will it take for the rocket to reach its maximum height in seconds?

A) 15

B) 20

C) 25

D) 30

29

What is the maximum height, in meters, of the rocket from the ground?

A) 1200

B) 1600

C) 2000

D) 2400

30

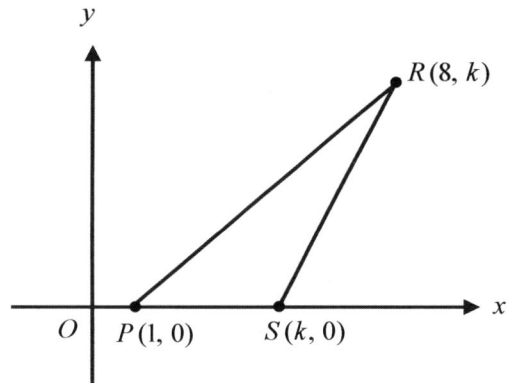

In the xy-plane above, the area of triangle PRS is 10. What is the value of k?

A) 4

B) 5

C) 6

D) 7

CONTINUE

DIRECTIONS

For questions 31-38, solve the problem and enter your answer in the grid, as described below, on the answer sheet.

Answer: $\dfrac{7}{12}$

Write answer in boxes.

Fraction line

Grid in result.

Answer: 2.5

Decimal point

1. Although not required, it is suggested that you write your answer in the boxes at the top of the columns to help you fill in the circles accurately. You will receive credit only if the circles are filled in correctly.
2. Mark no more than one circle in any column.
3. No question has a negative answer.
4. Some problems may have more than one answer.

5. **Mixed numbers** such as $3\dfrac{1}{2}$ must be gridded as 3.5 or 7/2. (If $\boxed{3\,|\,1\,|\,/\,|\,2}$ is entered into the grid, it will be interpreted as $\dfrac{31}{2}$, not $3\dfrac{1}{2}$.)

6. **Decimal answers:** If you obtain a decimal answer with more digits than the grid can accommodate, it may be either rounded or truncated, but it must fill the entire grid.

Acceptable ways to grid $\dfrac{2}{3}$ are:

Answer: 201
Either position is correct.

Note: You may start your answers in any column, space permitting. Columns you don't need to use should be left blank.

CONTINUE

31

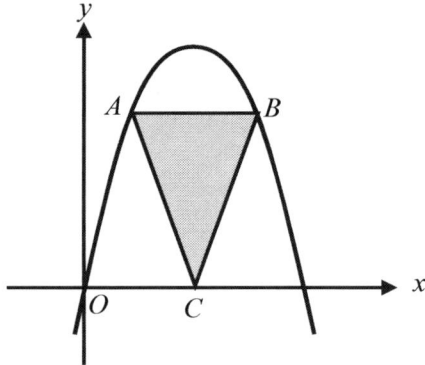

Note: Figure not drawn to scale.

The function $y = x(8 - x)$ is graphed in the xy-plane above. The length of $\overline{AB}$ of isosceles triangle ABC is 4 and $\overline{AB}$ is parallel to the x-axis. What is the area of the triangle?

32

The magnitude of a complex number is the length of a vector from the origin to the terminal point. What is the magnitude of $3 - 4i$?

33

For how many ordered pairs of positive integers (x, y) is $4x + 5y < 15$?

34

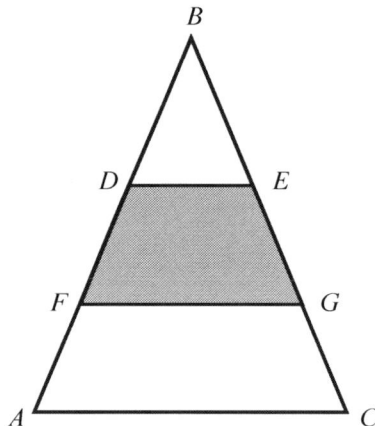

In the figure above, $\overline{DE} \parallel \overline{FG} \parallel \overline{AC}$ and $AF = FD = DB$. If the area of $AFGC$ is 20, what is the area of the shaded region?

CONTINUE

35

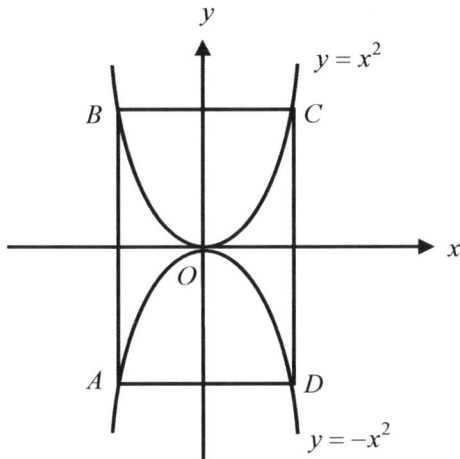

Note: Figure not drawn to scale.

The xy-plane above shows the graphs of two quadratic functions and a rectangle. Points A, B, C, and D lie on the graphs of $y = x^2$ and $y = -x^2$ respectively. If the area of rectangle $ABCD$ is 108, what is the length of $\overline{BC}$?

36

$$5s - 2t - 1 = -a$$
$$-8s + bt - 2 = 2$$

In the system of equations above, a and b are constants. If the system has infinitely many solutions, what is the value of a ?

37

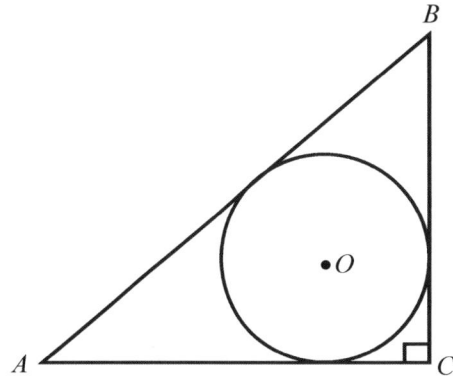

In the figure above, circle O is tangent to the sides of triangle ABC. If $AB = 10$ and $BC = 6$, what is the circumference of the circle to the nearest tenth?

38

$$f(x) = 3x^3 - 8x^2 + 5x - k$$

In the polynomial function above, k is a constant.

If $(x - 2)$ is a factor of $f(x)$, what is the value of k ?

STOP

**If you finish before time is called, you may check your work on this section only.
Do not turn to any other section in the test.**

Math Conversion Table

Raw Score	Scaled Score	Raw Score	Scaled Score
58	800	27	500
57	800	26	490
56	800	25	480
55	800	24	470
54	790	23	460
53	780	22	460
52	770	21	450
51	750	20	440
50	740	19	430
49	730	18	430
48	720	17	420
47	710	16	420
46	700	15	410
45	690	14	400
44	670	13	390
43	680	12	380
42	670	11	370
41	660	10	360
40	650	9	450
39	640	8	340
38	630	7	330
37	620	6	310
36	610	5	290
35	600	4	280
34	590	3	270
33	580	2	260
32	560	1	240
31	550	0	200
30	540		
29	530		
28	520		

Answer Explanations

Test 5 Answers and Explanations

SECTION 3	1	2	3	4	5	6	7	8	9	10
	C	B	C	A	A	B	D	A	C	D
	11	12	13	14	15	16	17	18	19	20
	A	D	D	B	D	10	100	3	20/3	56

SECTION 4	1	2	3	4	5	6	7	8	9	10
	C	D	D	B	C	B	B	A	B	D
	11	12	13	14	15	16	17	18	19	20
	C	B	B	A	C	A	C	A	A	C
	21	22	23	24	25	26	27	28	29	30
	A	D	D	A	A	B	C	B	C	B
	31	32	33	34	35	36	37	38		
	24	5	3	12	6	7/2	12.6	2		

SECTION 3

1. C

$$4r - 35 = 4s + 13 \;\rightarrow\; 4r - 4s = 35 + 13 \;\rightarrow\; 4(r - s) = 48 \;\rightarrow\; r - s = 12$$

2. B

$$x^2 - y^2 = 35 \;\rightarrow\; (x + y)(x - y) = 35 \;\rightarrow\; 5(x - y) = 35 \;\rightarrow\; x - y = 7$$
$$x + y = 5$$
Addition: $2x = 12 \;\rightarrow\; x = 6$

3. C

$$\left(x - \frac{1}{x}\right)^2 + 4 \;\rightarrow\; x^2 - 2 + \frac{1}{x^2} + 4 \;\rightarrow\; x^2 + 2 + \frac{1}{x^2} \;\rightarrow\; \left(x + \frac{1}{x}\right)^2$$

4. A

If $\cos a = \sin b$, then $a + b = 90$. $\;\rightarrow\; a + b = 3x + 20 + x - 10 = 4x + 10 \;\rightarrow\; 4x + 10 = 90 \;\rightarrow\; x = 20$

5. A

$BC = 10$ and $AC = 10\sqrt{3} \;\rightarrow\;$ Area of $\triangle ABC = \dfrac{10 \times 10\sqrt{3}}{2} = 50\sqrt{3}$

6. B

    $$\sqrt{-6} \cdot \sqrt{-24} = \left(i\sqrt{6}\right)\left(i\sqrt{24}\right) = i^2\sqrt{144} = (-1)(12) = -12$$

7. D

    $$0 = 750 + 10a \quad \rightarrow \quad a = \frac{-750}{10} = -75$$

8. A

    $$\frac{180}{75} = 2.4$$

9. C

    $$P = \frac{A-d}{B+d} \quad \rightarrow \quad PB + Pd = A - d \quad \rightarrow \quad Pd + d = A - PB \quad \rightarrow \quad d(P+1) = A - PB \quad \rightarrow \quad d = \frac{A-PB}{P+1}$$

10. D

    $$a = \text{length of an edge} \quad \rightarrow \quad \sqrt{a^2+a^2+a^2} = 12 \quad \rightarrow \quad 3a^2 = 144 \quad \rightarrow \quad a^2 = 48$$

    Surface area $= 6a^2 = 6 \times 48 = 288$

11. A

    $$A = \frac{1}{2}B + 50 \quad \rightarrow \quad A + B = \frac{1}{2}B + 50 + B = 500 \quad \rightarrow \quad \frac{3}{2}B = 450 \quad \rightarrow \quad B = 450 \times \frac{2}{3} = 300$$

    Therefore, $A = 500 - 300 = 200$.

12. D

    $$A = 25 + 0.2x \text{ and } B = 40 + 0.08x \quad \rightarrow \quad 25 + 0.2x = 40 + 0.08x \quad \rightarrow \quad 0.12x = 15 \quad \rightarrow \quad x = 125$$

13. D

    $$\frac{10x+5}{x-1} \quad \rightarrow \quad \frac{10(x-1)+15}{x-1} = 10 + \frac{15}{x-1}$$

14. B

    Tiger: $d = 20 \times 88 = 1760$

    Prey: $d = 500 + 20p = 1760 \quad \rightarrow \quad 20p = 1260 \quad \rightarrow \quad p = 63$

15. D

    $$P(x) = 5x^2 - 3x + 4 = (x-2)Q(x) + R \quad \rightarrow \quad P(2) = 20 - 6 + 4 = R \quad \rightarrow \quad 18 = R$$

16. 10

    $$R^2 - S^2 = 19 \quad \rightarrow \quad (R+S)(R-S) = 19 \quad \rightarrow \quad R+S = 19 \text{ and } R-S = 1 \quad \rightarrow \quad 2R = 20 \quad \rightarrow \quad R = 10$$

Answer Explanations

17. 100

In group A, there are 40 boys and 50 girls. In group B, there are $3k$ boys and $2k$ girls.

$$\frac{40+3k}{50+2k}=\frac{10}{9} \rightarrow 360+27k=500+20k \rightarrow 7k=140 \rightarrow k=20$$

Therefore, the number of students in group B is $5k=5(20)=100$.

18. 3

Identical equation: $a-1=0, b-2=0,$ and $c=0 \rightarrow a=1, b=2, c=0 \rightarrow a+b+c=3$

19. 20/3

$$\frac{3}{r}=\frac{p}{5}=\frac{12}{6}=\frac{2}{1} \rightarrow r=\frac{3}{2} \text{ and } p=10 \rightarrow \frac{p}{r}=\frac{10}{3/2}=\frac{20}{3}$$

20. 56

Ratio of areas of $\triangle DBE:\triangle ABC=1:4 \rightarrow k$ and $4k$

The area of the shaded region $=4k-k=3k=42 \rightarrow k=14$

Therefore, the area of $\triangle ABC=4k=4\times14=56$.

SECTION 4

1. C

2. D

$$\frac{P}{4}=\frac{S}{x} \rightarrow x=\frac{4S}{P}$$

3. D

$$f(7-5)=5(7)-14=21$$

4. B

5. C

$$\frac{L}{W}=k \rightarrow \frac{25}{10}=2.5=k \rightarrow L=kW=2.5W$$

6. B

$$33=2.5W \rightarrow W=\frac{33}{2.5}=13.2$$

7. B

$$20x\geq650-500 \rightarrow 20x\geq150 \rightarrow x\geq7.5 \rightarrow x=8$$

8. A

$4^{a+b} = 8$ and $9^{a-b} = 81$ → $2^{2(a+b)} = 2^3$ and $3^{2(a-b)} = 3^4$

$a+b = \dfrac{3}{2}$ and $a-b = \dfrac{4}{2} = 2$ → $a^2 - b^2 = (a+b)(a-b) = \dfrac{3}{2} \times 2 = 3$

9. B

$f(x+2-2) = 3(x+2)-5$ → $f(x) = 3x+1$

10. D

Slope of line m is $-\dfrac{2}{3}$. The equation of line ℓ is $y = \dfrac{3}{2}x$. $(6, 9)$ lies on this graph.

11. C

$5a+b = a-2b$ → $4a = -3b$ → $\dfrac{a}{b} = \dfrac{-3}{4}$

12. B

13. B

$ax+by = 5$ → $y = -\dfrac{a}{b}x + \dfrac{5}{b}$ → $y = \dfrac{-b}{b}x + \dfrac{5}{b}$ → $y = x + \dfrac{5}{b}$ $\left(\because a = -b\right)$

14. A

Four times as fast as she walked → Time will be $\dfrac{1}{4}$ of 14 minutes.

Therefore, $14 + \dfrac{14}{4} + \dfrac{14}{4} = 21$.

15. C

Simplify the inequalities. $x - 2y = 8$ and $x + 2y = 20$ → $x = 14$ and $y = 3$ → $x + y = 17$

16. A

Putting $(0,0)$ in the function to determine a. → $0 = a(-4)^2 + 8$ → $a = -\dfrac{1}{2}$

The linear function is $y = x$. Solve the equation. $-\dfrac{1}{2}(x-4)^2 + 8 = x$ → $(x-4)^2 - 16 + 2x = 0$ →

$x^2 - 8x + 16 - 16 + 2x = 0$ → $x^2 - 6x = 0$ → $x(x-6) = 0$ → $x = r = 6$

17. C

$\begin{cases} 4x - 5y - 10 > 0 & \rightarrow \quad 5y < 4x - 10 \quad \rightarrow \quad y < \dfrac{4}{5}x - 2 \text{ (dotted line)} \\ y \geq x^2 - 6x + 5 & \rightarrow \quad y \geq (x-3)^2 - 4 \quad \rightarrow \quad \text{positive axis of symmetry} \end{cases}$

18.   A

$$\frac{x-1}{3}=kx+2 \quad \rightarrow \quad x-1=3kx+6 \quad \rightarrow \quad x-3kx=7 \quad \rightarrow \quad x(1-3k)=7 \quad \rightarrow \quad k=\frac{1}{3}$$

If $k=\frac{1}{3}$, then $x\times 0=7$. There is no such a number to satisfy the equation.

19.   A

Area of the circle $=\pi\left(\dfrac{a}{2}\right)^2$ and area of the square $=\dfrac{a\times a}{2}=\dfrac{a^2}{2}$

Therefore, the area of the shaded region $=\dfrac{\pi a^2}{4}-\dfrac{a^2}{2}=\dfrac{\pi a^2-2a^2}{4}=\dfrac{a^2(\pi-2)}{4}$

20.   C

$$ax+by-2=0 \quad \rightarrow \quad y=-\frac{a}{b}x+\frac{2}{b} \quad \rightarrow \quad -\frac{a}{b}<0 \text{ and } \frac{2}{b}<0 \text{ (negative slope and negative } y\text{-intercept)}$$

Since $b<0,\ a$ must be negative.

21.   A

$$R=\frac{f(b)-f(a)}{b-a}=\frac{-3-5}{5-2}=-\frac{8}{3}$$

22.   D

Slope of $\overline{MT}$ is $\dfrac{11-1}{7-1}=\dfrac{5}{3}$. Slope of line ℓ is $-\dfrac{3}{5}$. Midpoint of $\overline{MT}$ is $\left(\dfrac{7+1}{2},\dfrac{11+1}{2}\right)=(4,6)$.

Therefore, the equation of $\overline{AH}$ is $y=-\dfrac{3}{5}x+b$. Putting $(4,6)$ in the equation $\quad \rightarrow \quad y=-\dfrac{3}{5}x+\dfrac{42}{5}$

23.   D

$$x^2+y^2-2x-2y=7 \quad \rightarrow \quad (x-1)^2+(y-1)^2=9 \quad \rightarrow \quad \text{center}(1,1)\text{ and radius}=3$$

$y=k \quad \rightarrow \quad$ In order to have two intersections: $\quad -2<k<4$

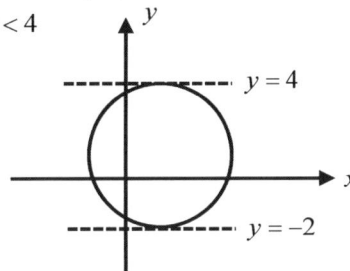

24.   A

$BC=10\sin A=10(0.4)=4$ and $\sin A=\cos B=0.4$

Therefore, $CD=4\cos C=4(0.4)=1.6$

25.   A

Radius $=\sqrt{(11-3)^2+(15-0)^2}=17 \quad \rightarrow \quad$ Therefore, $3-17=-14. \quad \rightarrow \quad (-14,0)$

26. B

$$(a+b)x^2 + (a-2b)x + k = (k-1)x^2 + 5x + 3$$

$$\begin{cases} a+b = k-1 & \to \quad a+b = 3 \quad \to \quad 2a+2b = 6 \\ a-2b = 5 & \hspace{4.5em} \to \quad a-2b = 5 \qquad \text{From the system of equations: } 3a = 9 \quad \to \quad a = 3 \\ k = 3 \end{cases}$$

27. C

Since the radius of the circle is 10, the area of the sector is $\dfrac{100\pi}{4} = 25\pi$. $OC = 5$ and $OA = 5\sqrt{3}$

Therefore, the area of the shaded region $= 25\pi - (5)\left(5\sqrt{3}\right) \approx 35.2$

28. B

$$h = v_0 t - \frac{1}{2}gt^2 + 40 \quad \to \quad h = -\frac{1}{2}(9.8)t^2 + 196t + 40 \quad \to \quad h = -4.9t^2 + 196t + 40$$

Axis of symmetry $t = \dfrac{-196}{2(-4.9)} = 20$

29. C

$$h(20) = -4.9(20)^2 + 196(20) + 40 = 2,000$$

30. B

$PS = k-1$ and the height is k. $\to$ Area $= \dfrac{(k-1)k}{2} = 10 \quad \to \quad k^2 - k - 20 = 0 \quad \to \quad (k-5)(k+4) = 0$

Therefore, $k = 5$.

31. 24

The x-coordinate of point B is 6 and y-coordinate is $y = 6(8-6) = 12$.

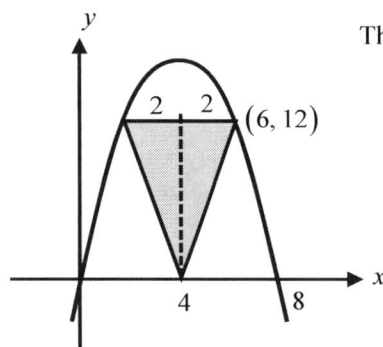

Therefore, the area is $\dfrac{4 \times 12}{2} = 24$.

32. 5

$$z = 3 - 4i \quad \to \quad |z| = \sqrt{3^2 + 4^2} = 5$$

33.　3

$$\begin{cases} \text{If } x=1 \;\to\; 4+5y<15, \;\; y=1,2 \;\to\; (1,1),(1,2) \\ \text{If } x=2 \;\to\; 8+5y<15, \;\; y=1 \;\to\; (2,1) \end{cases}$$

34.　12

Ratio of the lengths $=1:2:3$ $\to$ ratio of the areas $=1:4:9$ $\to$ $k, 4k, 9k$

$9k-4k=5k=20$ $\to$ $k=5$ $\to$ Therefore, the area of the shaded region is $4k-k=3k=12$.

35.　6

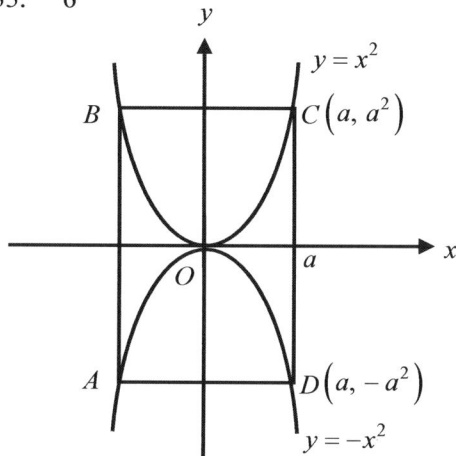

$CD=2a^2$ and $AD=2a$ $\to$ Area of the rectangle $=(2a)(2a^2)=108$

$a=3$ $\to$ $BC=2a=6$

36.　7/2 or 3.5

$$\frac{5}{-8}=\frac{-2}{b}=\frac{1-a}{4} \;\to\; -8+8a=20 \;\to\; a=\frac{7}{2}$$

37.　12.6

$AB=6-r+8-r=10$ $\to$ $r=2$

Circumference $=2\pi r=4\pi=12.6$.

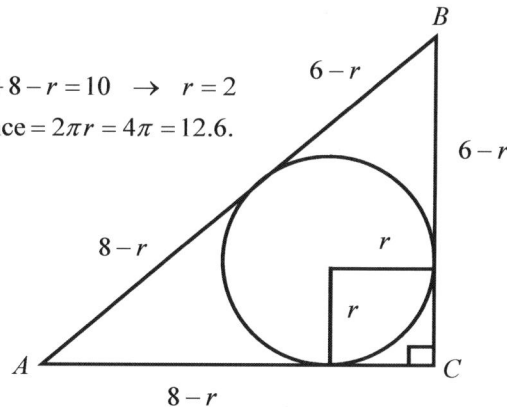

The length of a tangent from a exterior point is constant .

38.　2

$f(2)=0$ $\to$ $f(2)=3(8)-8(4)+5(2)-k=0$ $\to$ $k=2$

No Test Material on This Page

PRACTICE
TEST
6

Dr. John Chung's SAT Math

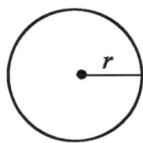

Math Test - No Calculator

25 MINUTES, 20 QUESTIONS

Turn to Section 3 of your answer sheet to answer the questions in this section.

DIRECTIONS

For questions 1–15, solve each problem, choose the best answer from the choices provided, and fill in the corresponding circle on your answer sheet. **For questions 16–20**, solve the problem and enter your answer in the grid on your answer sheet. Please refer to the directions before question 16 on how to enter your answers in the grid. You may use any available space in your test booklet for scratch work.

NOTE

1. The use of a calculator **is not permitted**.

2. All variables and expressions used represent real numbers unless otherwise indicated.

3. Figures provided in this test are drawn to scale unless otherwise indicated.

4. All figures lie in a plane unless otherwise indicated.

5. Unless otherwise indicated, the domain of a given function f is the set of all real numbers x for which $f(x)$ is a real number.

REFERENCE

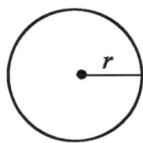 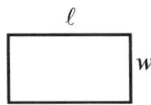 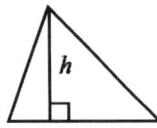 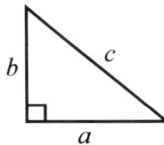 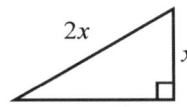 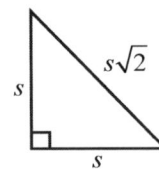

$A = \pi r^2$
$C = 2\pi r$

$A = \ell w$

$A = \dfrac{1}{2}bh$

$c^2 = a^2 + b^2$

Special Right Triangles

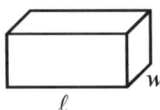 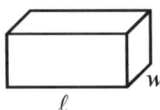   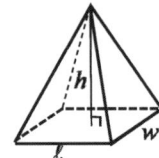

$V = \ell w h$

$V = \pi r^2 h$

$V = \dfrac{4}{3}\pi r^3$

$V = \dfrac{1}{3}\pi r^2 h$

$V = \dfrac{1}{3}\ell w h$

The number of degrees of arc in a circle is 360.
The number of radians of arc in a circle is 2π.
The number of the measures in degrees of the angles of a triangle is 180.

CONTINUE

1

If $10x - 5 = a$, what is the value of $2x - 1$?

A) $\dfrac{a}{5} - 1$

B) $\dfrac{a}{5}$

C) $\dfrac{a}{5} + 1$

D) $\dfrac{a}{5} + 5$

2

Claire is trying to get in shape for a town summer walking tour. She starts her exercise by walking on the treadmill for 20 minutes on the first day. She adds 5 minutes each day before the tour. At this rate how many minutes will she be walking on the treadmill on the 20th day?

A) 80

B) 100

C) 115

D) 120

3

$$|a - 1| < 3$$

In the absolute value inequality above, how many integers a satisfy the inequality?

A) 2

B) 3

C) 4

D) 5

4

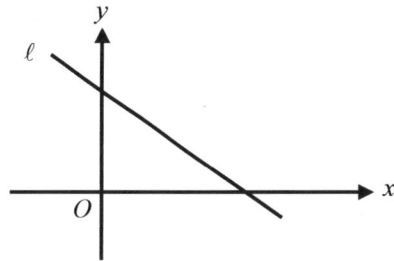

The graph of the line ℓ is shown in the xy-plane above. Which of the following could represent the graph of line ℓ?

A) $x - y + 2 = 0$

B) $x + y - 2 = 0$

C) $x + y + 2 = 0$

D) $x - y - 2 = 0$

CONTINUE

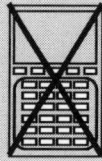

5

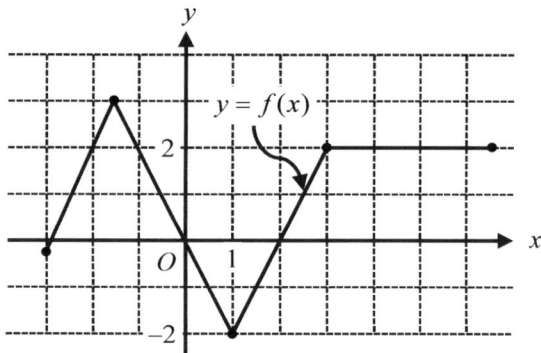

The complete graph of the function f is shown above. Which of the following are equal to 2?

I. $f(-2)$

II. $|f(1)|$

III. $f(4.7)$

A) I only

B) I and III only

C) II and III only

D) I, II, and III

6

$$\left(\sqrt[3]{x^{15}}\right)\left(\sqrt[2]{x^8}\right)$$

If x is positive, which of the following is equivalent to the expression above?

A) $\sqrt[6]{x^{23}}$

B) $\sqrt[5]{x^{23}}$

C) x^{20}

D) x^9

Questions 7 and 8 refer to the following information.

	Juniors	Seniors	Total
Physics	80		180
Statistics		100	
Total			300

The partially completed table gives the enrollment for Physics and Statistics at Jade High School. Only juniors and seniors take these classes.

7

According to the table, what is the number of juniors who take Statistics?

A) 20

B) 40

C) 60

D) 80

8

What percent of juniors is taking Statistics?

A) 6.7

B) 10

C) 20

D) 25

CONTINUE

9

Which of the following data sets appears to have the smallest standard deviation?

A)

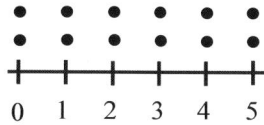

B)

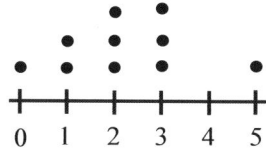

C)

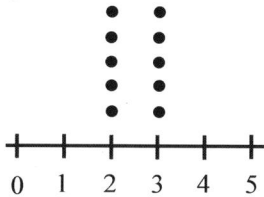

D)

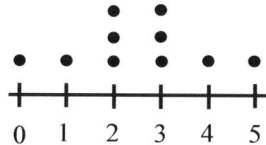

10

$$\sqrt{x+10} = x-2$$

What is the solution set for the equation above?

A) $\{-1\}$

B) $\{6\}$

C) $\{-1, 6\}$

D) No solution

11

$$x^2 - 4x + 5 = (x-1)(x-2) + ax + b$$

In the equation above, a and b are constants. If the equation is true for all values of x, what are the values of a and b?

A) $a = -3, b = -1$

B) $a = 3, b = -1$

C) $a = -1, b = 3$

D) $a = 3, b = 1$

12

$$x^4 - \frac{1}{81}$$

Which of the following is equivalent to the expression above?

A) $\left(x^2 - \frac{1}{9}\right)^2$

B) $\left(x - \frac{1}{3}\right)^4$

C) $\left(x^2 + x + \frac{1}{9}\right)^2$

D) $\left(x^2 + \frac{1}{9}\right)\left(x + \frac{1}{3}\right)\left(x - \frac{1}{3}\right)$

CONTINUE

13

$$4(x^2 - 5x) = 16$$

What is the sum of the solutions of the equation above?

A) 5

B) 10

C) $10 + \sqrt{41}$

D) $10 - \sqrt{41}$

Questions 14 and 15 refer to the following information.

STUDENTS' SALARIES

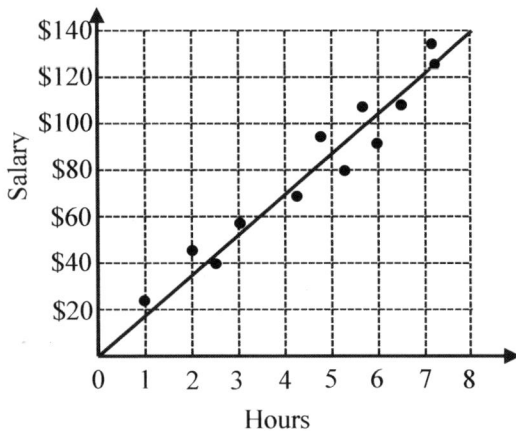

The scatterplot above shows the salary and hours worked by 12 students in the library after school, along with the line of best fit.

14

According to the line of best fit in the scatterplot, which of the following best approximates the average salary, in dollars per hour, of the 12 students?

A) 15

B) 16.2

C) 17.4

D) 20

15

Based on the information above, if a student works 20 hours, approximately how much will he be paid for the work?

A) 200

B) 300

C) 350

D) 500

CONTINUE

DIRECTIONS

For questions 16–20, solve the problem and enter your answer in the grid, as described below, on the answer sheet.

1. Although not required, it is suggested that you write your answer in the boxes at the top of the columns to help you fill in the circles accurately. You will receive credit only if the circles are filled in correctly.
2. Mark no more than one circle in any column.
3. No question has a negative answer.
4. Some problems may have more than one answer.
5. **Mixed numbers** such as $3\frac{1}{2}$ must be gridded as 3.5 or 7/2. (If $\boxed{3\ 1\ /\ 2}$ is entered into the grid, it will be interpreted as $\frac{31}{2}$, not $3\frac{1}{2}$.)
6. **Decimal answers:** If you obtain a decimal answer with more digits than the grid can accommodate, it may be either rounded or truncated, but it must fill the entire grid.

Answer: $\frac{7}{12}$

Write answer in boxes. ← Fraction line

Grid in result.

Answer: 2.5 ← Decimal point

Acceptable ways to grid $\frac{2}{3}$ are:

Answer: 201
Either position is correct.

Note: You may start your answers in any column, space permitting. Columns you don't need to use should be left blank.

CONTINUE

16

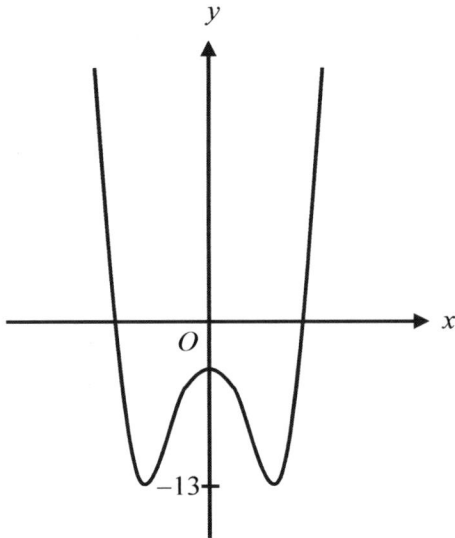

The function $f(x) = x^4 - 6x^2 - 4$ is graphed in the xy-plane as shown above. If the equation $y = -3$ is graphed in the plane, how many points of intersection with the function f are there?

17

$$f(x) = g(x) - k$$
$$g(x) = \sqrt{3x - 2}$$

In the system of equations above, k is a constant. If $f(2) = -3$, what is the value of k?

18

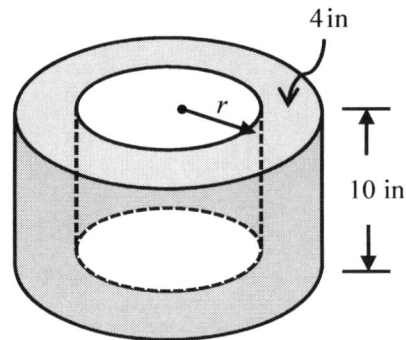

A water pipe is in the shape of a circular tube. The figure above shows the pipe with a portion cut out. The dimensions of the pipe above are height 10 inches with thickness 4 inches. If the volume of the figure above is 800π cubic inches, what is the radius r of the inner circle in inches?

CONTINUE

19

$$f(x) = g^2(x) - 7g(x) + 15$$

In the equation above, if $f(2) = 3$, what is one possible value of $g(2)$?

20

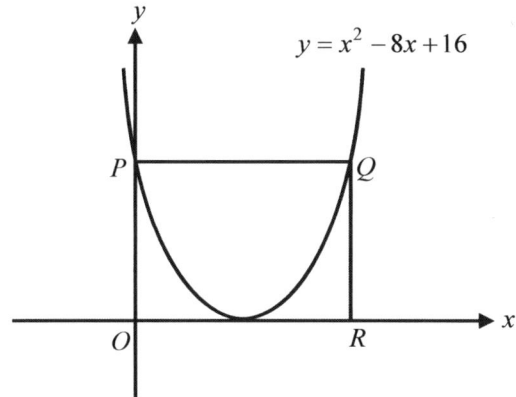

$y = x^2 - 8x + 16$

Figure not drawn to scale.

The graph of $y = x^2 - 8x + 16$ is shown in the xy-plane above. If point P is the y-intercept of the graph, what is the area of rectangle $OPQR$?

STOP

If you finish before time is called, you may check your work on this section only.
Do not turn to any other section in the test.

Math Test - Calculator

55 MINUTES, 38 QUESTIONS

Turn to Section 4 of your answer sheet to answer the questions in this section.

DIRECTIONS

For questions 1-30, solve each problem, choose the best answer from the choices provided, and fill in the corresponding circle on your answer sheet. **For questions 31-38**, solve the problem and enter your answer in the grid on your answer sheet. Please refer to the directions before question 31 on how to enter your answers in the grid. You may use any available space in your test booklet for scratch work.

NOTE

1. The use of a calculator **is permitted**.

2. All variables and expressions used represent real numbers unless otherwise indicated.

3. Figures provided in this test are drawn to scale unless otherwise indicated.

4. All figures lie in a plane unless otherwise indicated.

5. Unless otherwise indicated, the domain of a given function f is the set of all real numbers x for which $f(x)$ is a real number.

REFERENCE

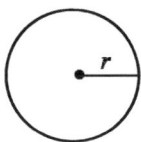

$A = \pi r^2$
$C = 2\pi r$

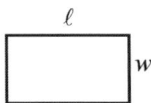

$A = \ell w$

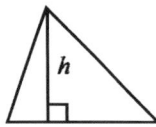

$A = \dfrac{1}{2}bh$

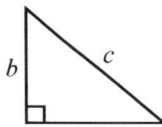

$c^2 = a^2 + b^2$

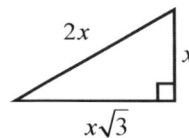

Special Right Triangles

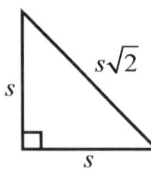

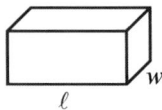

$V = \ell wh$

$V = \pi r^2 h$

$V = \dfrac{4}{3}\pi r^3$

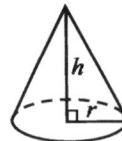

$V = \dfrac{1}{3}\pi r^2 h$

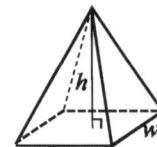

$V = \dfrac{1}{3}\ell wh$

The number of degrees of arc in a circle is 360.
The number of radians of arc in a circle is 2π.
The number of the measures in degrees of the angles of a triangle is 180.

CONTINUE

1

If $(x+3)y = x^2 - x + 12$, what is the value of y when $x = 3$?

A) 3

B) 4

C) 6

D) 8

2

The total cost of 10 equally priced notebooks is k dollars. If the cost per book is reduced by $1, how much will 2 of these notebooks cost at new rate?

A) $k-1$

B) $2x-2$

C) $\dfrac{k}{5} - 2$

D) $\dfrac{k}{10} - 2$

Questions 3 and 4 refer to the following information.

The cost C for maintenance on a heating system increases each year by 2.8%. If Mark paid $250 this year for maintenance, the cost t years from now can be given by the function $C(t) = 250P^t$.

3

What is the value of P?

A) 0.28

B) 0.028

C) 1.028

D) 1.28

4

What is the approximate cost in 4 years?

A) $265

B) $279

C) $310

D) $320

5

$$\frac{x}{4} - \frac{y}{4} = 1$$

In the xy-plane, which of the following could be the graph of the function above?

A)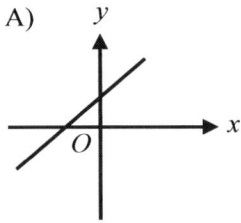

B)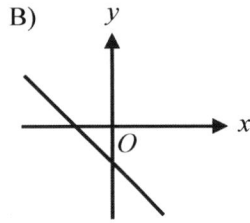

C)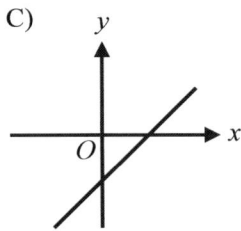

D)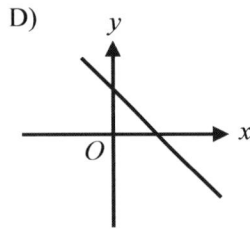

6

$$ax + 5 = 0.8x + b$$

In the equation above, a and b are constants. For which of the following values of a and b does the equation have no solution?

A) $a = 10,\ b = 5$

B) $a = 5,\ b = 0.8$

C) $a = 0.8,\ b = 5$

D) $a = 0.8,\ b = 0.8$

7

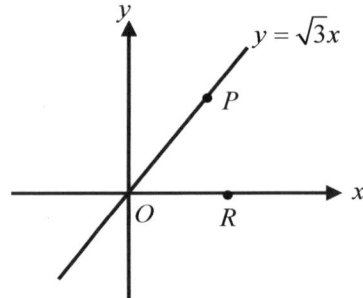

The graph of $y = \sqrt{3}x$ is shown in the xy-plane above. What is the measure, in radians, of angle POR?

A) $\dfrac{\pi}{6}$

B) $\dfrac{\pi}{4}$

C) $\dfrac{\pi}{3}$

D) $\dfrac{\pi}{2}$

8

If $i = \sqrt{-1}$, which of the following is equal to $\dfrac{1 - i^2}{i}$?

A) $-i$

B) $i + 2$

C) $-2i$

D) $i + 1$

CONTINUE

9

What is the remainder when $x^2 - 3x + 5$ is divided by $x - 1$?

A) 2

B) 3

C) 4

D) 5

10

$$\left(\sqrt[k]{16}\right)\left(\sqrt[k]{8}\right) = 2$$

In the equation above, what is the value of k?

A) 4

B) 5

C) 6

D) 7

11

$$ax - y = 1$$
$$x + 2y = 3$$

If the lines represented above are perpendicular, which of the following is the value of a?

A) 3

B) 2

C) −2

D) −3

12

Claire works one week and earns a dollars. If she had worked 5 more hours, she would have earned b dollars. If the hourly rate is constant, what is the hourly rate?

A) $\dfrac{b}{5}$ dollars

B) $\dfrac{a}{5}$ dollars

C) $\dfrac{a-b}{5}$ dollars

D) $\dfrac{b-a}{5}$ dollars

CONTINUE

Questions 13 and 14 refer to the following information.

| | | Holiday | | | |
		Thanksgiving	Memorial Day	Labor Day	Total
Gender	Males	40		35	125
	Females	63			
	Total		140	109	352

A community group responded to a survey that asked which holiday is their favorite. The incomplete survey data are shown in the table above.

13

How many females responded to the survey that Memorial Day is their favorite holiday?

A) 50

B) 75

C) 90

D) 105

14

Which of the following categories accounts for approximately 21 percent of all the survey respondents?

A) Females choosing Memorial Day

B) Males choosing Labor Day

C) Females choosing Thanksgiving

D) Females choosing Labor Day

CONTINUE

15

$$C = -1.5K + 300$$

The linear equation above shows the cost, C, of producing K toys. Based on the information, which of the following must be true?

   I. There is a positive correlation between C and K.

   II. When the company produces 20 toys, the cost is $270.

   III. As K increases by 10, C decreases by $15.

A) II only

B) I and II only

C) I and III only

D) II and III only

16

If $x^2 + kx + k + 1 = (x + p)(x + 2)$ for all values of x and k and p are constants, what is the value of k?

A) 5

B) 4

C) 3

D) 2

17

$$f(x) = 2x^2 - 3$$

In the equation above, if $\frac{1}{3}f\left(\sqrt{k}\right) = 3,$ what is the value of k?

A) 3

B) 4

C) 5

D) 6

18

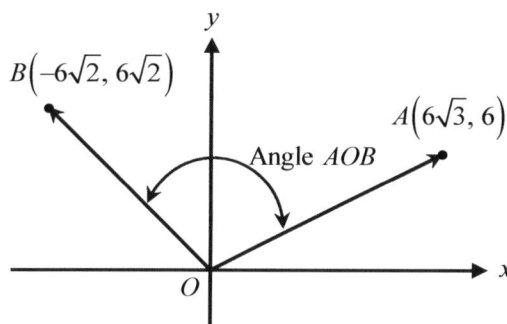

In the xy-plane above, what is the measure, in radians, of angle AOB?

A) $\dfrac{\pi}{3}$

B) $\dfrac{\pi}{2}$

C) $\dfrac{5\pi}{12}$

D) $\dfrac{7\pi}{12}$

CONTINUE

19

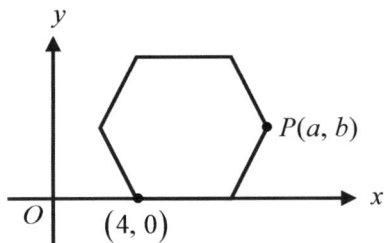

In the xy-plane above, the figure shows a regular hexagon with side length of 5. What is the value of a?

A) 10.5

B) 11

C) 11.5

D) 12

20

$$y > -2x + a$$
$$y < 3x + b$$

In the system of inequalities, a and b are constants. In the xy-plane, if $(0,1)$ is a solution to the system. Which of the following must be true?

A) $a > b$

B) $a = b$

C) $a < b$

D) $a = -b$

21

$$(x-1)(x^2 + 2x - 1) = 0$$

Which of the following is the solution set of the equation above?

A) $\left\{1, \dfrac{1 \pm \sqrt{5}}{2}\right\}$

B) $\left\{1, \dfrac{-2 \pm \sqrt{5}}{2}\right\}$

C) $\left\{1, -2 \pm 2\sqrt{2}\right\}$

D) $\left\{1, -1 \pm \sqrt{2}\right\}$

22

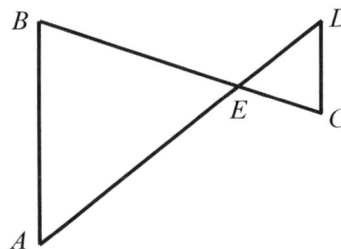

In the figure above, If $\overline{AB} \parallel \overline{CD}$, which of the following must be true?

A) $BE \cdot CE = AE \cdot DE$

B) $\angle ABE = \angle CDE$

C) $\overline{BC} \perp \overline{AD}$

D) $\dfrac{AB}{CD} = \dfrac{BE}{CE}$

CONTINUE

23

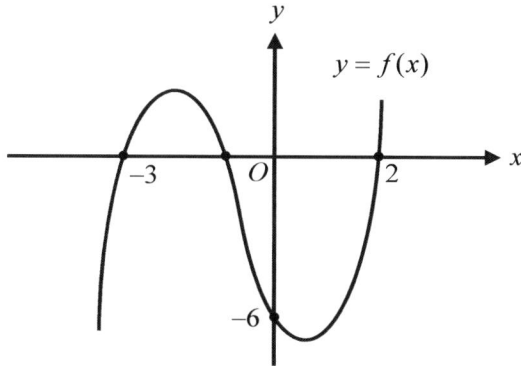

In the xy-plane above, the function f is defined by $f(x) = 2x^3 + 3x^2 + ax + b$, where a and b are constants. If the graph of f intersects the x-axis at three points, what is the value of a?

A) 11

B) 5

C) −5

D) −11

24

$$g(x) = f(x-3) - 10$$

If the slope of the linear function f is $\dfrac{2}{5}$, what is the slope of the function g shown above?

A) $-\dfrac{2}{5}$

B) $-\dfrac{5}{2}$

C) $\dfrac{2}{5}$

D) $\dfrac{5}{2}$

25

A certain dancing group does not receive applicants whose height is less than 5 feet or more than 6 feet. Which of the following inequalities can be used to determine the height h, in feet, of applicants who are not accepted in the group?

A) $|h - 5| > 6$

B) $|h - 6| > 5$

C) $|h - 5.5| < 0.5$

D) $|h - 5.5| > 0.5$

26

$$a^{-2} + 3a^{-1} - 10 = 0$$

In the equation above, $a > 0$. What is the value of a?

A) $\dfrac{1}{2}$

B) 2

C) 3

D) 4

27

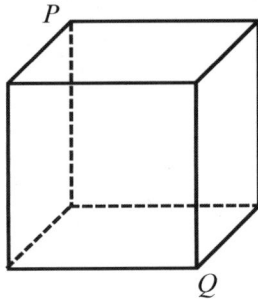

In the figure above, if the volume of the cube is 64, what is the length of diagonal $\overline{PQ}$?

A) 4

B) $4\sqrt{2}$

C) $4\sqrt{3}$

D) $8\sqrt{2}$

28

$$F = k\frac{v^2}{r}$$

In the equation above, k is a constant. If v is tripled and r is halved, which of the following is true?

A) F is tripled.

B) F is multiplied by 8.

C) F is multiplied by 12.

D) F is multiplied by 18.

29

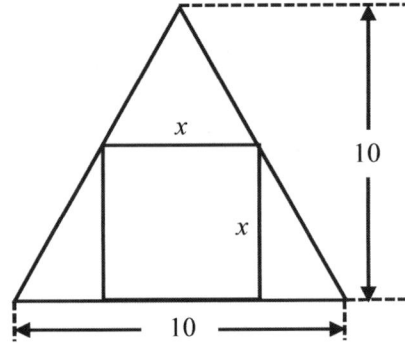

Note: Figure not drawn to scale.

In an isosceles triangle with a height 10 and a base 10, a square is inscribed with side x along the base of the triangle as shown above. What is the area of the square?

A) 16

B) 25

C) 26.25

D) 36

30

$$|k - 3| = 10$$
$$|m + 3| = 6$$

In the system of equation above, what is the greatest value of $k - m$?

A) 10

B) 16

C) 22

D) 24

CONTINUE

DIRECTIONS

For questions 31-38, solve the problem and enter your answer in the grid, as described below, on the answer sheet.

1. Although not required, it is suggested that you write your answer in the boxes at the top of the columns to help you fill in the circles accurately. You will receive credit only if the circles are filled in correctly.
2. Mark no more than one circle in any column.
3. No question has a negative answer.
4. Some problems may have more than one answer.
5. **Mixed numbers** such as $3\frac{1}{2}$ must be gridded as 3.5 or 7/2. (If $\boxed{3\ 1\ /\ 2}$ is entered into the grid, it will be interpreted as $\frac{31}{2}$, not $3\frac{1}{2}$.)
6. **Decimal answers:** If you obtain a decimal answer with more digits than the grid can accommodate, it may be either rounded or truncated, but it must fill the entire grid.

Answer: $\frac{7}{12}$

Write answer in boxes. ← → Fraction line

Grid in result.

Answer: 2.5

← Decimal point

Acceptable ways to grid $\frac{2}{3}$ are:

Answer: 201
Either position is correct.

Note: You may start your answers in any column, space permitting. Columns you don't need to use should be left blank.

CONTINUE

31

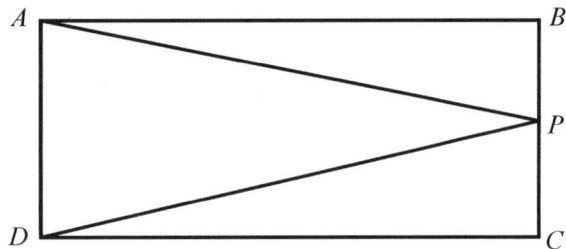

Note: Figure not drawn to scale.

In the rectangle above, $\tan \angle BAP = \dfrac{1}{3}$ and $\tan \angle CDP = \dfrac{2}{5}$. What is the value of $\dfrac{BP}{CP}$?

32

$$a^{(x+1)^2} = \left(\dfrac{1}{a}\right)^{-4x}$$

In the equation above, $a > 0$. What is the value of x?

33

$$\text{Kinetic energy} = \dfrac{1}{2}mv^2$$

In the equation above, kinetic energy is the energy of motion, where m is the mass and v is the speed of an object. If a k-kg roller coaster car is moving 16 meters per second and the other $2k$-kg roller coaster is moving 8 meters per second, what is the ratio of the kinetic energy of the k-kg roller coaster to the kinetic energy of the $2k$-kg roller coaster?

34

$$h = 3t(18 - t)$$

An arrow is shot upward on the moon with an initial velocity of 54 meters per second and returns to the surface. If the height is given by the equation above, what is the maximum height, in meters, that the arrow can reach?

CONTINUE

35

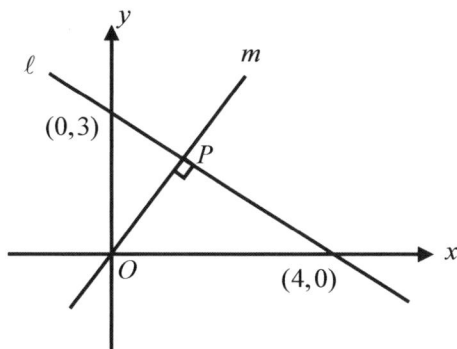

In the xy-plane above, the graphs of line ℓ and line m intersect at point P. If line ℓ is perpendicular to line m, what is the length of $\overline{OP}$?

36

$$15x + 9y = b$$
$$ax + by = 1$$

In the system of equations above, a and b are constants, where $b > 0$. If the system has infinitely many solutions, what is the value of a ?

37

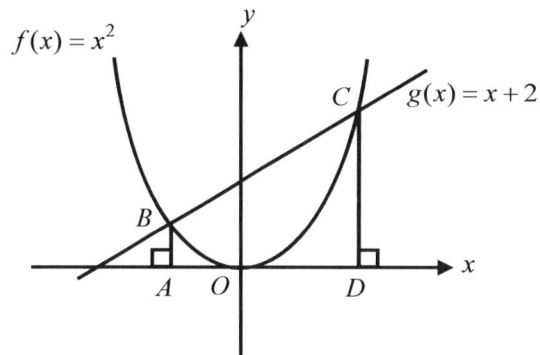

In the xy-plane above, the graphs of functions f and g intersect at points B and C. What is the area of quadrilateral $ABCD$?

38

$$g(x) = 2f(x) + k$$

In the equation above, $f(x)$ is a linear function and k is a constant. If $g(2) = 10$ and $g(5) = 18$, what is the slope of the function $f(x)$?

STOP

If you finish before time is called, you may check your work on this section only.
Do not turn to any other section in the test.

Math Conversion Table

Raw Score	Scaled Score	Raw Score	Scaled Score
58	800	27	500
57	800	26	490
56	800	25	480
55	800	24	470
54	790	23	460
53	780	22	460
52	770	21	450
51	750	20	440
50	740	19	430
49	730	18	430
48	720	17	420
47	710	16	420
46	700	15	410
45	690	14	400
44	670	13	390
43	680	12	380
42	670	11	370
41	660	10	360
40	650	9	450
39	640	8	340
38	630	7	330
37	620	6	310
36	610	5	290
35	600	4	280
34	590	3	270
33	580	2	260
32	560	1	240
31	550	0	200
30	540		
29	530		
28	520		

Answer Explanations

Test 6   Answers and Explanations

SECTION 3	1	2	3	4	5	6	7	8	9	10
	B	C	D	B	D	D	A	C	C	B
	11	12	13	14	15	16	17	18	19	20
	C	D	A	C	C	2	5	8	3, 4	128

SECTION 4	1	2	3	4	5	6	7	8	9	10
	A	C	C	B	C	D	C	C	B	D
	11	12	13	14	15	16	17	18	19	20
	B	D	C	D	D	A	D	D	C	C
	21	22	23	24	25	26	27	28	29	30
	D	D	D	C	D	A	C	D	B	C
	31	32	33	34	35	36	37	38		
	5/6	1	2	243	2.4	5	7.5	4/3		

SECTION 3

1.   B

$$10x - 5 = a \quad \rightarrow \quad 5(2x - 1) = a \quad \rightarrow \quad 2x - 1 = \frac{a}{5}$$

2.   C

$$20 + 19 \times 5 = 115$$

3.   D

$$|a - 1| < 3 \quad \rightarrow \quad -3 < a - 1 < 3 \quad \rightarrow \quad -2 < a < 4 \quad \rightarrow \quad 5 \text{ integers: } -1, 0, 1, 2, 3,$$

4.   B

Only B has a negative slope and positive y-intercept.

5.   D

$$f(-2) = 2, \ |f(1)| = |-2| = 2, \ f(4.7) = 2$$

6.   D

$$\left(\sqrt[3]{x^{15}}\right)\left(\sqrt[2]{x^8}\right) = x^{15/3} x^{8/2} = x^5 x^4 = x^9$$

Answer Explanations

7. A

8. C

$$\frac{20}{100} = 20\%$$

	Juniors	Seniors	Total
Physics	80	100	180
Statistics	20	100	120
Total	100	200	300

9. C

The data are not spread farther from the mean than any other data.

10. B

$$\sqrt{x+10} = x-2 \;\rightarrow\; x+10 = x^2 - 4x + 4 \;\rightarrow\; x^2 - 5x - 6 = 0 \;\rightarrow\; (x-6)(x+1) = 0 \;\rightarrow\; x = 6 \text{ or } -1$$

When $x = -1$, it's undefined. Therefore, $x = \{6\}$.

11. C

$$x^2 - 4x + 5 = (x-1)(x-2) + ax + b \;\rightarrow\; x^2 - 4x + 5 = x^2 + (a-3)x + 2 + b \;\rightarrow\; a - 3 = -4 \text{ and } 5 = 2 + b$$

Therefore, $a = -1$ and $b = 3$.

12. D

$$x^4 - \frac{1}{81} = \left(x^2 + \frac{1}{9}\right)\left(x^2 - \frac{1}{9}\right) = \left(x^2 + \frac{1}{9}\right)\left(x + \frac{1}{3}\right)\left(x - \frac{1}{3}\right)$$

13. A

$$4(x^2 - 5x) = 16 \;\rightarrow\; x^2 - 5x - 4 = 0 \;\rightarrow\; \text{Sum of the solutions is } \frac{-b}{a} = \frac{-(-5)}{1} = 5$$

14. C

$$\frac{140}{8} \approx 17.5 \;\rightarrow\; 17.4 \text{ is closest to the number.}$$

15. C

$$y = 17.5x \;\rightarrow\; y = 17.5 \times 20 = 350$$

16. 2

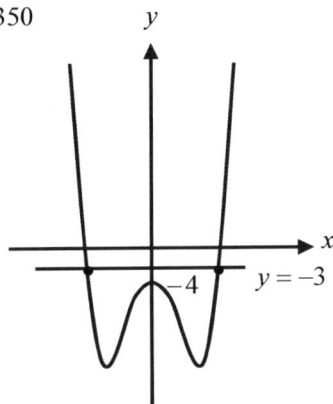

17. 5

$$f(2) = g(2) - k = -3 \;\rightarrow\; g(2) = k - 3$$

$$g(2) = \sqrt{3(2) - 2} = 2 \;\rightarrow\; \text{Therefore, } k - 3 = 2 \;\rightarrow\; k = 5.$$

Answer Explanations

18. 8

$$\text{Volume} = \pi\left(\left(r+4\right)^2 - r^2\right) \times 10 = 800\pi \;\rightarrow\; \left(r+4\right)^2 - r^2 = 80 \;\rightarrow\; 8r + 16 = 80 \;\rightarrow\; r = 8$$

19. 3 or 4

$$f(x) = g^2(x) - 7g(x) + 15 \;\rightarrow\; f(2) = g^2(2) - 7g(2) + 15 = 3 \;\rightarrow\; g^2(2) - 7g(2) + 12 = 0$$
$$\left(g(2) - 3\right)\left(g(2) - 4\right) = 0 \;\rightarrow\; g(2) = 3 \text{ or } 4$$

20. 128

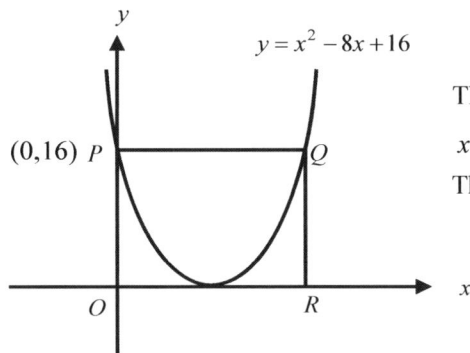

The equation of line $\overline{PQ}$ is $y = 16$.

$$x^2 - 8x + 16 = 16 \;\rightarrow\; x^2 - 8x = x(x-8) = 0$$

Therefore, $OR = 8$. $\text{Area} = 8 \times 16 = 128$

SECTION 4

1. A

$$(x+3)y = x^2 - x + 12 \;\rightarrow\; 6y = 9 - 3 + 12 \;\rightarrow\; y = 3$$

2. C

$$\text{Original price per book} = \frac{k}{10} \;\rightarrow\; \text{New price is } \frac{k}{10} - 1 \;\rightarrow\; \text{Price for 2 books} = 2\left(\frac{k}{10} - 1\right) = \frac{k}{5} - 2$$

3. C

$$C(t) = 250P^t = 250\left(1 + 0.028\right)^t \;\rightarrow\; P = 1.028$$

4. B

$$C = 250\left(1.028\right)^4 \approx 279$$

5. C

When $x = 0$, $y = -4$ and when $y = 0$, $x = 4$. $\rightarrow$ positive x-intercept and negetive y-intercept

Or, $y = x - 4$.

6. D

$$ax + 5 = 0.8x + b \;\rightarrow\; (a - 0.8)x = b - 5$$

Choice C) $a = 0.8$, $b = 5$ $\rightarrow$ $0(x) = 0$ $\rightarrow$ infinitely many solution

Choice D) $a = 0.8$, $b = 0.8$ $\rightarrow$ $0(x) = -4.2$ $\rightarrow$ no solution

7. C

Since slope is $\sqrt{3}$, $\angle POR = 60° = \dfrac{\pi}{3}$.

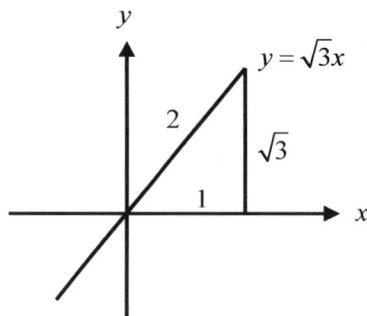

8. C

$$\dfrac{1-i^2}{i} = \dfrac{2 \times i}{i \times i} = \dfrac{2i}{-1} = -2i$$

9. B

$P(1) = 1 - 3 + 5 = 3$

10. D

$$\left(\sqrt[k]{16}\right)\left(\sqrt[k]{8}\right) = 2 \;\rightarrow\; 16^{\frac{1}{k}} \times 8^{\frac{1}{k}} = (128)^{\frac{1}{k}} = 2^{\frac{7}{k}} = 2^1 \;\rightarrow\; \dfrac{7}{k} = 1 \;\rightarrow\; k = 7$$

11. B

$ax - y = 1 \;\rightarrow\; y = ax - 1$

$x + 2y = 3 \;\rightarrow\; y = -\dfrac{1}{2}x + \dfrac{3}{2}$ $\rightarrow$ a is a negativ reciprocal of the other slope. $\rightarrow$ $a = 2$

12. D

$$a + 5x = b \;\rightarrow\; 5x = b - a \;\rightarrow\; x = \dfrac{b-a}{5}$$

13. C

		Holiday			
		Thanksgiving	Memorial Day	Labor Day	Total
Gender	Males	40	50	35	125
	Females	63	90	74	227
	Total	103	140	109	352

14. D

$352 \times 0.21 = 73.92$ $\rightarrow$ females choosing Labor Day

15. D

16. A

$f(-2) = 4 - 2k + k + 1 = 0 \;\rightarrow\; k = 5$

Or

$x^2 + kx + k + 1 = (x + p)(x + 2) \;\rightarrow\; x^2 + kx + k + 1 = x^2 + (2 + p)x + 2p \;\rightarrow\; k = 2 + p$ and $k + 1 = 2p$

$k + 1 = 2k - 4 \;\rightarrow\; k = 5$

17. **D**

$$\frac{1}{3}f\left(\sqrt{k}\right)=3 \;\rightarrow\; f\left(\sqrt{k}\right)=9 \;\rightarrow\; 2k-3=9 \;\rightarrow\; k=6$$

18. **D**

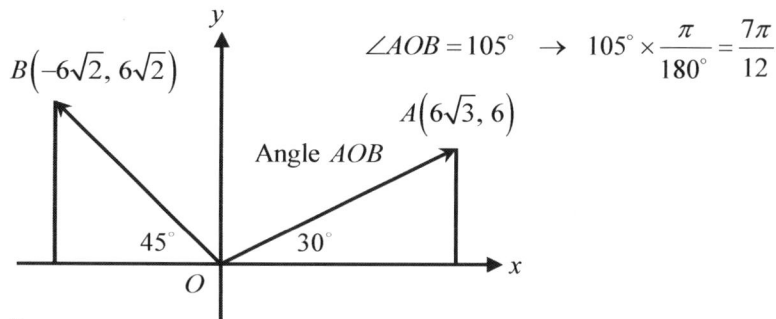

$$\angle AOB=105° \;\rightarrow\; 105°\times\frac{\pi}{180°}=\frac{7\pi}{12}$$

19. **C**

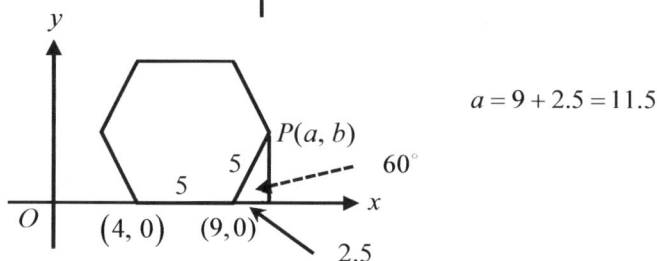

$$a=9+2.5=11.5$$

20. **C**

Putting $(0,1)$ in the inequalities.

$1>a$ and $1<b \;\rightarrow\; b>a$

21. **D**

Use quadratic formula.

22. **D**

The triangles are similar.

23. **D**

$$f(0)=b=-6 \;\rightarrow\; b=-6 \;\rightarrow\; f(2)=16+12+2a-6=0 \;\rightarrow\; 2a=-22 \;\rightarrow\; a=-11$$

24. **C**

Translate doesn't affect the slope.

25. **D**

If $h>6$ or $h<5$, Not accepted. $\;\rightarrow\; |h-5.5|>0.5$

26. **A**

$$a^{-2}+3a^{-1}-10=0 \;\rightarrow\; \frac{1}{a^2}+\frac{3}{a}-10=0 \;\rightarrow\; 10a^2-3a-1=0 \;\rightarrow\; (5a+1)(2a-1)=0 \;\rightarrow\; a=-\frac{1}{5} \text{ or } \frac{1}{2}$$

Therefore, $a=\frac{1}{2}$. $(a>0)$

27. C

The length of an edge is 4. The length of a diagonal $= \sqrt{4^2 + 4^2 + 4^2} = \sqrt{48} = 4\sqrt{3}$

28. D

$$F' = k\frac{(3v)^2}{\left(\frac{1}{2}r\right)} = k\frac{v^2(9)}{r^2(1/2)} = k\frac{v^2}{r}(18) = 18F$$

Or, use convenient number. $(k = 1, \ v = 1, \ \text{and} \ r = 1) \ \rightarrow \ F = 1 \ \rightarrow \ F' = \dfrac{3^2}{\dfrac{1}{2}} = 18$

29. B

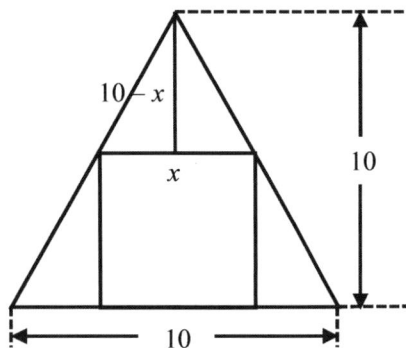

Similar: $\dfrac{10-x}{10} = \dfrac{x}{10} \ \rightarrow \ 10x = 100 - 10x$

$\rightarrow \ 20x = 100 \ \rightarrow \ x = 5$

Therefore, the area of the square is 25.

30. C

$|k - 3| = 10 \ \rightarrow \ k = 13, -7$

$|m + 3| = 6 \ \rightarrow \ m = 3, -9$

Therefore, the greatest value of $k - m$ is $13 - (-9) = 22$.

31. 5/6

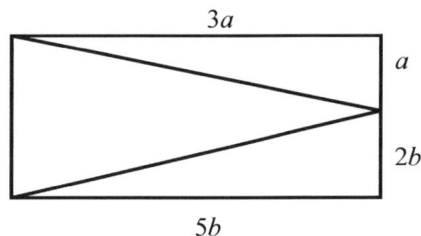

$3a = 5b \ \rightarrow \ $ Use $a = 5$ and $b = 3 \ \rightarrow \ \dfrac{BP}{CP} = \dfrac{5}{6}$

32. 1

$$a^{(x+1)^2} = \left(\frac{1}{a}\right)^{-4x} \ \rightarrow \ a^{x^2+2x+1} = \left(a^{-1}\right)^{-4x} \ \rightarrow \ a^{x^2+2x+1} = a^{4x} \ \rightarrow \ x^2 + 2x + 1 = 4x$$

$x^2 - 2x + 1 = 0 \ \rightarrow \ (x-1)^2 = 0 \ \rightarrow \ x = 1$

33. 2

$$\frac{KE_1}{KE_2} = \frac{\frac{1}{2}(k)(16)^2}{\frac{1}{2}(2k)(8)^2} = \frac{16 \times 16}{2 \times 8 \times 8} = 2$$

34. **243**

    When $t = 9$, it is the maximum height. $h = 3 \times 9(18-9) = 243$

35. **2.4**

    Area of the triangle:

    $$\frac{3 \times 4}{2} = \frac{5 \times OP}{2} \quad \rightarrow \quad OP = \frac{3 \times 4}{5} = 2.4$$

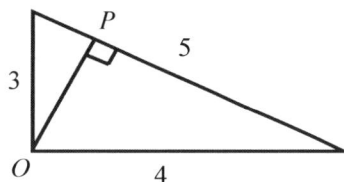

36. **5**

    $$\frac{15}{a} = \frac{9}{b} = \frac{b}{1} \quad \rightarrow \quad b^2 = 9 \quad \rightarrow \quad b = 3 \quad \rightarrow \quad \frac{15}{a} = 3 \quad \rightarrow \quad a = 5$$

37. **7.5**

    $$x^2 = x + 2 \quad \rightarrow \quad x^2 - x - 2 = 0 \quad \rightarrow \quad (x-2)(x+1) = 0$$

    $x = -1, 2$

    The area of trapezoid is $\dfrac{(1+4)3}{2} = 7.5$

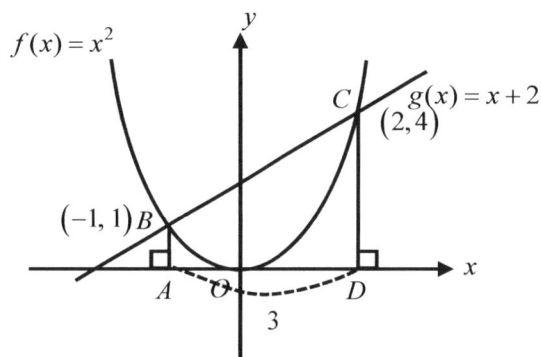

38. $\dfrac{4}{3}$

    $$g(5) = 2f(5) + k = 18 \quad \rightarrow \quad f(5) = \frac{18-k}{2}$$

    $$g(2) = 2f(2) + k = 10 \quad \rightarrow \quad f(2) = \frac{10-k}{2}$$

    $$\rightarrow \quad f(5) - f(2) = \frac{18-10}{2} = 4$$

    Slope of the function $f = \dfrac{f(5) - f(2)}{5 - 2} = \dfrac{4}{3}$

No Test Material on This Page

PRACTICE TEST
7

Dr. John Chung's SAT Math

Math Test - No Calculator

25 MINUTES, 20 QUESTIONS

Turn to Section 3 of your answer sheet to answer the questions in this section.

DIRECTIONS

For questions 1–15, solve each problem, choose the best answer from the choices provided, and fill in the corresponding circle on your answer sheet. **For questions 16–20,** solve the problem and enter your answer in the grid on your answer sheet. Please refer to the directions before question 16 on how to enter your answers in the grid. You may use any available space in your test booklet for scratch work.

NOTE

1. The use of a calculator **is not permitted**.

2. All variables and expressions used represent real numbers unless otherwise indicated.

3. Figures provided in this test are drawn to scale unless otherwise indicated.

4. All figures lie in a plane unless otherwise indicated.

5. Unless otherwise indicated, the domain of a given function f is the set of all real numbers x for which $f(x)$ is a real number.

REFERENCE

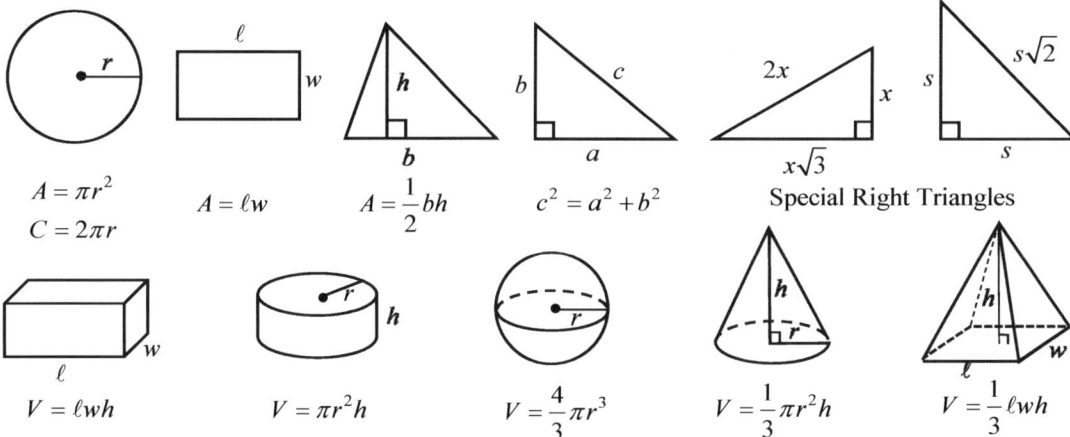

$A = \pi r^2$
$C = 2\pi r$

$A = \ell w$

$A = \frac{1}{2}bh$

$c^2 = a^2 + b^2$

Special Right Triangles

$V = \ell wh$

$V = \pi r^2 h$

$V = \frac{4}{3}\pi r^3$

$V = \frac{1}{3}\pi r^2 h$

$V = \frac{1}{3}\ell wh$

The number of degrees of arc in a circle is 360.
The number of radians of arc in a circle is 2π.
The number of the measures in degrees of the angles of a triangle is 180.

CONTINUE ➤

1

$$2 \le x \le 10$$

Which of the following is equivalent to the expression above?

A) $|x+6| \ge 4$

B) $|x+6| \le 4$

C) $|x-6| \ge 4$

D) $|x-6| \le 4$

2

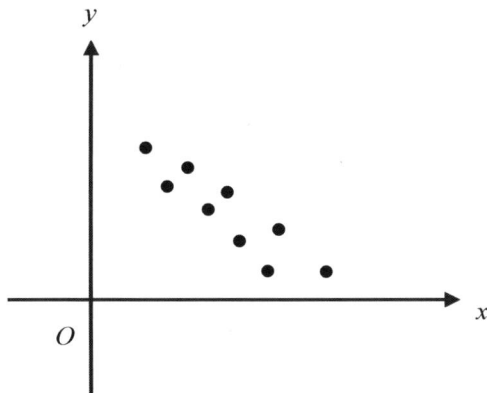

In the scatter plot above, what correlation coefficient best fits the data?

A) -1

B) -0.95

C) 0.95

D) 1

3

$$f(x) = -\frac{5}{3}x + b$$

In the function above, b is a constant. If $f(9) = 5$, what is the value of $f(3)$?

A) 15

B) 10

C) -15

D) -25

4

$$x^{-2}\left(\frac{1}{\sqrt{x}}\right)$$

Which of the following is equivalent to the expression shown above?

A) $\dfrac{1}{\sqrt{x^5}}$

B) $\dfrac{\sqrt{x^5}}{2}$

C) $\dfrac{1}{\sqrt{x^3}}$

D) $\dfrac{1}{x^3}$

CONTINUE

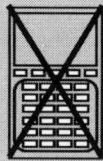

5

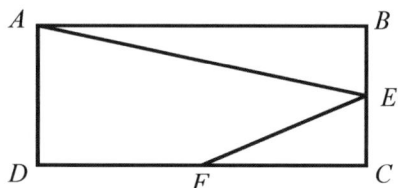

In the rectangle above, E and F are the midpoints of $\overline{BC}$ and $\overline{CD}$ respectively. If the value of $\sin \angle BAE$ is 0.6, what is the value of $\tan \angle EFC$?

A) $\dfrac{1}{2}$

B) $\dfrac{2}{3}$

C) $\dfrac{3}{5}$

D) $\dfrac{3}{2}$

6

If $f(2x) = 3x + 1$, which of the following represents $f(x)$?

A) $\dfrac{1}{2}(3x + 1)$

B) $\dfrac{1}{2}x + 1$

C) $\dfrac{3}{2}x + 1$

D) $\dfrac{3}{2}x + 2$

7

$$\left(a - \dfrac{1}{a}\right)^2$$

Which of the following is equivalent to the expression above?

A) $a^2 - \dfrac{1}{a^2}$

B) $a^2 + \dfrac{1}{a^2}$

C) $a^2 + \dfrac{1}{a^2} + 2$

D) $a^2 + \dfrac{1}{a^2} - 2$

CONTINUE

Questions 8 and 9 refer to the following information.

The price of a smart phone in 2015 is $300. The product will decrease in value at a rate of $20 per year. P is the dollar value of the smart phone and t $(0 \le t \le 10)$ is the number of years from 2015.

8

Based on the information above, which of the following represents the price, in dollars, in terms of t?

A) $P = 300(1+0.15)^t$

B) $P = 300 - 20t$

C) $P = 300 + 20t$

D) $P = 300(1-0.15)^t$

9

In how many years will the value of the smart phone be $200?

A) 3

B) 4

C) 5

D) 6

10

$$\frac{1}{3}x - \frac{1}{6}y = 10$$

Which of the following equations represents a line that is parallel to the graph of the equation above?

A) $4x + y = 5$

B) $2x + 4y = 9$

C) $5x - 10y = 9$

D) $10x - 5y = 11$

11

$$\sqrt{\frac{x-2}{x}} = 2$$

Which of the following is the solution to the equation above?

A) $\dfrac{2}{3}$

B) $\dfrac{3}{2}$

C) $-\dfrac{2}{3}$

D) Undefined

CONTINUE

12

$$y = k$$
$$y = (x+5)(x-5)$$

In the system of equations above, k is a constant. If the system has no solution, which of the following could be the value of k?

A)   50

B)   25

C) −25

D) −50

13

$$f(x) = 2(x-a)^2 + b$$

In the function above, a and b are constants. If $f(x) = 2x^2 - 4x + 27$ is equivalent to the expression above, what is the value of b?

A) 25

B) 26

C) 28

D) 32

14

If $a > 1$, which of the following is equivalent to

$$\dfrac{1 - \dfrac{2}{3a}}{a - \dfrac{4}{9a}}?$$

A) $\dfrac{3a-2}{3a+2}$

B) $\dfrac{3}{3a-2}$

C) $\dfrac{3}{3a+2}$

D) $\dfrac{3a-2}{3}$

15

If $4x - y = \dfrac{2}{3}$, what is the value of $\dfrac{81^{3x}}{27^y}$?

A) 3

B) 9

C) 27

D) 81

CONTINUE

DIRECTIONS

For questions 16–20, solve the problem and enter your answer in the grid, as described below, on the answer sheet.

Answer: $\frac{7}{12}$

Write answer in boxes. ← Fraction line

Grid in result.

Answer: 2.5

← Decimal point

1. Although not required, it is suggested that you write your answer in the boxes at the top of the columns to help you fill in the circles accurately. You will receive credit only if the circles are filled in correctly.

2. Mark no more than one circle in any column.

3. No question has a negative answer.

4. Some problems may have more than one answer.

5. **Mixed numbers** such as $3\frac{1}{2}$ must be gridded as 3.5 or 7/2. (If 3 1 / 2 is entered into the grid, it will be interpreted as $\frac{31}{2}$, not $3\frac{1}{2}$.)

6. **Decimal answers:** If you obtain a decimal answer with more digits than the grid can accommodate, it may be either rounded or truncated, but it must fill the entire grid.

Acceptable ways to grid $\frac{2}{3}$ are:

Answer: 201
Either position is correct.

Note: You may start your answers in any column, space permitting. Columns you don't need to use should be left blank.

CONTINUE

16

For the function f, $y = f(x)$ is inversely proportional to x. If $f(5) = 24$, what is the value of $f(10)$?

17

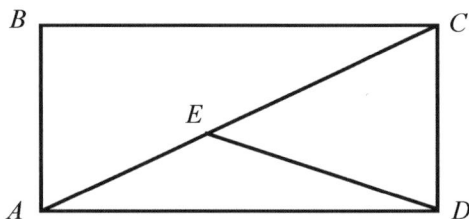

In the rectangle above, the length of $\overline{AE}$ is $\frac{2}{5}$ of the length of $\overline{AC}$. If the area of $\triangle AED$ is 18, what is the area of $\triangle CED$?

18

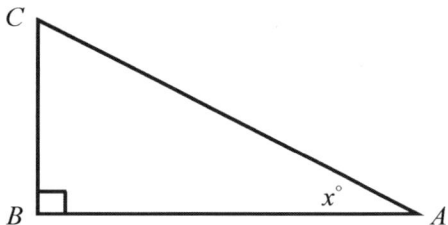

In the figure above, the value of $\sin x$ is 0.6 and the length of $\overline{AB}$ is 12. What is the area of $\triangle ABC$?

19

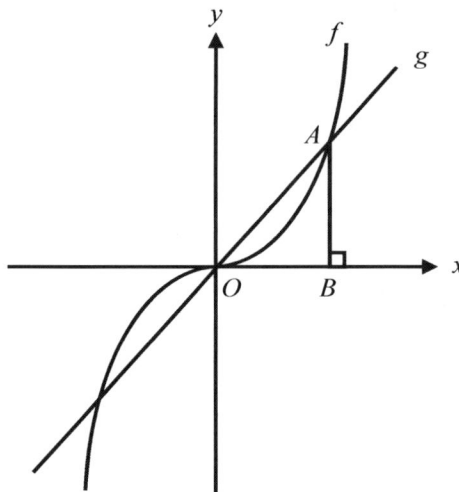

The graphs of $f(x) = ax^3$ and $g(x) = x$ are shown in the xy-plane above, where a is a constant. If the area of $\triangle AOB$ is $\frac{1}{8}$, what is the value of a?

20

$$\frac{14}{x^2 - 3x - 10} = \frac{a}{x-5} + \frac{b}{x+2}$$

In the equation above, a and b are constants. If the equation is true for all values of x except 5 and -2, what is the value of a?

STOP

If you finish before time is called, you may check your work on this section only.
Do not turn to any other section in the test.

No Test Material on This Page

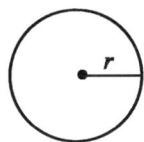
Math Test - Calculator

55 MINUTES, 38 QUESTIONS

Turn to Section 4 of your answer sheet to answer the questions in this section.

DIRECTIONS

For questions 1-30, solve each problem, choose the best answer from the choices provided, and fill in the corresponding circle on your answer sheet. **For questions 31-38,** solve the problem and enter your answer in the grid on your answer sheet. Please refer to the directions before question 31 on how to enter your answers in the grid. You may use any available space in your test booklet for scratch work.

NOTE

1. The use of a calculator **is permitted**.

2. All variables and expressions used represent real numbers unless otherwise indicated.

3. Figures provided in this test are drawn to scale unless otherwise indicated.

4. All figures lie in a plane unless otherwise indicated.

5. Unless otherwise indicated, the domain of a given function f is the set of all real numbers x for which $f(x)$ is a real number.

REFERENCE

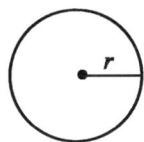 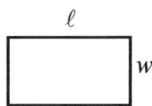 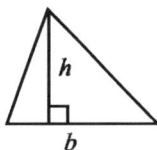 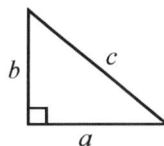 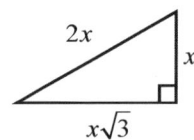 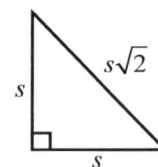

$A = \pi r^2$
$C = 2\pi r$
$\qquad A = \ell w \qquad A = \dfrac{1}{2}bh \qquad c^2 = a^2 + b^2 \qquad$ Special Right Triangles

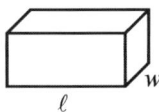 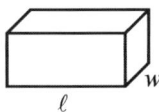   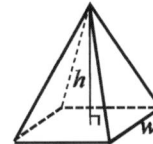

$V = \ell wh \qquad\qquad V = \pi r^2 h \qquad\qquad V = \dfrac{4}{3}\pi r^3 \qquad V = \dfrac{1}{3}\pi r^2 h \qquad V = \dfrac{1}{3}\ell wh$

The number of degrees of arc in a circle is 360.
The number of radians of arc in a circle is 2π.
The number of the measures in degrees of the angles of a triangle is 180.

CONTINUE

326

1

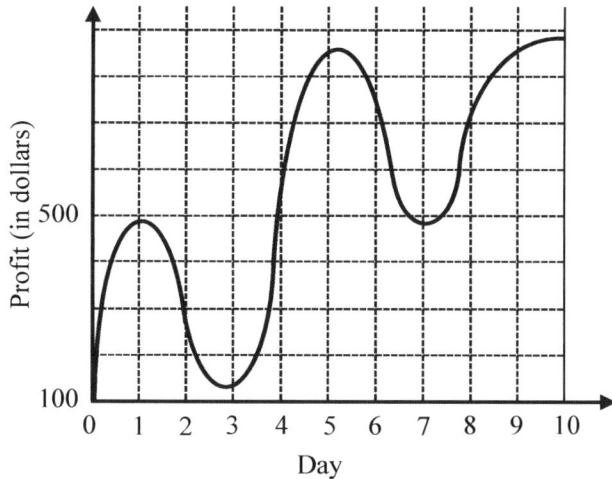

Peter opened a hardware store recently. The graph above shows the profit during the first 10 business days. On which interval is the profit strictly increasing?

A) Between day 1 and 3

B) Between day 3 and 4

C) Between day 4 and 6

D) Between day 6 and 10

2

If $\dfrac{x}{y} = 2$, what is the value of $6\left(\dfrac{x^2}{y}\right)\left(\dfrac{3}{2x}\right)$?

A) 18

B) 15

C) 12

D) 9

3

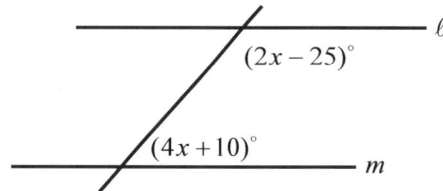

In the figure above, lines ℓ and m are parallel. What is the value of x?

A) 30

B) 32.5

C) 35

D) 37.5

4

If $x + y = 12$ and $x - y = 4$, what is the value of $\dfrac{x^2 - y^2}{2x}$?

A) 3

B) 6

C) 8

D) 9

CONTINUE

5

$$C(t) = 110t + 300$$

The cost C, in dollars, of renting a town party room is modeled by the function above, where t is the number of hours used. Claire rented the room for 5 hours, but she wants to add two more hours. How much more will she pay for using additional hours?

A) $200

B) $220

C) $500

D) $520

6

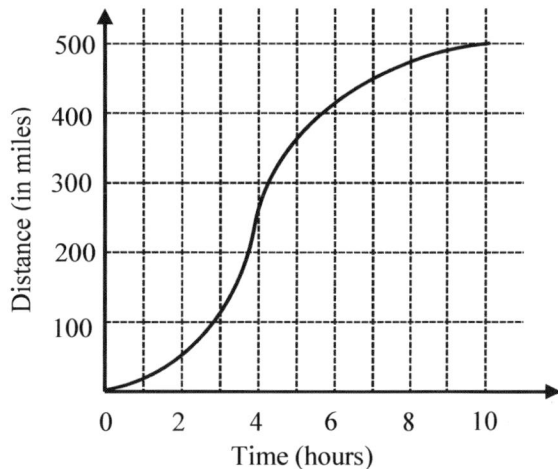

Jessica is travelling to Washington D.C. The graph above shows the distance she traveled during the first 10 hours. In which time interval did the graph show the greatest average rate of change?

A) Between 0 and 2

B) Between 2 and 4

C) Between 4 and 6

D) Between 8 and 10

7

For what value of n is $|n - 10| < 0$?

A) −5

B) 5

C) 15

D) There is no such value of n.

8

$$f(x) = x^2 + kx - 5$$

In the equation above, k is a constant. If $f(-2) = 3$, what is the value of $f(2)$?

A) −5

B) −3

C) 3

D) 5

CONTINUE

Questions 9 and 10 refer to the following information.

$$C(n) = 40n + 800$$
$$n(t) = 30t$$

A company produces a smartphone for which the weekly cost of producing n units is C, in dollars. The weekly cost C and the number of units n produced in t hours are given by the equations above.

9

What will be the increase in cost if the number of units increases by 100?

A) $800

B) $2000

C) $4000

D) $4800

10

If the weekly cost increases to $20,000, how many hours will it take to produce the units?

A) 16

B) 20

C) 40

D) 48

11

$$\left(\frac{3+i}{2-i}\right)(a+bi) = 1$$

In the equation above, a and b are constants. If $i = \sqrt{-1}$, what is the value of a?

A) $-\dfrac{1}{2}$

B) $-\dfrac{1}{5}$

C) $\dfrac{1}{5}$

D) $\dfrac{1}{2}$

12

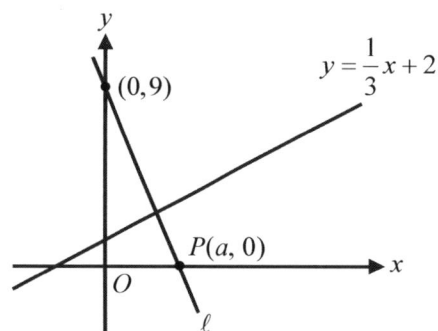

Note: Figure not drawn to scale.

In the xy-plane above, the graph of $y = \dfrac{1}{3}x + 2$ is perpendicular to the graph of line ℓ. What is the value of a?

A) 2

B) 3

C) 4

D) 4.5

CONTINUE

13

Boy's Shoe Size	7	7.5	8	8.5
Foot Length (in)	9.25	9.5	9.75	10

The table shows the relationship of a boy's shoe size and the length of a boy's foot, in inches. What is the correlation coefficient?

A) −1

B) −0.95

C) 0.95

D) 1

14

In triangle RST, if $\cos \angle R = \sin \angle T$, which of the following must be true?

A) Triangle RST is equilateral.

B) Triangle RST is isosceles.

C) Triangle RST is an obtuse triangle.

D) Triangle RST is a right triangle.

15

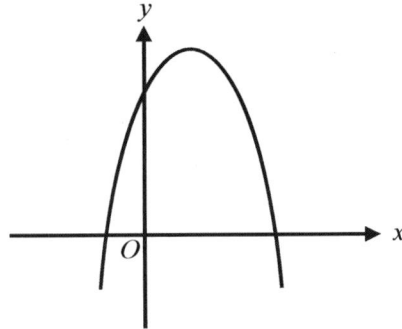

The graph of a quadratic function $f(x) = ax^2 + bx + c$ is shown in the xy-plane above. Which of the following must be true?

    I. $a > 0$

    II. $b > 0$

    III. $c > 0$

A) I only

B) I and II only

C) II and III only

D) I, II, and III

16

$$\frac{1}{12}x - \frac{1}{24}y = \frac{1}{8}$$

$$5x + 3y = 2$$

If (a, b) is the solution to the system of equations above, what is the value of a?

A) −2

B) −1

C) 1

D) 2

CONTINUE

Questions 17 and 18 refer to the following information

Survey Results

Number of pets	East Village	West Village
0	10	25
1	40	20
2	35	35
3	10	15
4	5	5

A statistician chose 100 families at random from each of two towns and asked how many pets they own. The results are shown in the table above. There is a total of 10,000 families in East village and 15,000 families in West village.

17

What is the median number of pets for all families surveyed?

A) 1

B) 1.5

C) 2

D) 3

18

What is the expected total number of families, who own 3 pets in the two villages?

A) 25

B) 1,000

C) 2,000

D) 3,125

19

$$x^2 + y^2 - 6x - 8y = 0$$

The equation of a circle in the xy-plane is shown above. What is the area of the circle?

A) 15.7

B) 31.4

C) 62.8

D) 78.5

20

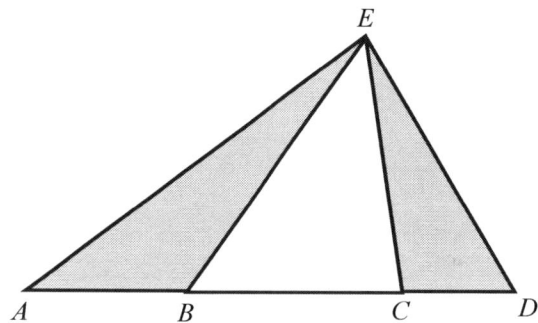

Note: Figure not drawn to scale.

In the figure above, $AB : BC : CD = 3 : 5 : 2$. If the sum of the areas of the shaded regions is 13, what is the area of $\triangle EBC$?

A) 13

B) 17

C) 20

D) 25

CONTINUE

21

$$K \le -6 \text{ or } K \ge 14$$

Which of the following is equivalent to the expression of inequalities above?

A) $|K + 4| \le 10$

B) $|K + 4| \ge 10$

C) $|K - 4| \le 10$

D) $|K - 4| \ge 10$

22

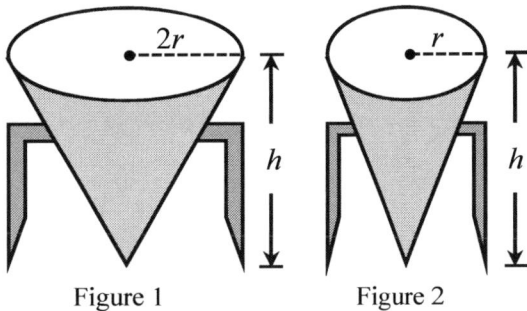

Figure 1   Figure 2

In the figures above, both of the water tanks are in the shape of a right circular cone. If the larger tank can hold 125 gallons of water, how many gallons of water can the smaller tank hold?

A) 5

B) 25

C) 31.25

D) 62.5

23

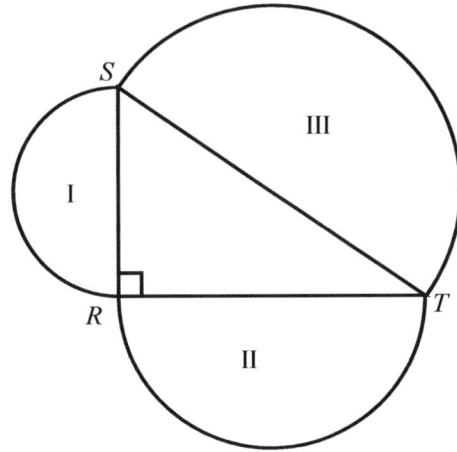

In the figure above, right triangle RST bordered by three semicircles on each side. If the area of semicircle I is 8 and the area of semicircle II is 24, what is the length of $\overline{ST}$?

A) $\dfrac{16}{\sqrt{\pi}}$

B) $\dfrac{16}{\pi}$

C) $\dfrac{8}{\sqrt{\pi}}$

D) $\dfrac{8}{\pi}$

CONTINUE

24

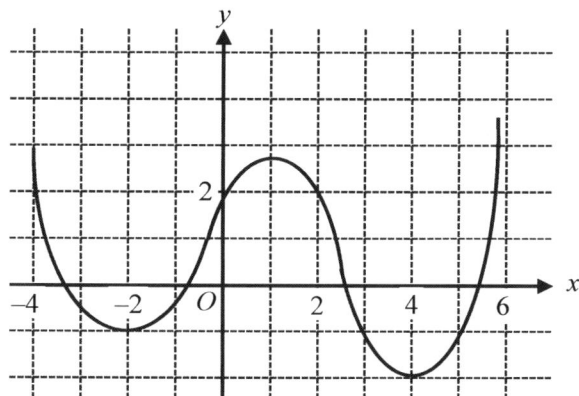

The graph of g is shown in the xy-plane above. If $f(x) = 2g(x) - 5$, what is the average rate of change of $f(x)$ between -2 and 4?

A) $\dfrac{1}{2}$

B) $\dfrac{1}{3}$

C) $-\dfrac{1}{4}$

D) $-\dfrac{1}{3}$

25

If the value of k^{-5} is twice the value of $4k^{-2}$, what is the value of k?

A) $\dfrac{1}{4}$

B) $\dfrac{1}{2}$

C) 2

D) 4

26

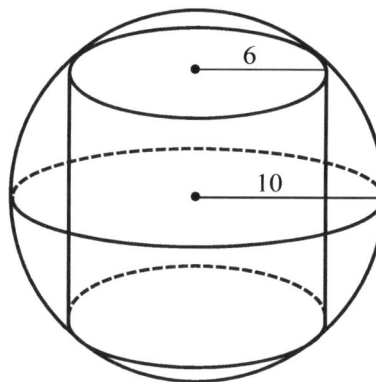

In the figure above, a right cylinder is inscribed in a sphere with radius 10. If the radius of the circular base of the cylinder is 6, what is the volume of the cylinder?

A) 144π

B) 288π

C) 576π

D) 720π

27

A carton contains k boxes of paper cups and each box contains 100 paper cups. If the carton cost d dollars, what is the cost per paper cup in cents?

A) kd

B) $\dfrac{d}{k}$

C) $\dfrac{d}{100k}$

D) $\dfrac{100k}{d}$

CONTINUE

28

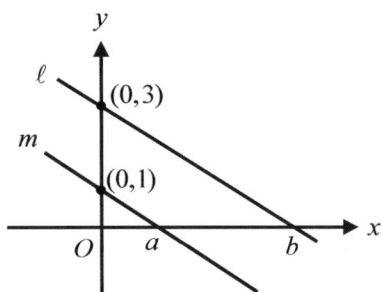

In the xy-plane above, lines ℓ and m are parallel and intersect the x-axis at $x = a$ and $x = b$ respectively. If $a + b = 8$, what is the value of a?

A) 1.5

B) 2

C) 2.5

D) 3

29

$$(k - 2a)x + k - 11 = 5$$

In the equation above, a and k are constants. If the equation has infinitely many solutions, what is the value of a?

A) 8

B) 16

C) 20

D) 24

30

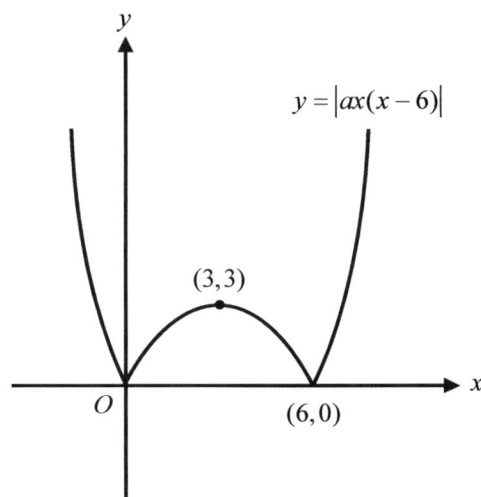

The graph of $y = |ax(x - 6)|$ is shown in the xy-plane above. Which of the following could be the value of a?

A) $-\dfrac{1}{2}$

B) $-\dfrac{1}{3}$

C) 2

D) 3

CONTINUE

DIRECTIONS

For questions 31-38, solve the problem and enter your answer in the grid, as described below, on the answer sheet.

Answer: $\frac{7}{12}$

Write answer in boxes. → 7 / 1 2 ← Fraction line

Grid in result. →

Answer: 2.5

2 . 5 ← Decimal point

1. Although not required, it is suggested that you write your answer in the boxes at the top of the columns to help you fill in the circles accurately. You will receive credit only if the circles are filled in correctly.
2. Mark no more than one circle in any column.
3. No question has a negative answer.
4. Some problems may have more than one answer.
5. **Mixed numbers** such as $3\frac{1}{2}$ must be gridded as 3.5 or 7/2. (If | 3 | 1 | / | 2 | is entered into the grid, it will be interpreted as $\frac{31}{2}$, not $3\frac{1}{2}$.)
6. **Decimal answers:** If you obtain a decimal answer with more digits than the grid can accommodate, it may be either rounded or truncated, but it must fill the entire grid.

Acceptable ways to grid $\frac{2}{3}$ are:

2 / 3   . 6 6 6   . 6 6 7

Answer: 201
Either position is correct.

2 0 1   2 0 1

Note: You may start your answers in any column, space permitting. Columns you don't need to use should be left blank.

CONTINUE →

31

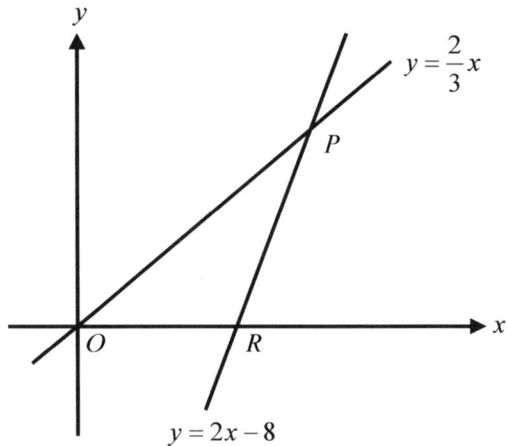

The lines with equations $y = \dfrac{2}{3}x$ and $y = 2x - 8$ are shown in the xy-plane above. What is the area of triangle OPR?

32

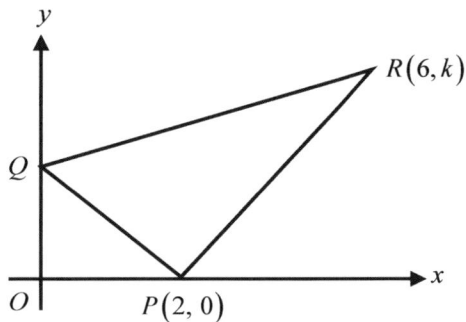

In the xy-plane above, the slope of $\overline{PQ}$ is -1 and the slope of $\overline{QR}$ is $\dfrac{1}{2}$. What is the slope of $\overline{PR}$?

33

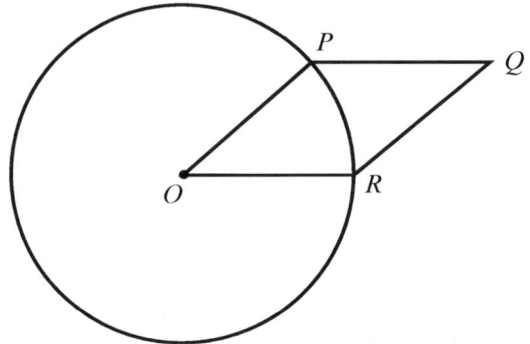

In the figure above, O is the center of the circle. If the area of the circle is 100π and measure of angle POR is $\dfrac{\pi}{6}$ radians, what is the area of parallelogram $OPQR$?

34

$$y \geq -3x + 1200$$
$$y \geq 15x + 300$$

In the xy-plane above, a point with coordinates (r, s) lies in the solution set of the system of inequalities above. What is the minimum possible value of s?

CONTINUE

35

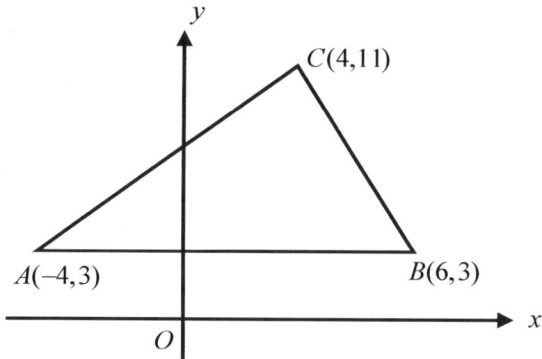

In the xy-plane above, what is the area of triangle ABC?

36

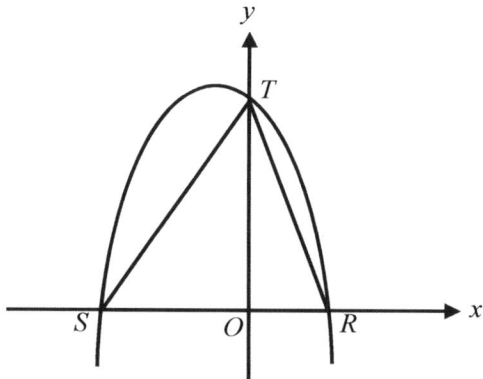

The graph of $y = -(x-2)(x+5)$ is shown in the xy-plane above. What is the area of triangle STR?

A consumer analyst believes that a new car will lose 18 percent of its value every year. After n years, the value of a new car that costs \$20,000 is modeled by $V(t) = 20,000 \cdot C^n$, where V is the value of the car after n years.

37

Based on the information above, what is the value of C?

38

To the nearest dollar, what is the value of the car 5 years after it was purchased? (Note: Disregard the \$ sign when gridding your answer.)

STOP

If you finish before time is called, you may check your work on this section only.
Do not turn to any other section in the test.

Math Conversion Table

Raw Score	Scaled Score	Raw Score	Scaled Score
58	800	27	500
57	800	26	490
56	800	25	480
55	800	24	470
54	790	23	460
53	780	22	460
52	770	21	450
51	750	20	440
50	740	19	430
49	730	18	430
48	720	17	420
47	710	16	420
46	700	15	410
45	690	14	400
44	670	13	390
43	680	12	380
42	670	11	370
41	660	10	360
40	650	9	450
39	640	8	340
38	630	7	330
37	620	6	310
36	610	5	290
35	600	4	280
34	590	3	270
33	580	2	260
32	560	1	240
31	550	0	200
30	540		
29	530		
28	520		

Answer Explanations

Test 7   Answers and Explanations

SECTION 3	1	2	3	4	5	6	7	8	9	10
	D	B	A	A	D	C	D	B	C	D
	11	12	13	14	15	16	17	18	19	20
	C	D	A	C	B	12	27	54	4	2

SECTION 4	1	2	3	4	5	6	7	8	9	10
	B	A	B	A	B	B	D	A	C	A
	11	12	13	14	15	16	17	18	19	20
	D	B	D	D	C	C	C	D	D	A
	21	22	23	24	25	26	27	28	29	30
	D	C	A	D	B	C	B	B	A	B
	31	32	33	34	35	36	37	38		
	8	5/4	50	1050	40	35	0.82	7415		

SECTION 3

1. D

Midpoint $= \dfrac{2+10}{2} = 6$ and distance $= 10 - 6 = 4$ $\rightarrow$ $|x - 6| \le 4$

2. B

3. A

$f(9) = -\dfrac{5}{3}(9) + b = 5$ $\rightarrow$ $b = 20$ $\rightarrow$ $f(3) = -\dfrac{5}{3}(3) + 20 = 15$

4. A

$x^{-2}\left(\dfrac{1}{\sqrt{x}}\right) = \dfrac{1}{x^2}\left(\dfrac{1}{x^{\frac{1}{2}}}\right) = \dfrac{1}{x^{\frac{5}{2}}} = \dfrac{1}{\sqrt{x^5}}$

5. D

$\tan \angle EFC = \dfrac{EC}{FC}$ and $\sin \angle BAE = \dfrac{6}{10}$ (Use these numbers)

$\tan \angle EFC = \dfrac{6}{4} = \dfrac{3}{2}$

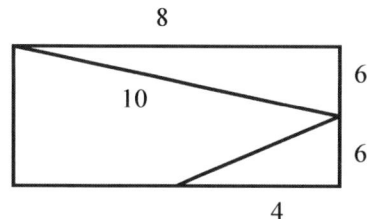

6. C

Putting $\dfrac{x}{2}$ in the equation $\rightarrow$ $f(x) = 3\left(\dfrac{x}{2}\right) + 1 = \dfrac{3}{2}x + 1$

7. D

8. B

9. C

$200 = 300 - 20t$ $\rightarrow$ $20t = 100$ $\rightarrow$ $t = 5$

10. D

$\dfrac{1}{3}x - \dfrac{1}{6}y = 10$ $\rightarrow$ $2x - y = 60$ is parallel to $10x - 5y = 11$, because $\dfrac{2}{10} = \dfrac{-1}{-5} \neq \dfrac{60}{11}$.

11. C

$\sqrt{\dfrac{x-2}{x}} = 2$ $\rightarrow$ $\dfrac{x-2}{x} = 4$ $\rightarrow$ $4x = x - 2$ $\rightarrow$ $x = -\dfrac{2}{3}$

12. D

Minimum of $y = (x+5)(x-5)$ is $f(0) = -25$. Therefore, if $k = -50$, no solution.

13. A

$f(x) = 2x^2 - 4x + 27$ $\rightarrow$ $f(x) = 2\left(x^2 - 2x + 1\right) + 25$ $\rightarrow$ $f(x) = 2(x-1)^2 + 25$

Therefore, $a = 1$ and $b = 25$.

14. C

$\dfrac{1 - \dfrac{2}{3a}}{a - \dfrac{4}{9a}} = \dfrac{\left(1 - \dfrac{2}{3a}\right)9a}{\left(a - \dfrac{4}{9a}\right)9a} = \dfrac{9a - 6}{9a^2 - 4} = \dfrac{3(3a-2)}{(3a+2)(3a-2)} = \dfrac{3}{3a+2}$

15. B

$\dfrac{81^{3x}}{27^y} = \dfrac{3^{12x}}{3^{3y}} = 3^{12x - 3y} = 3^2 = 9$, because $3(4x - y) = 3\left(\dfrac{2}{3}\right) = 2$.

16. 12

$5 \times 24 = 10 \times f(10)$ $\rightarrow$ $f(10) = \dfrac{120}{10} = 12$

17. 27

$\dfrac{AE}{EC} = \dfrac{2}{3}$ $\rightarrow$ $\dfrac{\text{area of } \triangle AED}{\text{area of } \triangle CED} = \dfrac{18}{x} = \dfrac{2}{3}$ $\rightarrow$ $x = 27$

18. 54

$$8k = 12 \quad \rightarrow \quad k = 1.5$$

Therefore, the area is $\dfrac{9 \times 12}{2} = 54$.

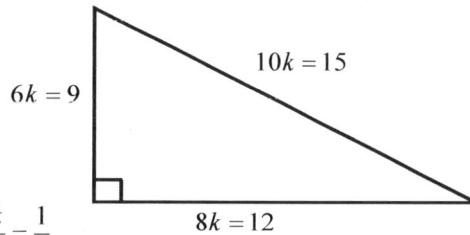

$10k = 15$

$6k = 9$

$8k = 12$

19. 4

Since $OA = AB = k$, the area of $\triangle OAB = \dfrac{k \times k}{2} = \dfrac{1}{8}$

$k^2 = \dfrac{1}{4} \quad \rightarrow \quad k = \dfrac{1}{2} \quad \rightarrow \quad$ The coordinates of point A is $\left(\dfrac{1}{2}, \dfrac{1}{2} \right)$.

Putting the ordered pair in $y = ax^3 \quad \rightarrow \quad \dfrac{1}{2} = a\left(\dfrac{1}{2} \right)^3 \quad \rightarrow \quad \dfrac{1}{2} = a\left(\dfrac{1}{8} \right) \quad \rightarrow \quad a = 4$

20. 2

$$\dfrac{14}{x^2 - 3x - 10} = \dfrac{a}{x-5} + \dfrac{b}{x+2} \quad \rightarrow \quad 14 = a(x+2) + b(x-5) \quad \rightarrow \quad \text{If } x = 5, \text{ then } 14 = 7a \quad \rightarrow \quad a = 2.$$

SECTION 4

1. B

2. A

$$6\left(\dfrac{x^2}{y} \right)\left(\dfrac{3}{2x} \right) = 9\left(\dfrac{x^2}{xy} \right) = 9\left(\dfrac{x}{y} \right) = 9(2) = 18$$

3. B

$$2x - 25 + 4x + 10 = 180 \quad \rightarrow \quad 6x = 195 \quad \rightarrow \quad x = 32.5$$

4. A

System of equations: $x = 8$ and $y = 4 \quad \rightarrow \quad \dfrac{x^2 - y^2}{2x} = \dfrac{64 - 16}{16} = 3$

5. B

Hourly rate is $110. \quad \rightarrow \quad 2 \times 110 = \$220$

6. B

Between 2 and 4, the graph has the greatest slope.

7. D

Absolute value CANNOT be negative.

Answer Explanations

8. A

$$f(-2) = 4 - 2k - 5 = 3 \;\rightarrow\; k = -2 \;\rightarrow\; f(x) = x^2 - 2x - 5 \;\rightarrow\; f(2) = 4 - 4 - 5 = -5$$

9. C

$$40 \times 100 = \$4,000$$

10. A

$$20,000 = 40n + 800 \;\rightarrow\; 40n = 19200 \;\rightarrow\; n = 480 \;\rightarrow\; 30t = 480 \;\rightarrow\; t = 16$$

11. D

$$a + bi = \frac{2-i}{3+i} = \frac{(2-i)(3-i)}{(3+i)(3-i)} = \frac{5-5i}{10} = \frac{1}{2} - \frac{1}{2}i \;\rightarrow\; \text{Therefore, } a = \frac{1}{2}.$$

12. B

Slope of line $\ell = \dfrac{9-0}{0-a} = -3$ (negative reciprocal) $\;\rightarrow\; a = 3$

13. D

All data lie exactly on the straight line that has a positive slope.

14. D

If $\cos \angle R = \sin \angle T$, $\angle R + \angle T = 90°$. $\;\rightarrow\;$ $\angle S = 90°$: Right triangle

15. C

1) y-intercept: $f(0) = c > 0$ $\;\rightarrow\;$ 2) Concave down(graph opens downward): $a < 0$

3) Axis of symmetry: $x = \dfrac{-b}{2a} > 0$ $\;\rightarrow\;$ $b > 0$, because $a < 0$.

16. C

$$\frac{1}{12}x - \frac{1}{24}y = \frac{1}{8} \;\rightarrow\; 2x - y = 3 \;\rightarrow\; 6x - 3y = 9$$

Solve the system equations using addition:
$$\begin{array}{r} 6x - 3y = 9 \\ 5x + 3y = 2 \\ \hline 11x \quad\;\; = 11 \end{array} \;\rightarrow\; x = a = 1$$

17. C

$$\frac{1 + 200}{2} = 100.5 \;\rightarrow\; \text{When you arrange the data from least to greatest, the median in betweeen 100 and 101.}$$

18. D

Probability (own 2 pets) $= \dfrac{10 + 15}{200} = \dfrac{1}{8}$ $\;\rightarrow\;$ Therefore, expected number $= (15,000 + 10,000) \times \dfrac{1}{8} = 3125$

Answer Explanations

19. D

$$x^2 + y^2 - 6x - 8y = 0 \;\rightarrow\; \left(x^2 - 6x + 9\right) + \left(x^2 - 8y + 16\right) = 9 + 16 \;\rightarrow\; \left(x^2 - 6x + 9\right) + \left(x^2 - 8y + 16\right) = 25$$

Area of the circle is $25\pi \approx 78.5$

20. A

$AB:BC:CD = 3:5:2. \;\rightarrow\;$ Ratio of the areas $= 3:5:2 = 3k, 5k, 2k$

Since $3k + 2k = 13 \;\rightarrow\; 5k = 13 \;\rightarrow\;$, area of $\triangle EBC$ is $5k = 13$.

21. D

Midpoint $= \dfrac{-6 + 14}{2} = 4$ and distance $= 14 - 4 = 10$. Therefore, $|K - 4| \geq 10$.

22. C

The ratio of the volumes $= \dfrac{\pi(2r)^2 h}{3} : \dfrac{\pi r^2 h}{3} = 4:1 \;\rightarrow\;$ Therefore, $\dfrac{1}{4} \times 125 = 31.25$

23. A

The area of semicircle III is $8 + 24 = 32$. $\;\rightarrow\; \pi\left(\dfrac{ST}{2}\right)^2 \times \dfrac{1}{2} = 32 \;\rightarrow\; \left(\dfrac{ST}{2}\right)^2 = \dfrac{64}{\pi} \;\rightarrow\; ST^2 = \dfrac{256}{\pi}$

$$ST = \sqrt{\dfrac{256}{\pi}} = \dfrac{16}{\sqrt{\pi}}$$

24. D

Average rate of change between -2 and $4 = \dfrac{f(4) - f(-2)}{4 - (-2)} = \dfrac{f(4) - f(-2)}{6}$

$f(4) = 2g(4) - 5 = 2(-2) - 5 = -9$ and $f(-2) = 2g(-2) - 5 = 2(-1) - 5 = -7$

Therefore, average rate of change is $\dfrac{-9 - (-7)}{6} = -\dfrac{1}{3}$.

25. B

$$k^{-5} = 2\left(4k^{-2}\right) \;\rightarrow\; \dfrac{1}{k^5} = \dfrac{8}{k^2} \;\rightarrow\; 8k^5 = k^2 \;\rightarrow\; k^3 = \dfrac{1}{8} \;\rightarrow\; k = \dfrac{1}{2} \;(k \neq 0)$$

26. C

The height of the cylinder is 16.

The volume is $\pi\left(6^2\right)(16) = 576\pi$

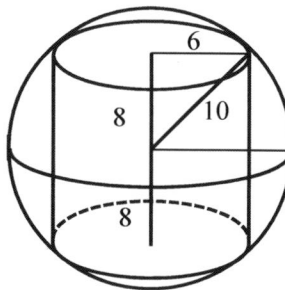

27. B

$$\dfrac{\$d}{100k \text{ paper cups}} = \dfrac{100d \text{ cents}}{100k \text{ cups}} = \dfrac{d}{k} \text{ cents/cup}$$

28. B

Same slope: $\dfrac{3-0}{0-b}=\dfrac{1-0}{0-a}$ → $-\dfrac{3}{b}=-\dfrac{1}{a}$ → $b=3a$: Putting $b=3a$ in the equation

$a+b=8$ → $a+(3a)=4a=8$ → $a=2$

29. A

$k-2a=0$ and $k-11=5$ → $k=16$ → $16-2a=0$ → $a=8$

30. B

$y=ax(x-6)$ has two possible graphs.

Case 1) $-3=a(3)(3-6)$ → $-3=-9a$ → $a=\dfrac{1}{3}$

Case 2) $3=a(3)(3-6)$ → $3=-9a$ → $a=-\dfrac{1}{3}$

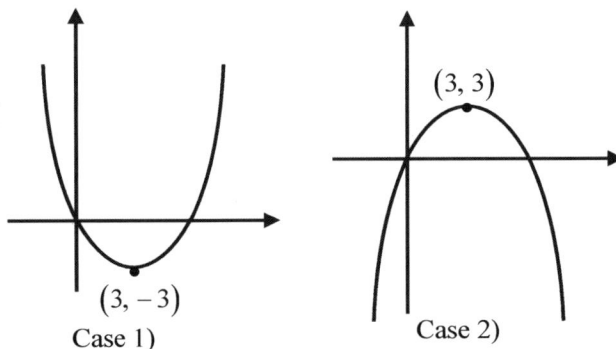

Case 1)   Case 2)

31. 8

Intersection of the two graphs: $\dfrac{2}{3}x=2x-8$ → $2x=6x-24$ → $24=4x$ → $x=6$ and $y=4$

x-intercept of $y=2x-8$ → $0=2x-8$ → $x=4$

Area $=\dfrac{4\times4}{2}=8$

32. $\dfrac{5}{4}$ or 1.25

$Q(0,2)$ and $R(6,k)$ → The slope of $\overline{QR}$ is $\dfrac{k-2}{6-0}=\dfrac{1}{2}$. → $k=5$.

Therefore, the slope of $\overline{PR}$ is $\dfrac{5-0}{6-2}=\dfrac{5}{4}$ or 1.25.

33. 50

Radius of the circle is 10. $\dfrac{\pi}{6}=\dfrac{180°}{6}=30°$

Area of the parallelogram is $10\times5=50$.

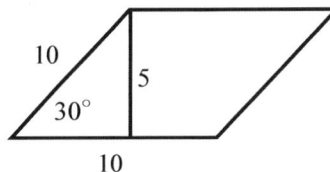

34. 1050

$15x+300=-3x+1200$

$18x=900$

$x=r=50$

$y=s=15(50)+300=1050$

At this point, the solution has minimum value of b.

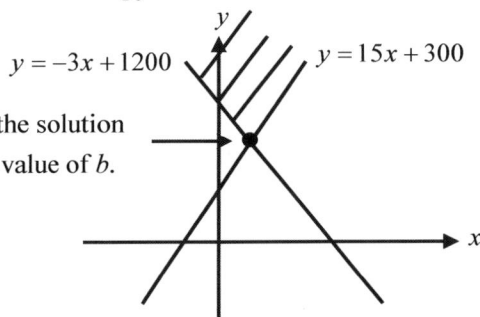

35. 40

$AB = 6 - (-4) = 10$ and the height is $11 - 3 = 8$.

$\text{Area} = \dfrac{10 \times 8}{2} = 40$

36. 35

$SR = 2 - (-5) = 7$ and $OT = f(0) = 10$

Therefore, the area of $\triangle STR = \dfrac{7 \times 10}{2} = 35$.

37. 0.82

$V(t) = 20,000(1 - 0.18)^n \quad \rightarrow \quad C = 0.82$

38. 7415

$c = 20,000 \times (0.82)^5 \approx \$7415$

Dr. John Chung's SAT Math Practice Test 7

No Test Material on This Page

PRACTICE TEST 8

Dr. John Chung's SAT Math

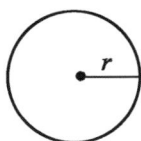
Math Test - No Calculator

25 MINUTES, 20 QUESTIONS

Turn to Section 3 of your answer sheet to answer the questions in this section.

DIRECTIONS

For questions 1–15, solve each problem, choose the best answer from the choices provided, and fill in the corresponding circle on your answer sheet. **For questions 16–20,** solve the problem and enter your answer in the grid on your answer sheet. Please refer to the directions before question 16 on how to enter your answers in the grid. You may use any available space in your test booklet for scratch work.

NOTE

1. The use of a calculator **is not permitted**.

2. All variables and expressions used represent real numbers unless otherwise indicated.

3. Figures provided in this test are drawn to scale unless otherwise indicated.

4. All figures lie in a plane unless otherwise indicated.

5. Unless otherwise indicated, the domain of a given function f is the set of all real numbers x for which $f(x)$ is a real number.

REFERENCE

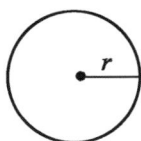 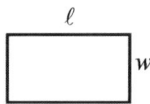 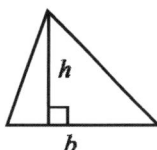 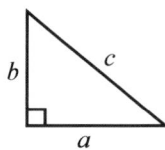 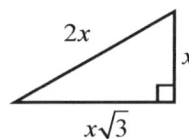 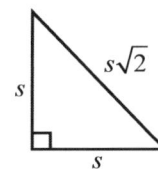

$A = \pi r^2$
$C = 2\pi r$ 　　　 $A = \ell w$ 　　　 $A = \dfrac{1}{2}bh$ 　　　 $c^2 = a^2 + b^2$ 　　　 Special Right Triangles

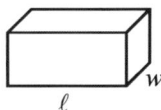 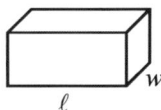   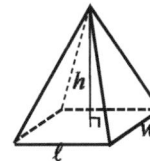

$V = \ell wh$ 　　 $V = \pi r^2 h$ 　　 $V = \dfrac{4}{3}\pi r^3$ 　　 $V = \dfrac{1}{3}\pi r^2 h$ 　　 $V = \dfrac{1}{3}\ell wh$

The number of degrees of arc in a circle is 360.
The number of radians of arc in a circle is 2π.
The number of the measures in degrees of the angles of a triangle is 180.

CONTINUE

348

1

If $\dfrac{3}{x-1} = x+1$ and $x \neq 1$, what is the value of x^2?

A) 1

B) 2

C) 4

D) 8

2

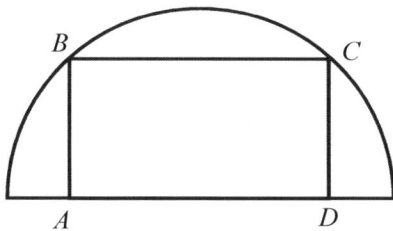

In the figure above, the diameter of the semicircle is 10 and the length of $\overline{CD}$ of rectangle $ABCD$ is 3. What is the length of $\overline{BC}$?

A) 4

B) 6

C) 7

D) 8

3

If $16 = \left(\dfrac{1}{4}\right)^{\frac{-1}{m}}$, what is the value of m?

A) $-\dfrac{1}{4}$

B) $-\dfrac{1}{2}$

C) $\dfrac{1}{4}$

D) $\dfrac{1}{2}$

4

Alex spends \$2.25 per gallon on gasoline. If Alex uses one gallon of gasoline to travel 30 miles, how many dollars will he spend to travel 240 miles?

A) 18

B) 20

C) 24

D) 28

CONTINUE

5

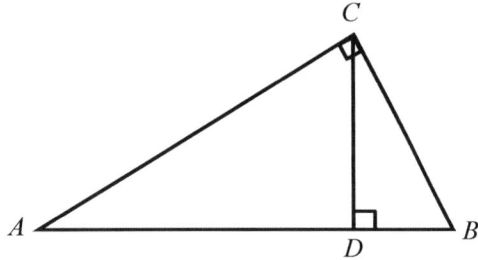

In the triangle above, the length of $\overline{DB}$ is 4 and the length of $\overline{CD}$ is 6. What is the area of triangle ACD?

A) 24

B) 27

C) 30

D) 39

6

If a and b are positive numbers, and 125 percent of a^2 is equal to 5 percent of b^2, what is the value of $\dfrac{a}{b}$?

A) $\dfrac{1}{5}$

B) $\dfrac{2}{5}$

C) 2

D) 5

Questions 7 and 8 refer to the following information.

$$C(t) = 15 + 0.15(t - K)$$

The cost of using a smart phone is \$15 for the first 200 minutes and \$0.15 for additional minute. The cost C is modeled by the equation above, where t is the length of time in minutes and K is a constant.

7

Based on the information above, what is the value of K?

A) 0.15

B) 1

C) 100

D) 200

8

If a customer paid \$36 for using his phone, how many minutes did he use?

A) 210

B) 340

C) 450

D) 500

CONTINUE

9

If $x+1$ is a factor of $x^4 - 3x^3 - ax + a$, where a is a constant, what is the value of a?

A) -4

B) -2

C) 2

D) 4

10

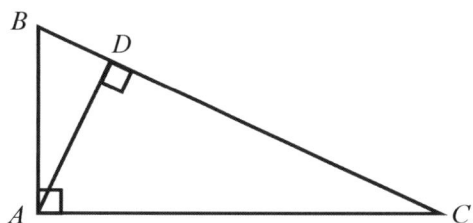

In the right triangle above, the value of $\sin C$ is 0.6 and the length of $\overline{BC}$ is 20. What is the length of $\overline{AD}$?

A) 7.2

B) 8.0

C) 9.6

D) 10

11

If $a(x+b) = 3x - 15$ for all real values of x, what is the value of b?

A) -5

B) -3

C) 3

D) 5

12

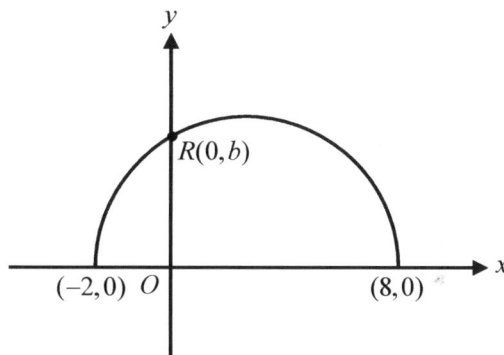

A semicircle is shown in the xy-plane above. If the semicircle intersects the y-axis at point R, what is the value of b?

A) 3

B) 4

C) 5

D) 6

CONTINUE

13

$$\frac{3+i}{3-i} = a + bi$$

In the equation above, a and b are real numbers. If $i = \sqrt{-1}$, what is the value of a?

A) 0.2

B) 0.4

C) 0.6

D) 0.8

Questions 14 and 15 refer to the following information.

Gender	Algebra	Geometry	Total
Male	80		
Female			90
Total	120		200

The incomplete table above shows the results of a survey about elective subject preferences given to 200 students.

14

What is the probability that a randomly selected student is a female who prefers geometry?

A) 0.2

B) 0.25

C) 0.3

D) 0.4

15

What fraction of male students prefer geometry?

A) $\dfrac{3}{20}$

B) $\dfrac{3}{11}$

C) $\dfrac{8}{11}$

D) $\dfrac{3}{8}$

CONTINUE

DIRECTIONS

For questions 16–20, solve the problem and enter your answer in the grid, as described below, on the answer sheet.

1. Although not required, it is suggested that you write your answer in the boxes at the top of the columns to help you fill in the circles accurately. You will receive credit only if the circles are filled in correctly.

2. Mark no more than one circle in any column.

3. No question has a negative answer.

4. Some problems may have more than one answer.

5. **Mixed numbers** such as $3\frac{1}{2}$ must be gridded as 3.5 or 7/2. (If $\boxed{3\ 1\ /\ 2}$ is entered into the grid, it will be interpreted as $\frac{31}{2}$, not $3\frac{1}{2}$.)

6. **Decimal answers:** If you obtain a decimal answer with more digits than the grid can accommodate, it may be either rounded or truncated, but it must fill the entire grid.

Answer: $\frac{7}{12}$

Write answer in boxes. →

Fraction line ←

Grid in result.

Answer: 2.5

Decimal point ←

Acceptable ways to grid $\frac{2}{3}$ are:

Answer: 201
Either position is correct.

Note: You may start your answers in any column, space permitting. Columns you don't need to use should be left blank.

CONTINUE

16

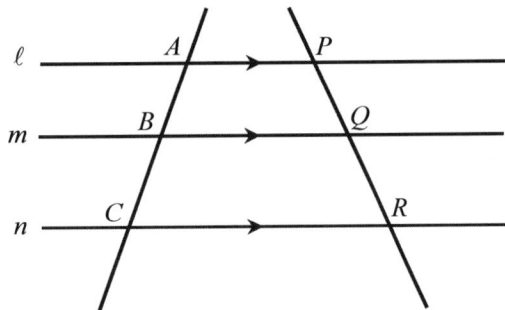

Note: Figure not drawn to scale.

In the figure above, line ℓ, m, and n are parallel. If $AB = 8$, $BC = 12$, and $PQ = 10$, what is the length of $\overline{QR}$?

17

Nigel drove from city A to city B at the speed of 60 miles per hour, and returned along the same route at the speed of 40 miles per hour. If it took $4\frac{1}{2}$ hours for the round trip, what is the distance, in miles, between city A and city B?

18

$$x^3 - 3x^2 + 5x = 15$$

For what real value of x is the equation above true?

19

$$x^2 + (k+1)x + 16 = (x+h)^2$$

In the equation above, k and h are positive constants. If the equation is true for all real numbers of x, what is the value of k?

20

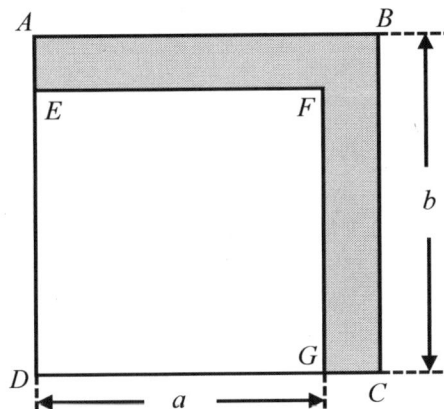

Squares $ABCD$ and $DEFG$ with integer-length sides of b and a respectively are shown in the figure above. If the area of the shaded region is 28, what is the area of square $ABCD$?

STOP

If you finish before time is called, you may check your work on this section only.
Do not turn to any other section in the test.

No Test Material on This Page

Math Test - Calculator

55 MINUTES, 38 QUESTIONS

Turn to Section 4 of your answer sheet to answer the questions in this section.

DIRECTIONS

For questions 1-30, solve each problem, choose the best answer from the choices provided, and fill in the corresponding circle on your answer sheet. **For questions 31-38**, solve the problem and enter your answer in the grid on your answer sheet. Please refer to the directions before question 31 on how to enter your answers in the grid. You may use any available space in your test booklet for scratch work.

NOTE

1. The use of a calculator **is permitted**.

2. All variables and expressions used represent real numbers unless otherwise indicated.

3. Figures provided in this test are drawn to scale unless otherwise indicated.

4. All figures lie in a plane unless otherwise indicated.

5. Unless otherwise indicated, the domain of a given function f is the set of all real numbers x for which $f(x)$ is a real number.

REFERENCE

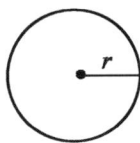
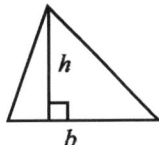
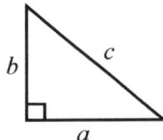
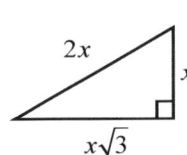
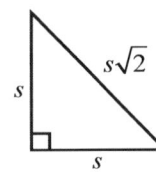

$A = \pi r^2$
$C = 2\pi r$

$A = \ell w$

$A = \dfrac{1}{2}bh$

$c^2 = a^2 + b^2$

Special Right Triangles

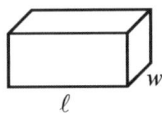

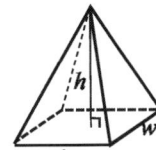

$V = \ell w h$

$V = \pi r^2 h$

$V = \dfrac{4}{3}\pi r^3$

$V = \dfrac{1}{3}\pi r^2 h$

$V = \dfrac{1}{3}\ell w h$

The number of degrees of arc in a circle is 360.
The number of radians of arc in a circle is 2π.
The number of the measures in degrees of the angles of a triangle is 180.

CONTINUE

1

If $ax + bx = 48$, what is the value of $2a + 2b$ when $x = 4$?

A) 12
B) 18
C) 24
D) 48

2

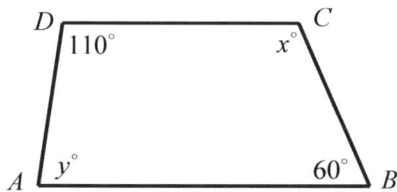

In the figure above, $\overline{AB}$ and $\overline{CD}$ are parallel. What is the value of $x - y$?

A) 50
B) 55
C) 60
D) 65

3

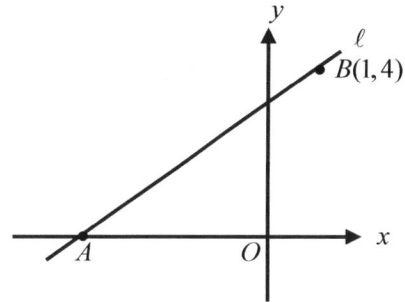

In the xy-plane above, points A and B lie on line ℓ. If $AB = 5$, what is the y-intercept of the line?

A) $\left(0, \dfrac{5}{2}\right)$

B) $\left(0, \dfrac{8}{3}\right)$

C) $(0, 3)$

D) $\left(0, \dfrac{10}{3}\right)$

4

In the xy-plane, line ℓ is parallel to the x-axis and passes through point $(-4, -8)$. What is the equation of the line?

A) $x = -4$
B) $y = -4$
C) $x = -8$
D) $y = -8$

CONTINUE

5

If $\left(\sqrt[4]{10}\right)\left(\sqrt[3]{10}\right) = 100^k$, what is the value of k?

A) $\dfrac{7}{24}$

B) $\dfrac{7}{12}$

C) 6

D) 12

6

$$F = \frac{9}{5}C + 32$$

The relationship between the temperature F, in degrees Fahrenheit, and the temperature C, in degrees Celsius, is given by the formula above. A student uses the approximate formula $F = 2C + 30$ to convert from degrees Celsius to degrees Fahrenheit. For what temperature in degrees Celsius does the student's approximate formula give the correct temperature in degrees Fahrenheit?

A) 0

B) 5

C) 10

D) 15

7

The graph of the function f in the xy-plane is a parabola that has a maximum at point $(5, 10)$. If the graph has one x-intercept at $(1, 0)$, what is the other x-intercept of f?

A) $(-1, 0)$

B) $(8, 0)$

C) $(9, 0)$

D) $(10, 0)$

8

$$a^2 = 4b^2$$
$$a = 1 + 2b$$

In the system of equations above, what is the value of a?

A) $-\dfrac{1}{2}$

B) $-\dfrac{1}{4}$

C) $\dfrac{1}{4}$

D) $\dfrac{1}{2}$

CONTINUE

9

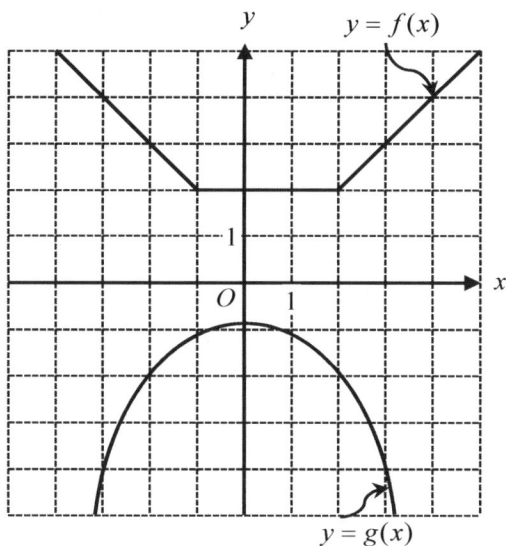

$y = f(x)$

$y = g(x)$

The graphs of the functions f and g are shown above in the xy-plane. If $f(1.5) = k$, what is the value of $g(k)$?

A) 2

B) −1

C) −2

D) −4

10

$$|2x - 6| < 10$$

Which of the following intervals is a subset of the values of x that satisfy the inequality above?

A) $-4 < x < 10$

B) $-2 < x < 9$

C) $-4 < x < 8$

D) $-1 < x < 7$

11

There are 20 black marbles and 16 white marbles in a container and no others. How many black marbles must be removed from the container so that the probability of randomly selecting a black marble from the container is $\frac{1}{3}$?

A) 8

B) 12

C) 16

D) 20

CONTINUE

12

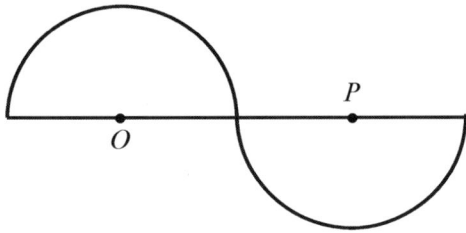

In the figure above, O and P are the centers of two semicircles of radius r. If the length of the perimeter is $8\pi + 16$, what is the value of r?

A) 2

B) 4

C) 6

D) 8

13

If $x^{12} = 5000$ and $\dfrac{x^{11}}{y} = 10$, what is the value of xy?

A) 500

B) 100

C) 50

D) 10

14

Test Scores for a class of 30 Students

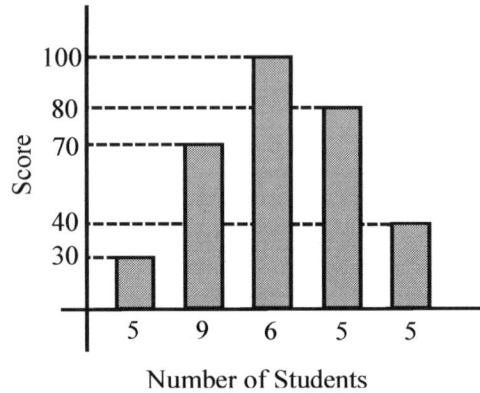

Number of Students

The graph above shows the test scores of 30 students. Based on the histogram above, what is the median score of the test?

A) 80

B) 70.5

C) 70

D) 40

CONTINUE

Question 15 and 16 refer to the following information.

An audio recording studio's fee consists of a setup charge of $100 plus a charge for session time at an hourly rate. The total fee for a session of 8 hours is $480.

15

Which of the following functions f gives the total fee, in dollars, for a session of t hours in the studio?

A) $100 + 60t$

B) $480 + 100t$

C) $100 + 48t$

D) $100 + 47.5t$

16

Jackson spent 10 hours recording his favorite pop songs in the studio. How much did he pay for his recording, in dollars?

A) 525.5

B) 565

C) 575

D) 600

17

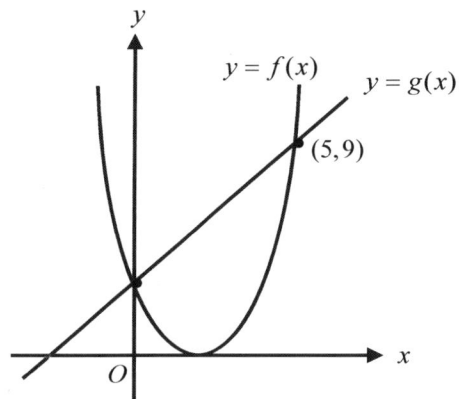

The graphs of $f(x) = (x-2)^2$ and $g(x)$ are shown in the xy-plane above. If the graphs intersect at point $(5, 9)$, which of the following is the equation of $g(x)$?

A) $y = \dfrac{1}{2}x + \dfrac{13}{2}$

B) $y = \dfrac{2}{5}x + 7$

C) $y = \dfrac{3}{5}x + 6$

D) $y = x + 4$

18

If $x^2 + y^2 = 85$ and $xy = 5$, what is the value of $\left(\dfrac{1}{x} - \dfrac{1}{y}\right)^2$?

A) 3

B) 5

C) 75

D) 80

CONTINUE

19

$$\frac{1}{x(x+1)} = \frac{a}{x} - \frac{b}{x+1}$$

In the equation above, a and b are constants. If the equation is true for all positive values of x, what is the value of b?

A) -1

B) 1

C) 2

D) 4

20

The linear function $2x + y = -5$ in the xy-plane is to be reflected about the y-axis. Which of the following ordered pairs CANNOT be the coordinates of the resulting graph?

A) $(0, -5)$

B) $(2, -1)$

C) $(3, 1)$

D) $(4, 0)$

21

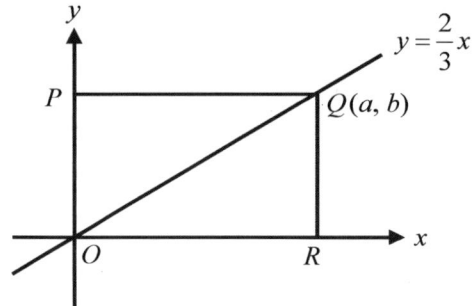

In the xy-plane above, the point Q is on the line $y = \frac{2}{3}x$. If the area of rectangle $OPQR$ is 54, what is the value of b?

A) 3

B) 6

C) 9

D) 12

22

Harry bought a 10 pound bag of flour for $80, a 25 pound bag of flour for $150, and a 50 pound bag of flour. If the average (arithmetic mean) cost per pound of all three bags is $6.00, what was the price of the 50 pound bag of flour?

A) $200

B) $240

C) $280

D) $320

CONTINUE

Questions 23 and 24 refer to the following information.

Test Scores

| | Test 1 |
| | Test 2 |

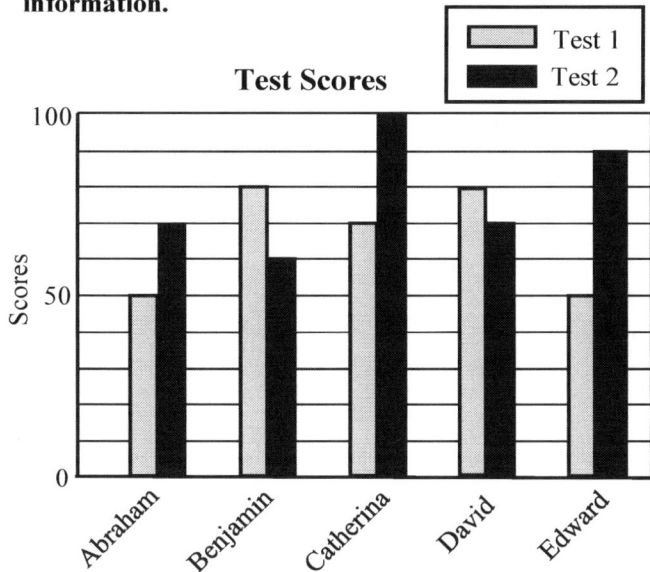

The bar graph shows the scores on the algebra tests for five students in Mrs. Lee's class.

23

Of the following, who has the greatest percent of increase in scores from test 1 to test 2?

A) Abraham

B) Benjamin

C) Catherina

D) Edward

24

Which of the following scatterplots represents the data on the bar graph?

A)

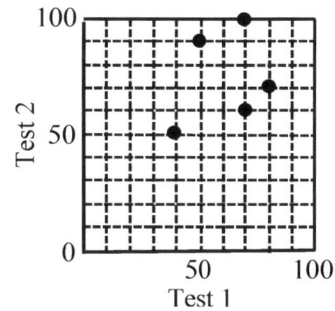

B)

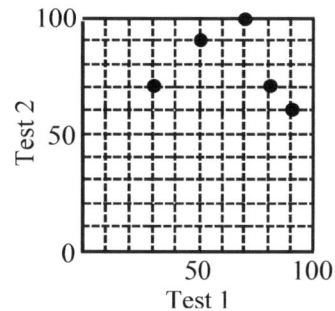

C)

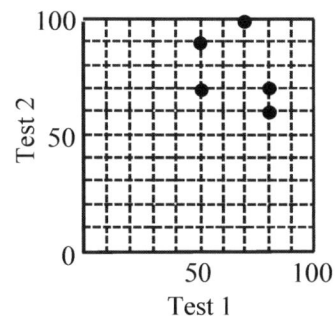

D)

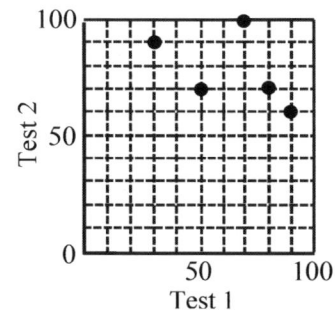

CONTINUE

25

City	Spring	Summer	Autumn	Winter
Amber	26	30	25	22
Buner	24	35	20	18

The table above gives high temperatures in degrees Celsius $(C°)$ for Amber City and Buner City over the four seasons. Which of the following is true about the data shown for the four seasons?

A) The standard deviation of high temperatures in Amber City is larger than Buner City.

B) The standard deviation of high temperatures in Buner City is larger than Amber City.

C) The standard deviation of high temperatures in Amber City is the same as that of Buner City.

D) Based on the data above, the standard deviation of high temperatures in these cities cannot be determined.

26

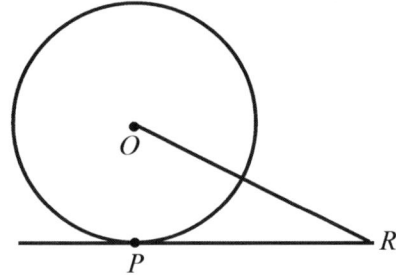

Note: Figure not drawn to scale.

In the figure above, $\overline{PR}$ is tangent to circle O at point P and the length of $\overline{PR}$ is 12. If the value of $\sin \angle R$ is 0.8, what is the radius of the circle?

A) 8

B) 12

C) 16

D) 20

27

$$2^{(x+2y)} = 16$$
$$3^{(2x+y)} = 81$$

In the system of equations, what is the value of x?

A) 4

B) 3

C) $\dfrac{4}{3}$

D) $\dfrac{3}{4}$

CONTINUE

28

$$(x+y)^2 - (x-y)^2 = 72$$

In the equation above, x and y are positive integers. Which of the following CANNOT be the value of $x+y$?

A) 9

B) 11

C) 19

D) 24

29

P = The average of a and b

Q = The average of b and c

R = The average of c and a

The various averages (arithmetic mean) of two of the three numbers a, b, and c are calculated and arranged as shown above. If $P > Q > R$, which of the following is true?

A) $a = b = c$

B) $a > b > c$

C) $c > b > a$

D) $b > a > c$

30

Peter sets up a lemonade stand. He paid a set-up cost of $120 and each cup of lemonade costs him $0.30 to make. He sells each cup of lemonade for $0.75. Which of the following represents the profit P as a function of the number of cups n of lemonade sold?

A) $P(n) = 0.75n - 120$

B) $P(n) = 0.75n + 120$

C) $P(n) = 0.45n - 120$

D) $P(n) = 0.45n + 120$

CONTINUE

Answer: $\frac{7}{12}$

Write answer in boxes. → Fraction line

Answer: 2.5 → Decimal point

Grid in result.

DIRECTIONS

For questions 31-38, solve the problem and enter your answer in the grid, as described below, on the answer sheet.

1. Although not required, it is suggested that you write your answer in the boxes at the top of the columns to help you fill in the circles accurately. You will receive credit only if the circles are filled in correctly.
2. Mark no more than one circle in any column.
3. No question has a negative answer.
4. Some problems may have more than one answer.
5. **Mixed numbers** such as $3\frac{1}{2}$ must be gridded as 3.5 or 7/2. (If $3 1 / 2$ is entered into the grid, it will be interpreted as $\frac{31}{2}$, not $3\frac{1}{2}$.)
6. **Decimal answers:** If you obtain a decimal answer with more digits than the grid can accommodate, it may be either rounded or truncated, but it must fill the entire grid.

Acceptable ways to grid $\frac{2}{3}$ are:

Answer: 201
Either position is correct.

Note: You may start your answers in any column, space permitting. Columns you don't need to use should be left blank.

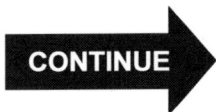

CONTINUE

31

$$x^2 - 4x + y^2 + 4y = 0$$

The equation of a circle in the xy-plane is shown above. To the nearest tenth, what is the area of the circle?

32

The cost C, in dollars, of producing x units of a certain product can be modeled by the equation $C = \dfrac{198.4x + 1097}{16}$. According to the model, for every increase of 1 unit, by how many dollars will the cost increase? (Disregard the $ sign when gridding your answer.)

33

If f is a linear function for which $f(10) - f(5) = 10$, what is the value of $f(20) - f(8)$?

34

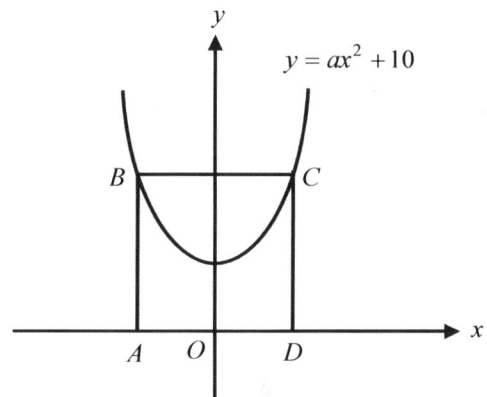

The graph of $y = ax^2 + 10$ is shown in the xy-plane above. If the area of square $ABCD$ is 400, what is the value of a?

CONTINUE

35

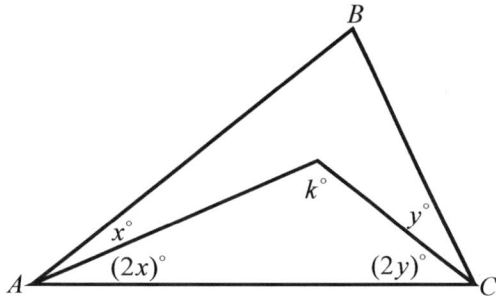

Note: Figure not drawn to scale.

In the figure above, if the measure of $\angle ABC$ is $30°$, what is the value of k?

36

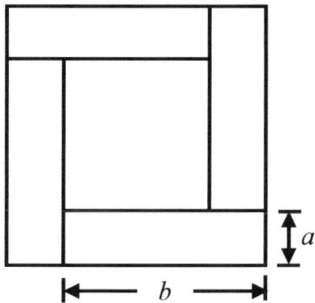

In the figure above, four congruent rectangles and a square are put together to form a larger square. The perimeter of each rectangle is 24, and the area of the smaller square is 36. If the dimensions of each rectangle are a and b as shown above, what is the value of b?

37

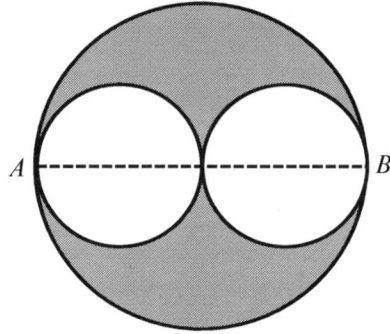

In the figure above, two congruent circles are tangent to each other and are internally tangent to the larger circle. Line segment AB is the diameter of the larger circle. If the area of each smaller circle is 10, what is the area of the shaded region?

38

$$f(x) = x^2 - 6x + 12$$
$$g(x) = k$$

In the equations above, $f(x) \geq g(x)$ for all real numbers x. If k is a constant, what is the maximum value of k?

STOP

If you finish before time is called, you may check your work on this section only.
Do not turn to any other section in the test.

No Test Material on This Page

Math Conversion Table

Raw Score	Scaled Score	Raw Score	Scaled Score
58	800	27	500
57	800	26	490
56	800	25	480
55	800	24	470
54	790	23	460
53	780	22	460
52	770	21	450
51	750	20	440
50	740	19	430
49	730	18	430
48	720	17	420
47	710	16	420
46	700	15	410
45	690	14	400
44	670	13	390
43	680	12	380
42	670	11	370
41	660	10	360
40	650	9	450
39	640	8	340
38	630	7	330
37	620	6	310
36	610	5	290
35	600	4	280
34	590	3	270
33	580	2	260
32	560	1	240
31	550	0	200
30	540		
29	530		
28	520		

Test 8   Answers and Explanations

SECTION 3	1	2	3	4	5	6	7	8	9	10
	C	D	D	A	B	A	D	B	B	C
	11	12	13	14	15	16	17	18	19	20
	A	B	D	B	B	15	108	3	7	64

SECTION 4	1	2	3	4	5	6	7	8	9	10
	C	A	B	D	A	C	C	D	C	D
	11	12	13	14	15	16	17	18	19	20
	B	B	A	C	D	C	D	A	B	D
	21	22	23	24	25	26	27	28	29	30
	B	C	D	C	B	C	C	D	D	C
	31	32	33	34	35	36	37	38		
	25.1	12.4	24	0.1	80	9	20	3		

SECTION 3

1.   C

$$\frac{3}{x-1} = x+1 \quad \rightarrow \quad 3 = x^2 - 1 \quad \rightarrow \quad x^2 = 4$$

2.   D

Radius $OC = 5$, $OD = 4$ $\rightarrow$ $BC = 2 \times OD = 8$

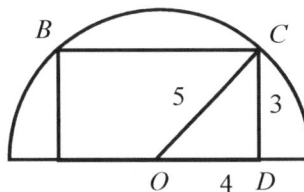

3.   D

$$16 = \left(\frac{1}{4}\right)^{\frac{-1}{m}} \quad \rightarrow \quad 4^2 = 4^{\frac{1}{m}} \quad \rightarrow \quad 2 = \frac{1}{m} \quad \rightarrow \quad m = \frac{1}{2}$$

4.   A

$$\left(\frac{240}{30}\right) \times 2.25 = \$18$$

5.   B

$$AD \times DB = CD^2 \quad \rightarrow \quad AD = \frac{36}{4} = 9 \quad \rightarrow \quad \text{Area of } \triangle ACD = \frac{9 \times 6}{2} = 27$$

6. A

$$1.25a^2 = 0.05b^2 \;\rightarrow\; \frac{a^2}{b^2} = \frac{0.05}{1.25} = \frac{1}{25} \;\rightarrow\; \frac{a}{b} = \frac{1}{5}$$

7. D

K must be 200.

8. B

$$36 = 15 + 0.15(T - 200) \;\rightarrow\; T - 200 = \frac{36 - 15}{0.15} = 140 \;\rightarrow\; T = 200 + 140 = 340 \text{ minutes}$$

9. B

$$f(-1) = 1 + 3 + a + a = 0 \;\rightarrow\; 2a = -4 \;\rightarrow\; a = -2$$

10. C

$BC = 20$, $AB = 20 \times \sin C = 20 \times 0.6 = 12$ and $AC = 16$.

Therefore, $\dfrac{16 \times 12}{2} = \dfrac{20 \times AD}{2} \;\rightarrow\; AD = \dfrac{16 \times 12}{20} = 9.6$

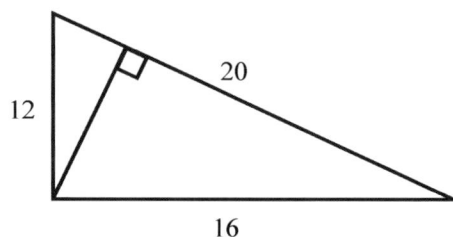

11. A

$$a(x + b) = 3x - 15 \;\rightarrow\; ax + ab = 3x - 15 \;\rightarrow\; a = 3 \text{ and } ab = -15 \;\rightarrow\; b = -5$$

12. B

Center is at $(3, 0)$. Radius $= 5$ and $OR = 4$.

Therefore, $b = 4$.

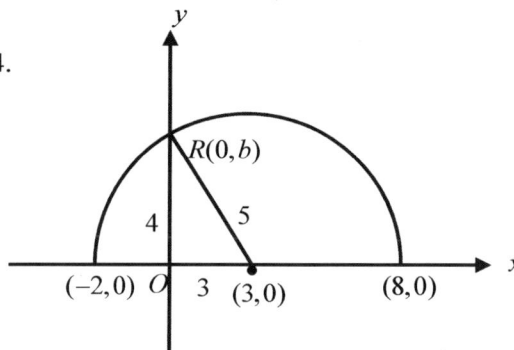

13. D

$$\frac{3 + i}{3 - i} = \frac{(3 + i)(3 + i)}{(3 - i)(3 + i)} = \frac{8 + 6i}{10} = \frac{4}{5} + \frac{3}{5}i = a + bi \;\rightarrow\; a = \frac{4}{5}$$

14. B

$$P = \frac{50}{200} = 0.25$$

Gender	Algebra	Geometry	Total
Male	80	30	110
Female	40	50	90
Total	120	80	200

15. B

$$P = \frac{30}{110} = \frac{3}{11}$$

16. 15 (splitting theorem)

$$\frac{8}{10} = \frac{12}{x} \quad \rightarrow \quad x = \frac{12 \times 10}{8} = 15$$

17. 108

$$\frac{d}{60} + \frac{d}{40} = 4.5 \quad \rightarrow \quad 2d + 3d = 540 \quad \rightarrow \quad 5d = 540 \quad \rightarrow \quad d = 108 \, \text{miles}$$

18. 3

$$x^3 - 3x^2 + 5x = 15 \quad \rightarrow \quad x^2(x-3) + 5(x-3) = 0 \quad \rightarrow \quad (x-3)(x^2+5) = 0 \quad \rightarrow \quad x = 3 : (x^2 + 5 \neq 0)$$

19. 7

$$x^2 + (k+1)x + 16 = (x+h)^2 \quad \rightarrow \quad x^2 + (k+1)x + 16 = x^2 + 2hx + h^2 \quad \rightarrow \quad h^2 = 16 \text{ and } k+1 = 2h$$
$$\rightarrow \quad h = 4 \text{ and } k = 2h - 1 = 7$$

20. 64

$$b^2 - a^2 = 28 \quad \rightarrow \quad (b+a)(b-a) = 28 \quad \rightarrow \quad b+a = 14 \text{ and } b-a = 2 \quad \rightarrow \quad 2b = 16 \quad \rightarrow \quad b = 8$$

Therefore, the area of the square is 64.

CF) If you chose $b + a = 28$ and $b - a = 1$, or $b + a = 7$ and $b - a = 4$, it's not working, because b and a are not integers.

SECTION 4

1. C

$$4a + 4b = 48 \quad \rightarrow \quad 2(2a + 2b) = 48 \quad \rightarrow \quad 2a + 2b = 24$$

2. A

$$x = 180 - 60 = 120 \text{ and } y = 180 - 110 = 70 \quad \rightarrow \quad x - y = 120 - 70 = 50$$

3. B

$$\frac{2}{3} = \frac{b}{4} \quad \rightarrow \quad b = \frac{8}{3}$$

Or, you can find the equation of line ℓ. $\quad y = \frac{4}{3}x + \frac{8}{3}$

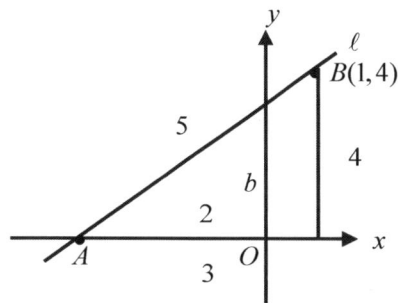

4. D

5. A

$$\left(\sqrt[4]{10}\right)\left(\sqrt[3]{10}\right) = 100^k \quad \rightarrow \quad 10^{\frac{1}{4}} \times 10^{\frac{1}{3}} = 10^{\frac{7}{12}} = 10^{2k} \quad \rightarrow \quad 2k = \frac{7}{12} \quad \rightarrow \quad k = \frac{7}{24}$$

6. C

$$\frac{9}{5}C + 32 = 2C + 30 \quad \rightarrow \quad 2 = \frac{1}{5}C \quad \rightarrow \quad C = 10$$

7. C

In the vertex form: $y = a(x-5)^2 + 10$, where $a < 0$. Therefore, $\dfrac{1+x}{2} = 5 \rightarrow x = 9 \rightarrow (9, 0)$ is the other x-intercept.

8. D

Putting $a = 1 + 2b$ in $a^2 = 4b^2$. $\rightarrow (1+2b)^2 = 4b^2 \rightarrow 1 + 4b + 4b^2 = 4b^2 \rightarrow 4b = -1 \rightarrow b = -\dfrac{1}{4}$

Therefore, $a = 1 + 2\left(-\dfrac{1}{4}\right) = \dfrac{1}{2}$

9. C

$$f(1.5) = k \quad \rightarrow \quad k = 2 \quad \rightarrow \quad g(2) = -2$$

10. D

$$|2x - 6| < 10 \quad \rightarrow \quad -10 < 2x - 6 < 10 \quad \rightarrow \quad -4 < 2x < 16 \quad \rightarrow \quad -2 < x < 8$$

$$\{-1 < x < 7\} \subset \{-2 < x < 8\}$$

11. B

Let x = number of black marbles removed from the container.

$$P = \frac{20 - x}{(20 - x) + 16} = \frac{1}{3} \quad \rightarrow \quad 60 - 3x = 36 - x \quad \rightarrow \quad 24 = 2x \quad \rightarrow \quad x = 12$$

12. B

Perimeter $= 2\pi r + 4r = 8\pi + 16 \quad \rightarrow \quad (2\pi + 4)r = (2\pi + 4)4 \quad \rightarrow \quad r = 4$

13. A

$$x^{12} = 5000 \text{ and } \frac{x^{11}}{y} = 10 \quad \rightarrow \quad \frac{x^{11} \times x}{y \times x} = \frac{x^{12}}{xy} = 10 \quad \rightarrow \quad \frac{5000}{xy} = 10 \quad \rightarrow \quad xy = \frac{5000}{10} = 500$$

14. C

$$\frac{1 + 30}{2} = 15.5 \quad \rightarrow \quad \text{median score must between 15th and 16th students.} \quad \rightarrow \quad 70$$

30 30 30 30 30 40 40 40 40 40 70 70 70 70 **70 70** 70 70 70 $\cdots$

15. D

$C = 100 + kt \quad \rightarrow \quad 480 = 100 + k(8) \quad \rightarrow \quad 8k = 380 \quad \rightarrow \quad k = 47.5$

Therefore, the modeled equation is $C = 100 + 47.5t$.

16. C

$$C = 100 + 47.5 \times 10 = \$575$$

17. D

$f(0) = 4 \quad \rightarrow \quad$ The line passes through points $(0, 4)$ and $(5, 9)$.

Slope: $\dfrac{9-4}{5-0} = 1 \quad \rightarrow \quad$ Therefore, the equation is $y = x + 4$.

18. A

$$\left(\frac{1}{x} - \frac{1}{y}\right)^2 = \left(\frac{y-x}{xy}\right)^2 = \frac{x^2 + y^2 - 2xy}{(xy)^2} = \frac{85 - 10}{25} = 3$$

19. B

$$\frac{1}{x(x+1)} = \frac{a}{x} - \frac{b}{x+1} \quad \rightarrow \quad \frac{1}{x(x+1)} = \frac{a(x+1) - bx}{x(x+1)} \quad \rightarrow \quad \frac{1}{x(x+1)} = \frac{(a-b)x + a}{x(x+1)} \quad \rightarrow \quad a = 1 \text{ and } b = 1$$

20. D

Reflected function must be $-2x + y = -5 \quad \rightarrow \quad y = 2x - 5$

$(4, 0)$ is not on the resulting graph of $y = 2x - 5$.

21. B

$$b = \frac{2}{3}a \quad \rightarrow \quad a\left(\frac{2}{3}a\right) = 54 \quad \rightarrow \quad a^2 = 81 \quad \rightarrow \quad a = 9 \quad \rightarrow \quad \text{Therefore, } b = \frac{2}{3}(9) = 6.$$

22. C

Price for the 50 pound bag $= x \quad \rightarrow \quad \dfrac{80 + 150 + x}{10 + 25 + 50} = 6 \quad \rightarrow \quad \dfrac{230 + x}{85} = 6 \quad \rightarrow \quad x = \$280$

23. D

Abraham: $\dfrac{70 - 50}{50} \times 100 = 40\%$ Benjamin: Decreased   Catherina: $\dfrac{100 - 70}{70} \times 100 \approx 42.9\%$

Edward: $\dfrac{90 - 50}{50} \times 100 = 80\%$

24. C

Ordered pairs $(\text{test }1,\ \text{test }2) \rightarrow (50, 70), (80, 60), (70, 100), (80, 70), (50, 90)$

25. B

Data of Burner City are more widely spread.

26. C

Since $\sin \angle R = 0.8$, $OP : OP : PR = 10 : 8 : 6$ or $5 : 4 : 3$.

Therefore, $\dfrac{OP}{PR} = \dfrac{4}{3} \quad \rightarrow \quad \dfrac{OP}{12} = \dfrac{4}{3} \quad \rightarrow \quad OP = 16$.

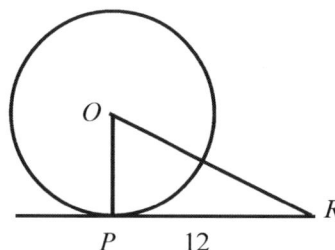

27. C

$$2^{(x+2y)} = 16 = 2^4 \rightarrow x+2y = 4 \qquad \begin{array}{r} x+2y = 4 \\ -4x-2y = -8 \\ \hline -3x = -4 \end{array} \rightarrow x = \frac{4}{3}$$
$$3^{(2x+y)} = 81 = 3^4 \rightarrow 2x+y = 4$$

28. D

$$(x+y)^2 - (x-y)^2 = 72 \rightarrow x^2 + 2xy + y^2 - x^2 + 2xy - y^2 = 72 \rightarrow 4xy = 72 \rightarrow xy = 18$$

Because x and y are positive integers, $(x, y) \rightarrow (1, 18), (2, 9), (3, 6), (6, 3), (9, 2), (18, 1)$

Therefore, possible sum of the numbers are $19, 11, 9$.

29. D

$$P > Q > R \rightarrow \frac{a+b}{2} > \frac{b+c}{2} > \frac{c+a}{2} \rightarrow \begin{cases} a+b > b+c \rightarrow a > c \\ b+c > c+a \rightarrow b > a \end{cases} \rightarrow \text{Therefore, } b > a > c.$$

30. C

Cost $= 120 + 0.3n$ and selling amount $= 0.75n \rightarrow$ Therefore, the profit $= 0.75n - (120 + 0.3n) = 0.45n - 120$

31. 25.1

$$x^2 - 4x + y^2 + 4x = 0 \rightarrow (x - 4x + 4) + (y + 4y + 4) = 4 + 4 \rightarrow (x-2)^2 + (y+2)^2 = 8$$

The area of the circle $= \pi r^2 = \pi(8) \approx 25.1$

32. 12.4

$$C = \frac{198.4x + 1097}{16} \rightarrow C = \frac{198.4}{16}x + \frac{1097}{16} \rightarrow C = 12.4x + 68.5625$$

For every 1 unit increase, $12.4 increases.

33. 24

Constant slope: $\text{slope} = \dfrac{f(10) - f(5)}{10-5} = \dfrac{10}{5} = 2 \rightarrow \dfrac{f(20)-f(8)}{20-8} = 2 \rightarrow f(20)-f(8) = 12 \times 2 = 24$

34. $\dfrac{1}{10}$ or 0.1

$OD = 10$ and $BC = 20$

Putting $(10, 20)$ in the equation $y = ax^2 + 10 \rightarrow 20 = a(100) + 10$

$100a = 10 \rightarrow a = \dfrac{1}{10}$.

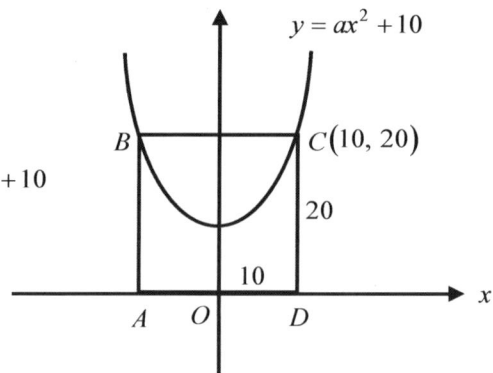

35. 80

$3x + 3y = 180 - 30 = 150 \rightarrow x + y = 50 \rightarrow 2x + 2y = 100$

Therefore, $k = 180 - 100 = 80$

36. 9

$$2(a+b) = 24 \text{ and } (b-a)^2 = 36 \quad \rightarrow \quad b-a = 6$$

$$\begin{cases} b+a = 12 \\ b-a = 6 \end{cases} \quad \rightarrow \quad 2b = 18 \quad \rightarrow \quad b = 9$$

37. 20

Ratio of corresponding sides $= 1:1:2 \quad \rightarrow \quad$ Ratio of their areas $= 1:1:4$.

Define their areas as k, k, and $4k$. $\quad \rightarrow \quad k = 10$ and $4k = 40$

Therefore, the area of the shaded region is $40 - (10 + 10) = 20$.

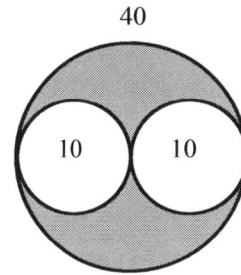

38. 3

$$f(x) = x^2 - 6x + 12 = (x-3)^2 + 3$$

$$g(x) = k \quad \rightarrow \quad \text{The possible maximum value of } k \text{ is } 3.$$

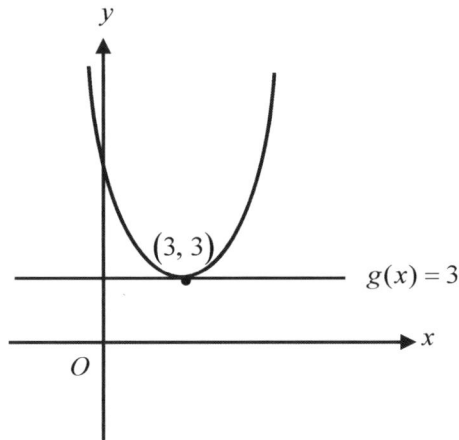

No Test Material on This Page

PRACTICE TEST 9

Dr. John Chung's SAT Math

Math Test - No Calculator

25 MINUTES, 20 QUESTIONS

Turn to Section 3 of your answer sheet to answer the questions in this section.

DIRECTIONS

For questions 1–15, solve each problem, choose the best answer from the choices provided, and fill in the corresponding circle on your answer sheet. **For questions 16–20,** solve the problem and enter your answer in the grid on your answer sheet. Please refer to the directions before question 16 on how to enter your answers in the grid. You may use any available space in your test booklet for scratch work.

NOTE

1. The use of a calculator **is not permitted**.

2. All variables and expressions used represent real numbers unless otherwise indicated.

3. Figures provided in this test are drawn to scale unless otherwise indicated.

4. All figures lie in a plane unless otherwise indicated.

5. Unless otherwise indicated, the domain of a given function f is the set of all real numbers x for which $f(x)$ is a real number.

REFERENCE

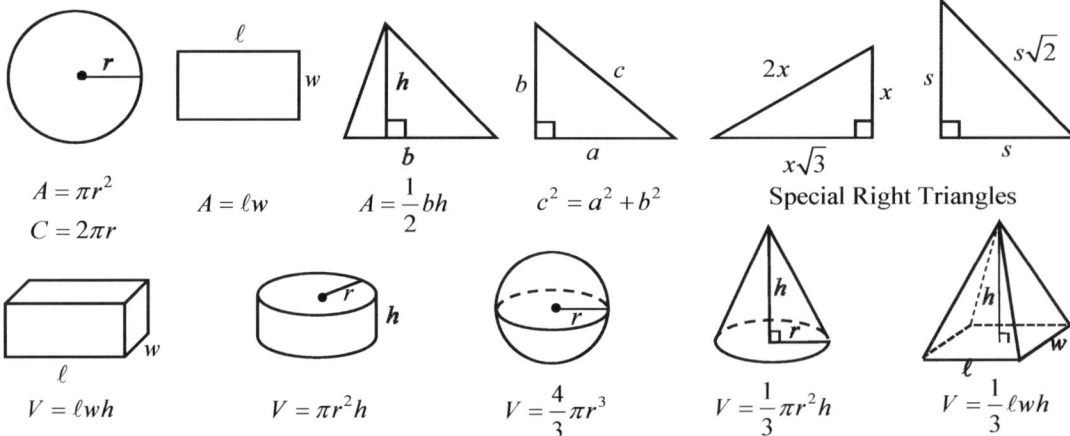

$A = \pi r^2$
$C = 2\pi r$

$A = \ell w$

$A = \dfrac{1}{2}bh$

$c^2 = a^2 + b^2$

Special Right Triangles

$V = \ell w h$

$V = \pi r^2 h$

$V = \dfrac{4}{3}\pi r^3$

$V = \dfrac{1}{3}\pi r^2 h$

$V = \dfrac{1}{3}\ell w h$

The number of degrees of arc in a circle is 360.
The number of radians of arc in a circle is 2π.
The number of the measures in degrees of the angles of a triangle is 180.

CONTINUE

1

If $|k-5| \le 8,$ which of the following CANNOT be the value of k?

A) 8

B) 2

C) -3

D) -4

2

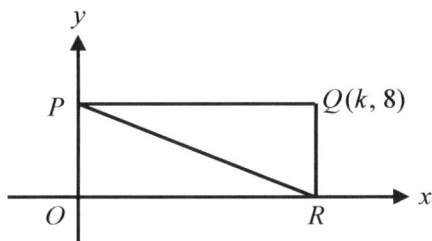

In the xy-plane above, $\overline{PR}$ is the diagonal of rectangle $OPQR$. If the length of $\overline{PR}$ is 17, what is the value of k?

A) 12

B) 15

C) 16

D) 20

3

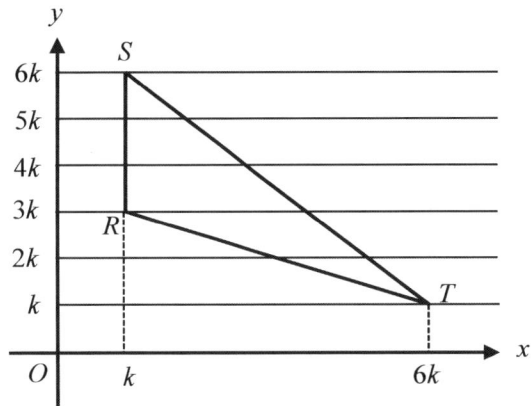

In the xy-plane above, the area of triangle RST is 30. What is the value of k?

A) -2

B) 2

C) 3.5

D) 4

4

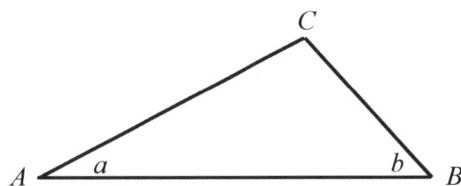

Note: Figure not drawn to scale.

In the triangle above, $\cos a = \sin b$. If the length of $\overline{AB}$ is 20 and the measure of angle b is $\dfrac{\pi}{3}$ radians, what is the area of the triangle?

A) $40\sqrt{2}$

B) $40\sqrt{3}$

C) $50\sqrt{3}$

D) $100\sqrt{3}$

CONTINUE

5

$$f(x) = x - p$$

In the function above, p is a constant. If $f(2) = 5$, what is the value of $f(2p)$?

A) -3

B) -1

C) 3

D) 6

6

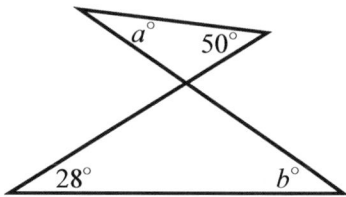

Note: Figure not drawn to scale.

In the figure above, what is the value of $|a - b|$?

A) 20

B) 22

C) 24

D) 26

Questions 7 and 8 refer to the following information.

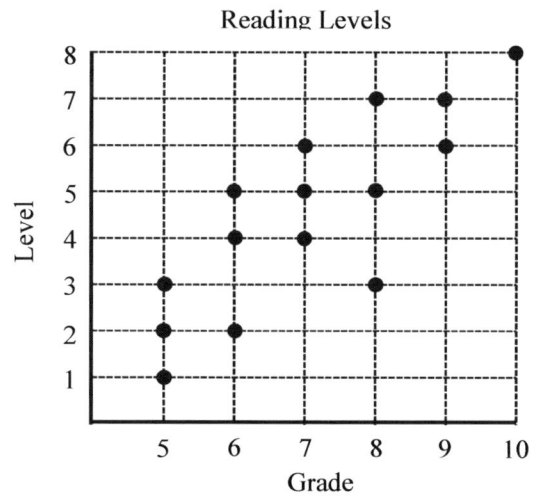

Reading Levels

The scatterplot above shows the reading levels by grade for 15 students in the J.H book-reading club.

7

Based on the data above, what is the median reading level for the 15 students?

A) 4

B) 5

C) 5.5

D) 6

8

What is the average reading level of 7th and 8th grade students?

A) 4

B) 5

C) 6

D) 6.5

CONTINUE

9

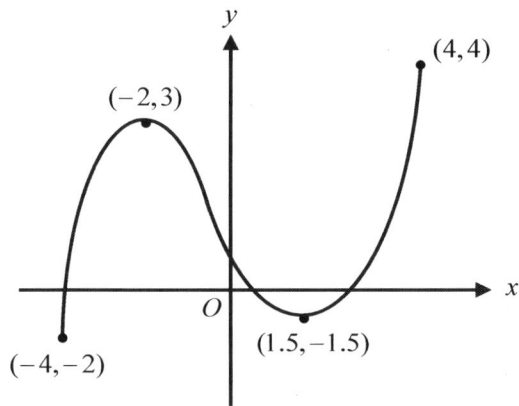

In the xy-plane above, the graph of f is shown in $-4 \le x \le 4$. If $f(k) = 3$, how many values of k are there in the interval?

A) 1

B) 2

C) 3

D) 4

10

$$P = \frac{50n - 200}{n} + k$$

The profit P from a car wash is modeled by the equation above, where n is the number of cars and k is a constant. Which of the following expressions represents n?

A) $n = \dfrac{200 + k}{50 - p}$

B) $n = \dfrac{200 - k}{50 + p}$

C) $n = \dfrac{200}{50 + k - p}$

D) $n = \dfrac{200}{k + p - 50}$

11

$$P(x) = 3x^3 + ax - 2$$

In the function above, a is a constant. If the remainder when $P(x)$ is divided by $x + 1$ is 2, what is the value of a?

A) -7

B) -5

C) 5

D) 7

CONTINUE

12

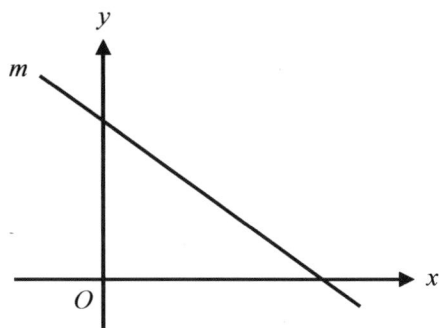

The graph of $ax + by = 5$ is shown in the xy-plane above. Which of the following must be true?

A) $a < 0$ and $b < 0$

B) $a > 0$ and $b < 0$

C) $a < 0$ and $b > 0$

D) $a > 0$ and $b > 0$

13

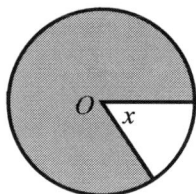

In the figure above, the center of the circle is O.

The area of the shaded region is 80π and the measure of x is $\dfrac{2\pi}{5}$ radians. What is the radius of the circle?

A) 6

B) 8

C) 9

D) 10

14

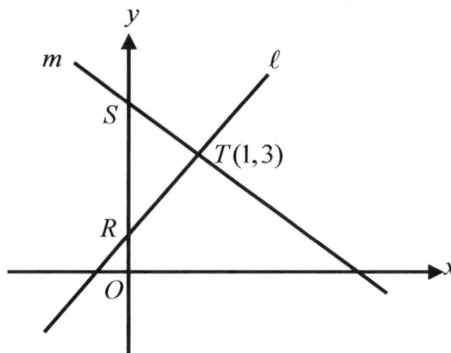

Note: Figure not drawn to scale.

Lines ℓ and m are perpendicular and intersect at point $T(1,3)$ as shown in the xy-plane above. If the slope of line ℓ is 1, what is the area of $\triangle RST$?

A) 1

B) 1.5

C) 2

D) 2.5

15

$$\frac{x^2 + 3}{x - 1}$$

Which of the following is equivalent to the expression above?

A) $x + 1$

B) $x(x+1) + 3$

C) $x + 1 + \dfrac{4}{x-1}$

D) $\dfrac{x}{x-1} + x$

CONTINUE

DIRECTIONS

For questions 16–20, solve the problem and enter your answer in the grid, as described below, on the answer sheet.

1. Although not required, it is suggested that you write your answer in the boxes at the top of the columns to help you fill in the circles accurately. You will receive credit only if the circles are filled in correctly.

2. Mark no more than one circle in any column.

3. No question has a negative answer.

4. Some problems may have more than one answer.

5. **Mixed numbers** such as $3\frac{1}{2}$ must be gridded as 3.5 or 7/2. (If $\boxed{3\,|\,1\,|\,/\,|\,2}$ is entered into the grid, it will be interpreted as $\frac{31}{2}$, not $3\frac{1}{2}$.)

6. **Decimal answers:** If you obtain a decimal answer with more digits than the grid can accommodate, it may be either rounded or truncated, but it must fill the entire grid.

Answer: $\frac{7}{12}$

Write answer in boxes. → Fraction line

Grid in result. →

Answer: 2.5

Decimal point

Acceptable ways to grid $\frac{2}{3}$ are:

Answer: 201
Either position is correct.

Note: You may start your answers in any column, space permitting. Columns you don't need to use should be left blank.

CONTINUE

16

If $\dfrac{1}{x-1} + \dfrac{1}{2x-2} = \dfrac{1}{4}$, what is the value of $x-1$?

17

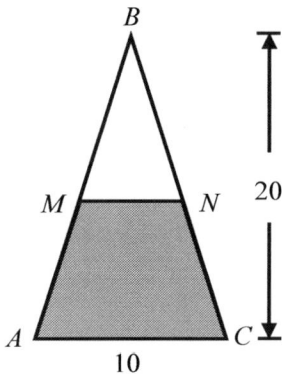

The figure above shows triangle ABC. The length of $\overline{AC}$ is 10 and the altitude of the triangle is 20. If M and N are the midpoints of $\overline{AB}$ and $\overline{BC}$ respectively, what is the area of the shaded region?

18

$$(a-1)x^2 + (b-2)x + ab = 4x^2 + 5x + k$$

In the equation above, a, b, and k are constants. If the equation is true for all values of x, what is the value of k?

19

If $P(x) = 2\sqrt{x-5} + 3x$, what is the minimum value of P?

20

$$x^2 + y^2 = 8$$
$$y = \sqrt{2x}$$

In the system of equations above, what is the value of y?

STOP

If you finish before time is called, you may check your work on this section only.
Do not turn to any other section in the test.

No Test Material on This Page

Math Test - Calculator

55 MINUTES, 38 QUESTIONS

Turn to Section 4 of your answer sheet to answer the questions in this section.

DIRECTIONS

For questions 1-30, solve each problem, choose the best answer from the choices provided, and fill in the corresponding circle on your answer sheet. **For questions 31-38**, solve the problem and enter your answer in the grid on your answer sheet. Please refer to the directions before question 31 on how to enter your answers in the grid. You may use any available space in your test booklet for scratch work.

NOTE

1. The use of a calculator **is permitted**.

2. All variables and expressions used represent real numbers unless otherwise indicated.

3. Figures provided in this test are drawn to scale unless otherwise indicated.

4. All figures lie in a plane unless otherwise indicated.

5. Unless otherwise indicated, the domain of a given function f is the set of all real numbers x for which $f(x)$ is a real number.

REFERENCE

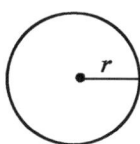 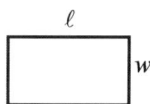 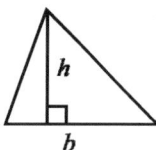 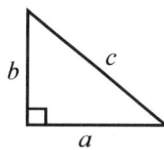 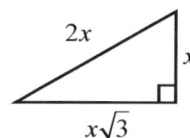 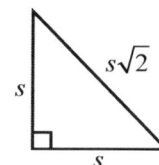

$A = \pi r^2$
$C = 2\pi r$

$A = \ell w$

$A = \dfrac{1}{2}bh$

$c^2 = a^2 + b^2$

Special Right Triangles

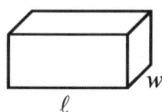  

$V = \ell w h$

$V = \pi r^2 h$

$V = \dfrac{4}{3}\pi r^3$

$V = \dfrac{1}{3}\pi r^2 h$

$V = \dfrac{1}{3}\ell w h$

The number of degrees of arc in a circle is 360.

The number of radians of arc in a circle is 2π.

The number of the measures in degrees of the angles of a triangle is 180.

CONTINUE

1

The Sky Telephone Company charges a cents for the first 3 minutes of a call and charges at the rate of r cents for each additional minute. If Jackson uses t minutes, where $t > 3$, how much, in dollars, is his call?

A) $a + rt$

B) $a + r(t-3)$

C) $0.01(a + r(t-3))$

D) $0.01(a + rt - 3)$

2

If Sally drives m miles from her house to her office in f hours, and drives back to her house in g hours, what is her average speed of the entire trip, in miles per hour?

A) $\dfrac{f+g}{2}$

B) $\dfrac{m}{f+g}$

C) $\dfrac{2m}{f+g}$

D) $\dfrac{m}{f} + \dfrac{m}{g}$

3

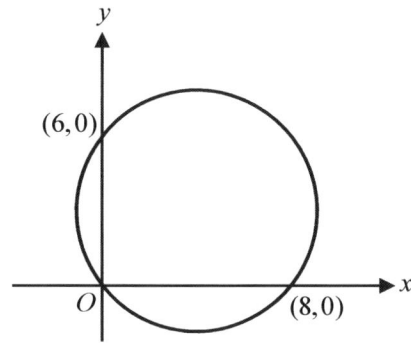

The graph of a circle shown in the xy-plane above, intersects the x-axis and y-axis at three points. What is the radius of the circle?

A) 4

B) 5

C) 6

D) 7

4

$$f(x) = (x+a)^2 + 5$$

In the equation above, a is a constant. If a is increased by 4 units, which of the following is true about the resulting graph?

A) The resulting graph would be shifted by 4 units right.

B) The resulting graph would be shifted by 4 units left.

C) The resulting graph would be shifted by 4 units up.

D) The resulting graph would be shifted by 4 units down

CONTINUE

Questions 5 and 6 refer to the following information.

Claire has $40 in her own savings jar and puts in $10 every week. David has $80 in his own savings jar and puts in $8 every week. Each of the graphs below shows the amount in the jar over time.

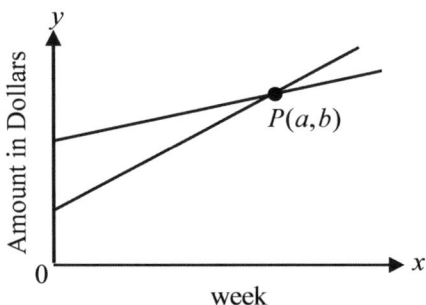

5

If the graphs intersect at the point $P(a,b)$, what is the value of b?

A) 200

B) 220

C) 240

D) 260

6

When Claire has $200 in her savings jar, how many dollars does David have in his savings jar?

A) 208

B) 220

C) 246

D) 252

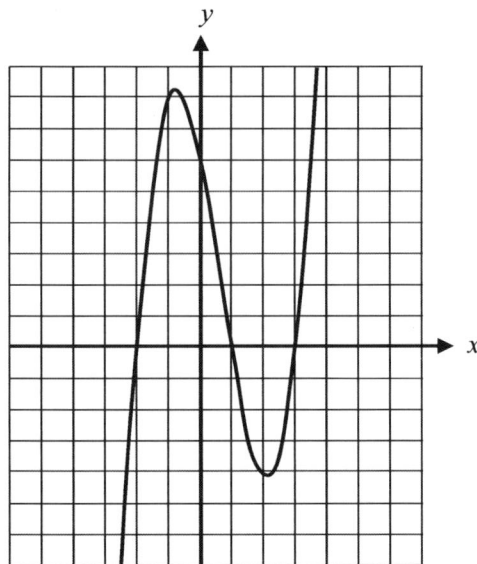

Which of the following functions could represent the graph of $f(x)$ shown in the xy-plane above?

A) $f(x) = (x-2)(x^2 + 4x - 3)$

B) $f(x) = (x-2)(x^2 - 4x - 3)$

C) $f(x) = (x+2)(x^2 + 4x + 3)$

D) $f(x) = (x+2)(x^2 - 4x + 3)$

8

The cost of a notebook is $1.25. The cost of a pencil is $0.30. Grace has $35.00 to spend on notebooks and pencils for her study club. If she must buy fifteen notebooks, what is the maximum number of pencils she can buy?

A) 54

B) 55

C) 70

D) 116

CONTINUE

9

$$x + y > 3$$
$$ax + 2y < -2$$

In the system of inequalities, a is a constant. If the system has no solution, what is the value of a?

A) -2

B) -1

C) 1

D) 2

10

$$f(x) = ax^2 - 4x + b$$

In the function f above, a and b are constants. If the zeros of the function $f(x)$ are -2 and 3, what is the value of b?

A) -24

B) -6

C) 6

D) 24

11

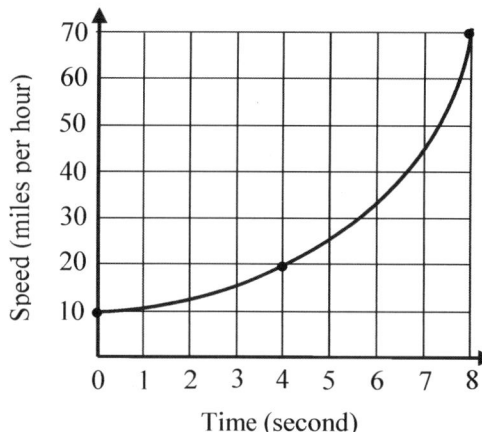

Jackie is driving a car at 10 miles per hour. The graph above shows the speed of his car over 8 seconds. During which of the following time intervals did the speed show the greatest average rate of change?

A) 0 to 2 seconds

B) 2 to 4 seconds

C) 4 to 6 seconds

D) 6 to 8 seconds

12

$$P(x) = x^2 + ax + b$$

In the function above, if the value of $P(0)$ is 1, which of the following must be true?

A) x is a factor of $P(x)$.

B) $x - 1$ is a factor of $P(x)$.

C) The remainder when $P(x)$ is divided by $x - 1$ is 0.

D) The remainder when $P(x)$ is divided by x is 1.

CONTINUE

13

$$f(x) = 2x^2 - 16x + 18$$

If $f(x) = a(x+b)^2 + c$ is equivalent to the function above, what is the value of c for $f(x)$?

A) -48

B) -14

C)   2

D)   50

14

$$\frac{5 - 10x}{2x + 1}$$

Which of the following is equivalent to the expression above?

A) -5

B) $\dfrac{5}{2x+1} - 5$

C) $\dfrac{10}{2x+1} + 5$

D) $\dfrac{10}{2x+1} - 5$

15

City A

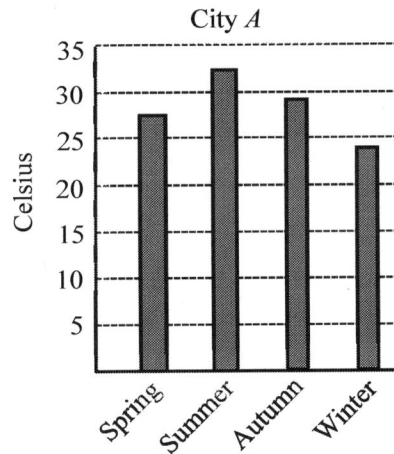

City B

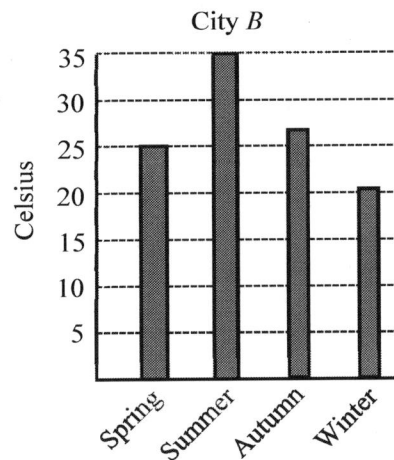

The bar graphs above show the average temperatures in degrees Celsius for City A and City B over the four seasons. Based on the graphs above, which of the following is true?

A) The standard deviation of the average temperatures in City A is larger than City B.

B) The standard deviation of the average temperatures in City B is larger than City A.

C) The standard deviation of the average temperatures in City B is the same as that of City A.

D) Based on graphs above, the standard deviation of the average temperatures in these cities cannot be determined.

CONTINUE

16

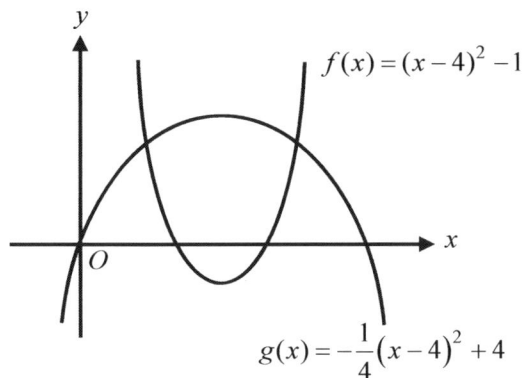

$$f(x) = (x-4)^2 - 1$$
$$g(x) = -\frac{1}{4}(x-4)^2 + 4$$

The graphs of the functions f and g are shown in the xy-plane above. For which of the following values of x does $f(x) - g(x) = 0$?

A) 1

B) 3

C) 6

D) 8

17

$$x^2 + y^2 - 4x + 4y = k$$

The equation of a circle in the xy-plane, where k is a constant, is shown above. If the radius of the circle is 6, what is the value of k?

A) -28

B) -2

C) 14

D) 28

Questions 18 and 19 refer to the following information.

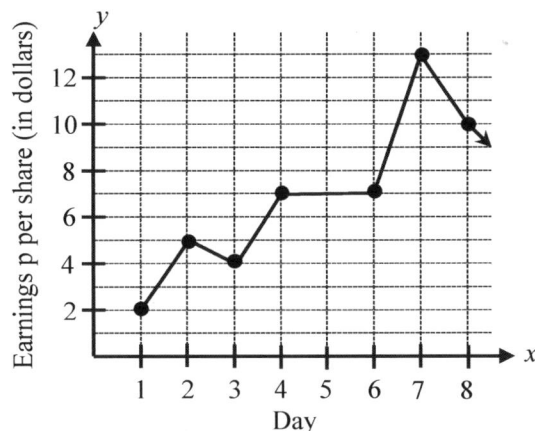

The graph above shows the earnings per share of stock for Milly Electronics for the first 8 days in March this year.

18

What is the average rate of change between days 1 and 8?

A) $\frac{3}{4}$ dollars per day

B) $\frac{7}{8}$ dollars per day

C) $\frac{8}{7}$ dollars per day

D) It cannot be determined based on the information given.

19

What is the equation of the line between day 2 and day 7?

A) $y = x + 3$

B) $y = 1.6x + 1.8$

C) $y = 2.4x + 0.2$

D) $y = 3x - 1$

CONTINUE

20

$$f(x) = 2^{x+2}$$

In the function above, which of the following is equivalent to $f(a+b)$?

A) $2\left(2^a + 2^b\right)$

B) $4\left(2^a + 2^b\right)$

C) $4\left(2^a \times 2^b\right)$

D) $4 + \left(2^a \times 2^b\right)$

21

The sum of four numbers is 783. One of the numbers, a, is 25% more than the sum of the other three numbers. What is the value of a?

A) 348

B) 435

C) 520

D) 585

22

Mr. Lee brought reading books to his class. If each student takes 5 books, there will be 15 books left. If 5 students do not take a book and the rest of the students take 7 books each, there will be no books left. How many books were brought to the class?

A) 140

B) 125

C) 120

D) 104

23

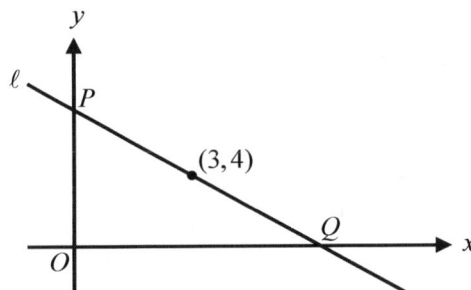

In the xy-plane above, if line ℓ has a slope of $-\dfrac{1}{3}$, what is the area of triangle OPQ?

A) 37.5

B) 60

C) 62.5

D) 75

CONTINUE

24

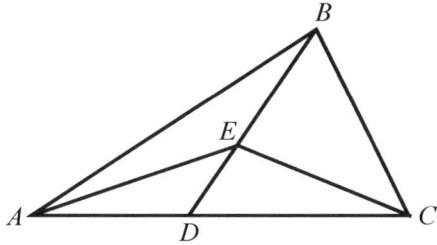

In the figure above, $BE : ED = 3 : 2$. The area of $\triangle BEC$ is 15 and the area of $\triangle BEA$ is 12. What is the area of $\triangle AEC$?

A) 15

B) 18

C) 20

D) 24

25

$$f(x) = a(x - b)^2 + k$$

In the function above, $a, b,$ and k are constants. If a and k are negative numbers, which of the following CANNOT be true?

A) $f(5) = -1$

B) $f(1) = k$

C) $f(2) = b$

D) $f(3) = 1$

26

$$(a - 2)x + (b + 2)y = 8$$
$$bx + ay = 4$$

In the system of equations above, a and b are constants. If the system has infinitely many solutions, what is the value of a?

A) $-\dfrac{4}{3}$

B) $-\dfrac{2}{3}$

C) $\dfrac{2}{3}$

D) $\dfrac{4}{3}$

27

A cylinder was altered by increasing the radius of its circular base by 10 percent and decreasing its height by k percent. If the volume of the resulting cylinder is 8.9% greater than the volume of the original cylinder, what is the value of k?

A) 8.9

B) 10

C) 12

D) 15

CONTINUE

▼

Questions 28 and 29 refer to the following information.

When Albert starts walking, Kimberly is 60 yards ahead of him. They are moving in the same direction on the same straight path. Albert walks 8 yards for every 4 yards that Kimberly walks. Albert walks 3 yards per second.

28

At these relative rates, in how many seconds will Albert catch up with Kimberly?

A) 20

B) 25

C) 30

D) 40

29

How many yards will Albert have to walk in order to catch up with Kimberly?

A) 100

B) 120

C) 240

D) 320

30

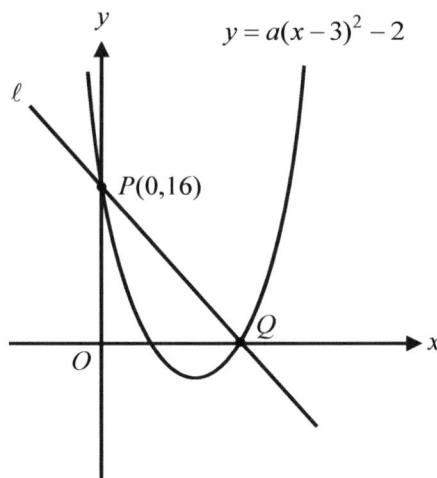

$$y = a(x-3)^2 - 2$$

$P(0,16)$

Note: Figure not drawn to scale.

In the xy-plane above, the graph of

$y = a(x-3)^2 - 2,$ where a is a constant, intersects line ℓ at points $P(0,16)$ and Q. What is the equation of line ℓ?

A) $y = -8x + 16$

B) $y = -4x + 16$

C) $y = -3x + 16$

D) $y = -2x + 16$

▲

CONTINUE

DIRECTION

For questions 31-38, solve the problem and enter your answer in the grid, as described below, on the answer sheet.

1. Although not required, it is suggested that you write your answer in the boxes at the top of the columns to help you fill in the circles accurately. You will receive credit only if the circles are filled in correctly.
2. Mark no more than one circle in any column.
3. No question has a negative answer.
4. Some problems may have more than one answer.
5. **Mixed numbers** such as $3\frac{1}{2}$ must be gridded as 3.5 or 7/2. (If $\boxed{3\,1\,/\,2}$ is entered into the grid, it will be interpreted as $\frac{31}{2}$, not $3\frac{1}{2}$.)
6. **Decimal answers:** If you obtain a decimal answer with more digits than the grid can accommodate, it may be either rounded or truncated, but it must fill the entire grid.

Answer: $\frac{7}{12}$

Write answer in boxes.

Fraction line

Grid in result.

Answer: 2.5

Decimal point

Acceptable ways to grid $\frac{2}{3}$ are:

Answer: 201
Either position is correct.

Note: You may start your answers in any column, space permitting. Columns you don't need to use should be left blank.

CONTINUE

31

$$h(x) = \frac{x-3}{x^3 - 3x^2 + x - 3}$$

For what value of x is the function h above undefined?

32

The state income tax where Alison lives is levied at the rate of $k\%$ of the first $\$30,000$ of annual income plus $(k+4)\%$ of any amount above $\$30,000$. This year Alison's income was $\$65,000$ and she paid $\$9,850$ for the income tax. What is the value of k?

33

$$b - 3 + (a - 5)i = a + 8i$$

In the equation above, a and b are real numbers. If $i = \sqrt{-1}$, what is the value of b?

34

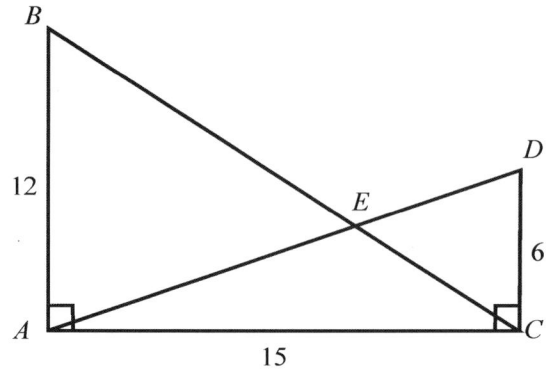

In the figure above, $AB = 12$, $AC = 15$, and $CD = 6$. Both $\overline{AB}$ and $\overline{CD}$ are perpendicular to $\overline{AC}$. If the area of $\triangle ABE$ is p and the area of $\triangle CDE$ is q, then what is the value of $p - q$?

35

$$h = -16t^2 + at$$

A football game begins with a kickoff. The formula for the kickoff is modeled by the equation above, where h is the height in feet of the football at t seconds and a is a constant. If the kickoff is in the air for 5 seconds, what is the value of a?

CONTINUE

36

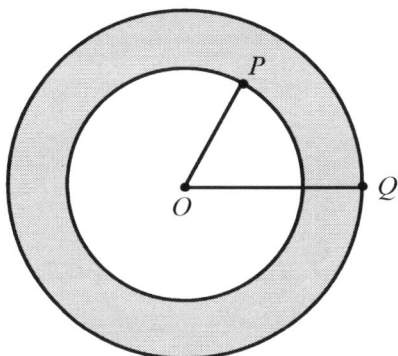

In the figure above, each of two circles has the same center O. If $OP : OQ = 3 : 5$ and the area of the shaded region is 40, what is the area of the larger circle?

37

$$f(x) = ax^2 + bx + c$$

In the function f above, $f(0) = 10$ and $f(-1) = 18$. What is the value of $a - b$?

38

In the xy-plane, the graphs of $y = -x^2 + 12$ and line ℓ intersect at points $P(p, 3)$ and $Q(q, -4)$. What is the greatest possible value of the slope of line ℓ?

STOP

If you finish before time is called, you may check your work on this section only.
Do not turn to any other section in the test.

Math Conversion Table

Raw Score	Scaled Score	Raw Score	Scaled Score
58	800	27	500
57	800	26	490
56	800	25	480
55	800	24	470
54	790	23	460
53	780	22	460
52	770	21	450
51	750	20	440
50	740	19	430
49	730	18	430
48	720	17	420
47	710	16	420
46	700	15	410
45	690	14	400
44	670	13	390
43	680	12	380
42	670	11	370
41	660	10	360
40	650	9	450
39	640	8	340
38	630	7	330
37	620	6	310
36	610	5	290
35	600	4	280
34	590	3	270
33	580	2	260
32	560	1	240
31	550	0	200
30	540		
29	530		
28	520		

Answer Explanations

Test 9 Answers and Explanations

<table>
<tr><th rowspan="3">SECTION
3</th><th>1</th><th>2</th><th>3</th><th>4</th><th>5</th><th>6</th><th>7</th><th>8</th><th>9</th><th>10</th></tr>
<tr><td>D</td><td>B</td><td>B</td><td>C</td><td>A</td><td>B</td><td>B</td><td>B</td><td>B</td><td>C</td></tr>
<tr><td>11</td><td>12</td><td>13</td><td>14</td><td>15</td><td>16</td><td>17</td><td>18</td><td>19</td><td>20</td></tr>
<tr><td></td><td>A</td><td>D</td><td>D</td><td>A</td><td>C</td><td>6</td><td>75</td><td>35</td><td>15</td><td>2</td></tr>
<tr><th rowspan="6">SECTION
4</th><th>1</th><th>2</th><th>3</th><th>4</th><th>5</th><th>6</th><th>7</th><th>8</th><th>9</th><th>10</th></tr>
<tr><td>C</td><td>C</td><td>B</td><td>B</td><td>C</td><td>A</td><td>D</td><td>A</td><td>D</td><td>A</td></tr>
<tr><td>11</td><td>12</td><td>13</td><td>14</td><td>15</td><td>16</td><td>17</td><td>18</td><td>19</td><td>20</td></tr>
<tr><td>D</td><td>D</td><td>B</td><td>D</td><td>B</td><td>C</td><td>D</td><td>C</td><td>B</td><td>C</td></tr>
<tr><td>21</td><td>22</td><td>23</td><td>24</td><td>25</td><td>26</td><td>27</td><td>28</td><td>29</td><td>30</td></tr>
<tr><td>B</td><td>A</td><td>A</td><td>B</td><td>D</td><td>C</td><td>B</td><td>D</td><td>B</td><td>B</td></tr>
<tr><td></td><td>31</td><td>32</td><td>33</td><td>34</td><td>35</td><td>36</td><td>37</td><td>38</td><td></td><td></td></tr>
<tr><td></td><td>3</td><td>13</td><td>16</td><td>45</td><td>80</td><td>62.5</td><td>8</td><td>7</td><td></td><td></td></tr>
</table>

SECTION 3

1. D

$|-4-5| = 9 \le 8 \text{(False)}$

2. B

$k = \sqrt{17^2 - 8^2} = \sqrt{225} = 15$

3. B

$SR = 3k$ and height $= 6k - k = 5k \rightarrow \text{Area} = \dfrac{3k \times 5k}{2} = 30 \rightarrow k^2 = 4 \rightarrow k = 2$ (k = positive)

4. C

Since $\sin a = \cos b$, $b = 30°$ and $a = 60°$. $BC = 10$ and $AC = 10\sqrt{3} \rightarrow$ Area is $\dfrac{10 \times 10\sqrt{3}}{2} = 50\sqrt{3}$.

5. A

$f(2) = 2 - p = 5 \rightarrow p = -3 \rightarrow f(2p) = f(-6) = -6 - (-3) = -3$

6. B

$a + 50 = b + 28 \rightarrow a - b = 28 - 50 = -28 \rightarrow |a - b| = 28$

7. B

1 2 2 3 3 4 4 **5** 5 5 6 6 7 7 8 $\rightarrow$ 5 is in the middle.

8. B

Average reading level is $\dfrac{4+5+6+3+5+7}{6} = \dfrac{30}{6} = 5$

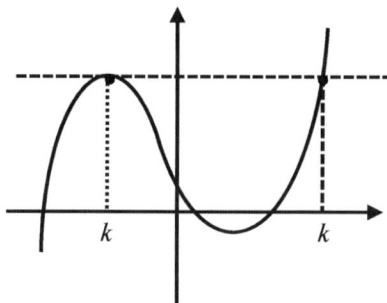

9. B

10. C

$p = \dfrac{50n-200}{n} + k \quad \rightarrow \quad np - nk = 50n - 200 \quad \rightarrow \quad 200 = 50n + nk - np \quad \rightarrow \quad 200 = n(50 + k - p)$

$\rightarrow \quad n = \dfrac{200}{50 + k - p}$

11. A

$P(-1) = 2 \quad \rightarrow \quad P(-1) = 3(-1)^3 + a(-1) - 2 = 2 \quad \rightarrow \quad -3 - a - 2 = 2 \quad \rightarrow \quad a = -7$

12. D

$by = -ax + 5 \quad \rightarrow \quad y = -\dfrac{a}{b}x + \dfrac{5}{b} \quad \rightarrow \quad$ negative slope and positive y-intercept $\rightarrow b > 0$ and $a > 0$

13. D

$\dfrac{2\pi}{5} = \dfrac{2(180)}{5} = 72°$, which is $\dfrac{1}{5}$ of $360°$. $\rightarrow \quad \dfrac{80\pi}{4} = \dfrac{A}{5} \quad \rightarrow \quad A = 100\pi \quad \rightarrow \quad r = 10 \; : \; A = $ area of the circle

14. A

line $\ell: y = x + b \quad \rightarrow \quad$ putting (1, 3) in the equation. $\rightarrow \quad 3 = 1 + b \rightarrow b = 2 \quad \rightarrow \quad y = x + 2$

line $m: y = -x + b \quad \rightarrow \quad$ putting (1, 3) in the equation. $\rightarrow \quad 3 = -1 + b \rightarrow b = 4$

Area of the triangle $= \dfrac{2 \times 1}{2} = 1$

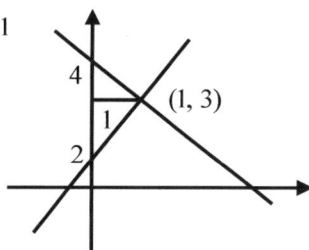

15. C

Long division: $\quad x-1\overline{)x^2+3}$ with quotient $x+1$ R 4 $\quad \rightarrow \quad \dfrac{x^2+3}{x-1} = x + 1 + \dfrac{4}{x-1}$

16. 6

$$\frac{1}{x-1}+\frac{1}{2x-2}=\frac{1}{4} \;\;\rightarrow\;\; \frac{1}{(x-1)}+\frac{1}{2(x-1)}=\frac{3}{2(x-1)}=\frac{1}{4} \;\;\rightarrow\;\; 2(x-1)=12 \;\;\rightarrow\;\; (x-1)=6$$

17. 75

The area of $\triangle ABC = \dfrac{10\times 20}{2}=100$

$\rightarrow$ The ratio of corresponding sides of $\triangle BMN:\triangle BAC=1:2$ $\rightarrow$ The ratio of their areas is 1:4.

Let their areas be k and $4k$. $\rightarrow$ $4k=100$ $\rightarrow$ $k=25$ $\rightarrow$ Area of the shaded region $=3k=75$

18. 35

$a-1=4$, $b-2=5$, and $ab=k$ $\rightarrow$ $a=5$ and $b=7$ $\rightarrow$ Therefore, $k=35$.

19. 15

At $x=5$, it has a minimum. $P(5)=2\sqrt{5-5}+3(5)=15$

20. 2

Substitution: $x^2+2x-8=0$ $\rightarrow$ $(x+4)(x-2)=0$ $\rightarrow$ $x=-4$ or $2\;(x\geq 0)$

Therefore, $y=\sqrt{2\times 2}=2$.

SECTION 4

1. C

For the first 3 minutes: $\$0.01a$, and for additional time $(t-3)$ minutes : $\$0.01r(t-3)$

2. C

3. B

$AB=\text{diameter}=\sqrt{6^2+8^2}=10$
Therefore, radius is 5.

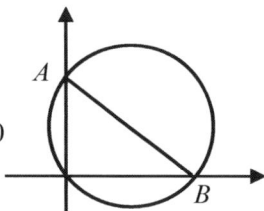

4. B

$f(x+4)=(x+a+4)^2+5$ $\rightarrow$ The graph was shifted by 4 left.

5. C

Claire: $y=40+10x$, and David: $y=80+8x$

$40+10x=80+8x$ $\rightarrow$ $2x=40$ $\rightarrow$ $x=a=20$ and $b=40+10(20)=240$

6. A

$200=40+10x$ $\rightarrow$ $x=16$, David: $y=80+8(16)=\$208$

7. D

Only choice D has three zeros such as $x = -2, 1,$ and 3.

8. A

Let x = number of pencils. $1.25 \times 15 + 0.3x \leq 35$ → $0.3x \leq 16.25$ → $x \leq 54.16$

maximum number is 54.

9. D

$$\begin{cases} x + y > 3 \;\rightarrow\; y > -x + 3 \\ ax + 2y < -2 \;\rightarrow\; y < -\dfrac{a}{2}x - 1 \end{cases} \;\rightarrow\; -1 = -\dfrac{a}{2} \;\rightarrow\; a = 2 \;\; : \text{Slopes must be same.}$$

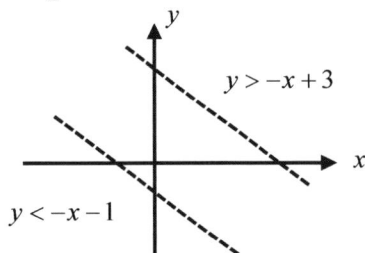

10. A

$$(x+2)(x-3) = 0 \;\rightarrow\; x^2 - x + 6 = 0 \;\rightarrow\; \text{Multiply by 4} \;\rightarrow\; 4x^2 - 4x - 24 = 0$$

Therefore, $a = 4$ and $b = -24$.

11. D

For 6 to 8 seconds, the average slope is the greatest.

12. D

Remainder Theorem

13. B

$$f(x) = 2x^2 - 16x + 18 \;\rightarrow\; f(x) = 2(x^2 - 8x + 16) + 18 - 32 \;\rightarrow\; f(x) = 2(x-4)^2 - 14 \;\rightarrow\; c = -14$$

14. D

$$\frac{5 - 10x}{2x + 1} = \frac{-5(2x+1) + 10}{2x + 1} = \frac{10}{2x + 1} - 5 \quad \text{Or, use long division.}$$

15. B

Temperatures in city B are spread widely.

16. C

$$f(x) = g(x) \;\rightarrow\; (x-4)^2 - 1 = -\frac{1}{4}(x-4)^2 + 4 \;\rightarrow\; \frac{5}{4}(x-4)^2 = 5 \;\rightarrow\; (x-4)^2 = 4$$

$$x - 4 = 2, -2 \;\rightarrow\; x = 6 \text{ or } 2$$

Answer Explanations

17. D

$$x^2 + y^2 - 4x + 4y = k \quad \to \quad (x^2 - 4x + 4) + (y^2 + 4y + 4) = k + 4 + 4 \quad \to \quad (x-2)^2 + (y+2)^2 = k + 8$$

$$r^2 = k + 8 = 6^2 \quad \to \quad k = 28$$

18. C

Average rate of change $= \dfrac{10 - 2}{8 - 1} = \dfrac{8}{7}$

19. B

Slope between $(7, 13)$ and $(2, 5)$ is $\dfrac{13 - 5}{7 - 2} = 1.6 \quad \to \quad y = 1.6x + b \quad \to \quad$ putting $(2, 5)$ in the equation

$5 = 1.6(2) + b \quad \to \quad b = 1.8 \quad \to \quad$ Therefore, $y = 1.6x + 1.8$.

20. C

$$f(a+b) = 2^{a+b+2} = 2^a \times 2^b \times 2^2 = 4\left(2^a \times 2^b\right)$$

21. B

$s =$ sum of the other three numbers $\quad \to \quad a + s = 783 \quad \to \quad a = 1.25s \quad \to \quad 1.25s + 5 = 783$

$\to \quad 2.25s = 783 \quad \to \quad s = 348 \quad \to \quad$ Therefore, $a = 1.25s = 1.25 \times 348 = 435$

22. A

$n =$ the number of students $\quad \to \quad$ the number of books $= 5n + 15 = 7(n - 5) \quad \to \quad 50 = 2n \quad \to \quad n = 25$

Therefore, the number of books $= 5(25) + 15 = 140$.

23. A

$y = -\dfrac{1}{3}x + 5 \quad \to \quad OP = 5$ and $OQ = 15 \quad \to \quad$ The area of the triangle $= \dfrac{5 \times 15}{2} = 37.5$

24. B

Since $\triangle BEA$ and $\triangle AED$ have the same height, the ratio of their aresa is also 3:2.

$\dfrac{\text{area of } \triangle BEA}{\text{area of } \triangle AED} = \dfrac{3}{2} = \dfrac{15}{x} \quad \to \quad x = 10$ and $\dfrac{\text{area of } \triangle BEC}{\text{area of } \triangle CED} = \dfrac{3}{2} = \dfrac{12}{y} \quad \to \quad y = 8$

Therefore, the area of $\triangle AEC$ is $10 + 8 = 18$.

25. D

Since a and k are negative, $y < 0$ for all value s of x.

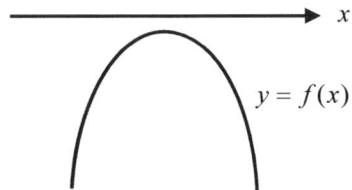

26. C

$\dfrac{a-2}{b} = \dfrac{b+2}{a} = \dfrac{8}{4} \quad \to \quad a - 2 = 2b$ and $b + 2 = 2a \quad \to \quad a = \dfrac{2}{3}$ and $b = -\dfrac{2}{3}$

27. B

$$V_1 = \pi r^2 h \text{ and } V_2 = \pi (1.1r)^2 \left(1 - \frac{k}{100}\right)h \rightarrow V_2 = 1.089V_1 \rightarrow \pi(1.1r)^2\left(1 - \frac{k}{100}\right)h = 1.089\pi r^2 h$$

$$1.21\left(1 - \frac{k}{100}\right) = 1.089 \rightarrow 1 - \frac{k}{100} = \frac{1.089}{1.21} = 0.9 \rightarrow k = 10$$

28. D

Kimberly walks 1.5 yards per second, because Albert walks 8 yards for every 4 yards that Kimberly walks.

For every second Albert walks 1.5 yard more than Kimberly. Therefore, $\frac{60}{1.5} = 40$ seconds.

29. B

$3 \times 40 = 120$ yards

30. B

First determine the value of a by putting $(0, 16)$ in the equation. $16 = a(0-3)^2 - 2 \rightarrow 18 = 9a \rightarrow a = 2$

Now find x-intercepts. $0 = 2(x-3)^2 - 2 \rightarrow (x-3)^2 = 1 \rightarrow x-3 = 1, -1 \rightarrow x = 4 \text{ or } 2 \rightarrow Q(4, 0)$

The equation of line $\ell \rightarrow$ slope $= \frac{16-0}{0-4} = -4 \rightarrow$ Therefore, $y = -4x + 16$

31. 3

Denominator: $x^3 - 3x^2 + x - 3 = 0 \rightarrow x^2(x-3) + (x-3) = (x-3)(x^2+1) = 0 \rightarrow x = 3 \ (x^2+1 \neq 0)$

32. 13

$$\frac{k}{100}(30,000) + \frac{k+4}{100}(65,000 - 30,000) = 9,850 \rightarrow 300k + (k+4)350 = 9,850 \rightarrow 650k = 8,450$$

$$k = \frac{8450}{650} = 13$$

33. 16

Since $b - 3 = a$ and $a - 5 = 8$, $a = 13$ and $b = 16$.

34. 45

Since $\triangle ABE$ and $\triangle CDE$ are similar, the ratio of corresponding sides are 12:6 or 2:1.

$$\frac{h_1}{h_2} = \frac{2}{1} \rightarrow h_1 + h_2 = 15 \rightarrow h_1 = 10 \text{ and } h_2 = 5$$

Therefore, $p - q = \frac{12 \times 10}{2} - \frac{6 \times 5}{2} = 60 - 15 = 45$

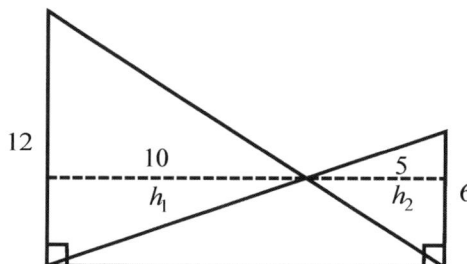

Answer Explanations

35. 80

Since $h(5) = 0$, $-16\left(5^2\right) + a(5) = 0$ → $400 = 5a$ → $a = 80$

36. 62.5

Since $OP:OQ = 3:5$, the ratio of their areas is 9:25. Let their areas $9k$ and $25k$. The area of the shaded region is $25k - 9k = 16k = 40$. → $k = 2.5$ → Therefore, the area of the larger circle is $25(2.5) = 62.5$.

37. 8

$f(0) = c = 10$ and $f(-1) = a(-1)^2 + b(-1) + c = a - b + c = 18$ → Therefore, $a - b = 18 - 10 = 8$.

38. 7

Point P: $-x^2 + 12 = 3$ → $x^2 = 9$ → $x = -3$ or 3 (two possible points)

Point Q: $-x^2 + 12 = -4$ → $x^2 = 16$ → $x = -4$ or 4 (two possible points)

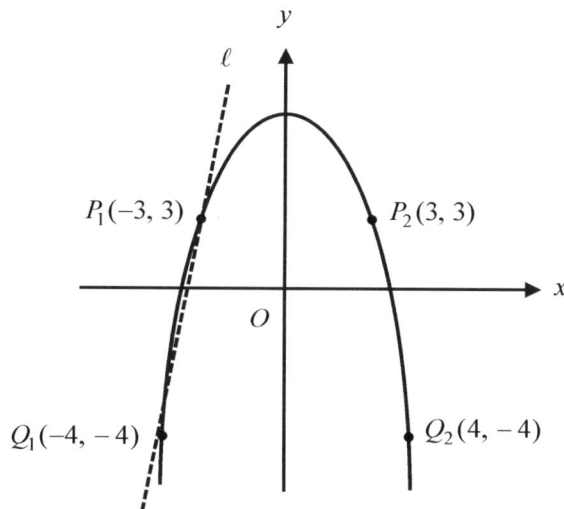

In the xy-plane above, line ℓ has the greatest slope of 7. $\dfrac{3 - (-4)}{-3 - (-4)} = 7$.

No Test Material on This Page

PRACTICE TEST 10

Dr. John Chung's SAT Math

Math Test - No Calculator

25 MINUTES, 20 QUESTIONS

Turn to Section 3 of your answer sheet to answer the questions in this section.

DIRECTIONS

For questions 1–15, solve each problem, choose the best answer from the choices provided, and fill in the corresponding circle on your answer sheet. **For questions 16–20,** solve the problem and enter your answer in the grid on your answer sheet. Please refer to the directions before question 16 on how to enter your answers in the grid. You may use any available space in your test booklet for scratch work.

NOTE

1. The use of a calculator **is not permitted**.

2. All variables and expressions used represent real numbers unless otherwise indicated.

3. Figures provided in this test are drawn to scale unless otherwise indicated.

4. All figures lie in a plane unless otherwise indicated.

5. Unless otherwise indicated, the domain of a given function f is the set of all real numbers x for which $f(x)$ is a real number.

REFERENCE

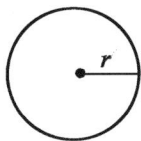 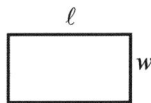 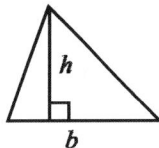 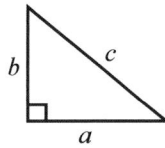 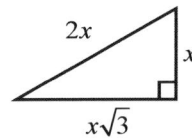 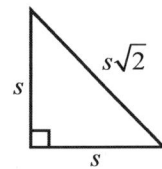

$A = \pi r^2$
$C = 2\pi r$

$A = \ell w$

$A = \dfrac{1}{2}bh$

$c^2 = a^2 + b^2$

Special Right Triangles

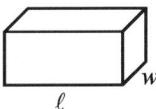 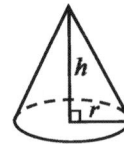 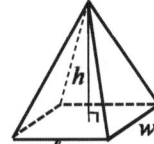

$V = \ell w h$

$V = \pi r^2 h$

$V = \dfrac{4}{3}\pi r^3$

$V = \dfrac{1}{3}\pi r^2 h$

$V = \dfrac{1}{3}\ell w h$

The number of degrees of arc in a circle is 360.
The number of radians of arc in a circle is 2π.
The number of the measures in degrees of the angles of a triangle is 180.

CONTINUE

1

If $\dfrac{5}{2x-3} = \dfrac{5}{x}$, what is the value of $2x-3$?

A) 0

B) 1

C) 3

D) 5

2

$$2x + y \le 3$$
$$x - y < -3$$

Which of the following ordered pairs (x, y) satisfies the system of inequalities above?

A) $(-1, 0)$

B) $(-1, -1)$

C) $(-2, 4)$

D) $(-3, 0)$

3

A salesman's commission is k percent of the selling price of a car. This week Peter, a salesman, sold 10 cars for $20,000 each. Which of the following represents the commission this week?

A) $200k$

B) $2,000k$

C) $\dfrac{20,000}{k}$

D) $\dfrac{20,000k}{100+k}$

4

Emily is walking a trail. After walking k percent of the length of the trail, she has 10 km left to go. Which of the following represents the length of the trail?

A) $10(100 - k)$

B) $\dfrac{100 - k}{10}$

C) $\dfrac{10k}{100 - k}$

D) $\dfrac{1000}{100 - k}$

CONTINUE

5

$$C(x) = 140,000 + 85x$$

A company that produces smart phones pays a start-up cost and a certain amount of money to produce each smart phone. The cost of producing x smart phones is given by the function above. What is the meaning of the value 85 in the function?

A) the start-up cost

B) the selling price of one smart phone

C) the amount spent to produce each smart phone

D) the profit earned from the sale of one smart phone

6

Which of the following equations has the same solution as $2x^2 + 12x - 32 = 0$?

A) $2(x+3)^2 = 32$

B) $2(x-3)^2 = 25$

C) $(x+3)^2 = 25$

D) $(x+3)^2 = 32$

7

The marketing department of a company estimates the price P, in dollars, of a smart phone by the equation $P = 500 - 25x$ over 10 years, where x is the number of years. What is the estimated decrease, in dollars, each year?

A) 20

B) 25

C) 100

D) 500

8

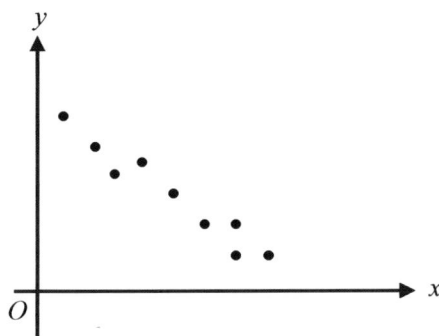

Which of the following best represents the correlation coefficient of the linear fit of the data shown above?

A) 0.95

B) −0.95

C) −1.00

D) −1.05

CONTINUE

9

$$2x - 3y = 6$$
$$y = x - 4$$

What is the solution (x, y) to the system of equations above?

A) $(3, -1)$

B) $(4, 0)$

C) $(5, 1)$

D) $(6, 2)$

10

Which of the following equations has a graph in the xy-plane for which y is always greater than 0?

A) $y = x + 2$

B) $y = (x - 2)^2$

C) $y = x^3 + 2$

D) $y = |x| + 2$

11

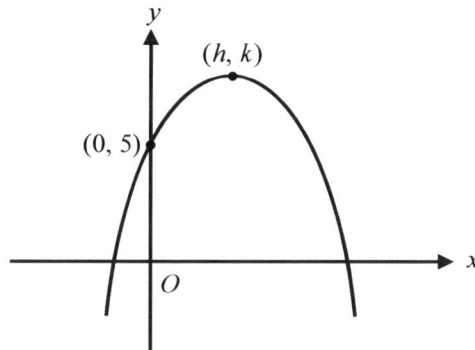

Note: Figure not drawn to scale.

The graph of $y = a(x + 1)(x - 5)$ is shown in the xy-plane above, where a is a constant. If the graph with vertex (h, k) intersects the y-axis at point $(0, 5)$, which of the following is equal to k?

A) 7

B) 8

C) 9

D) 10

12

If $k = \dfrac{(x + 1)(x - 1)}{3}$ and $k \neq 0$, what does $3x^2$ equal in terms of k?

A) $3k$

B) $9k$

C) $3k + 1$

D) $9k + 3$

CONTINUE

13

The average (arithmetic mean) of three positive numbers, a, b, and c is 15. When the greatest of these numbers is subtracted from the sum of the other two, the result is 5. If $a < b < c,$ what is the value of $a + b$?

A) 20

B) 25

C) 30

D) 40

14

$$\frac{a(x+1) + b(x-1)}{x-2} = 2 + \frac{1}{x-2}$$

The equation above is true for all values of $x \neq 2$, where a and b are constants. What is the value of a?

A) $-\dfrac{1}{2}$

B) 2

C) 3

D) 4

15

What are the solutions to $3(x-3)^2 - 6 = 14$?

A) $x = 3 \pm \sqrt{20}$

B) $x = \dfrac{3 \pm \sqrt{20}}{3}$

C) $x = 3 \pm \dfrac{\sqrt{20}}{3}$

D) $x = 3 \pm \dfrac{\sqrt{60}}{3}$

CONTINUE

DIRECTIONS

For questions 16–20, solve the problem and enter your answer in the grid, as described below, on the answer sheet.

Answer: $\dfrac{7}{12}$

Write answer in boxes. →

Answer: 2.5

Fraction line ←

Decimal point ←

Grid in result.

1. Although not required, it is suggested that you write your answer in the boxes at the top of the columns to help you fill in the circles accurately. You will receive credit only if the circles are filled in correctly.

2. Mark no more than one circle in any column.

3. No question has a negative answer.

4. Some problems may have more than one answer.

5. **Mixed numbers** such as $3\frac{1}{2}$ must be gridded as 3.5 or 7/2. (If `3 1 / 2` is entered into the grid, it will be interpreted as $\dfrac{31}{2}$, not $3\frac{1}{2}$.)

6. **Decimal answers:** If you obtain a decimal answer with more digits than the grid can accommodate, it may be either rounded or truncated, but it must fill the entire grid.

Acceptable ways to grid $\dfrac{2}{3}$ are:

Answer: 201
Either position is correct.

Note: You may start your answers in any column, space permitting. Columns you don't need to use should be left blank.

16

$$|x - 5| \le \frac{1}{2}$$

What is the least value of x that satisfies the inequality above?

17

If the diameter of a cylindrical jar is increased by 100% without altering the volume, by what percent must the height be decreased? (Note: Disregard the % sign when gridding your answer.)

18

$$f(x) = \frac{x^2}{2} - 20x + k$$

In the function f above, k is a constant. In the xy-plane, for what value of x does $f(x)$ have the same value of $f(10)$?

19

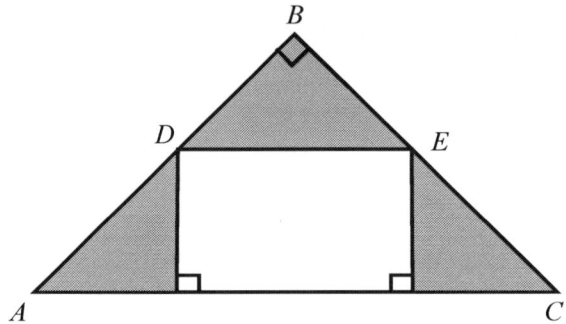

In the isosceles right triangle above, $AB = BC = 10\sqrt{2}$. Points D and E are the midpoints of $\overline{AB}$ and $\overline{BC}$ respectively. What is the area of the shaded region?

20

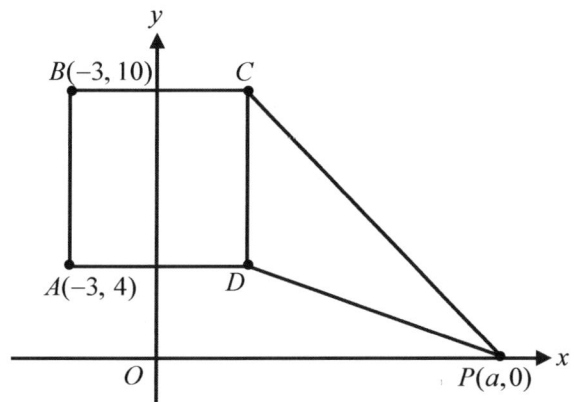

In the xy-plane above, the area of square $ABCD$ is equal to the area of triangle CDP. What is the value of a?

STOP

If you finish before time is called, you may check your work on this section only.
Do not turn to any other section in the test.

No Test Material on This Page

Math Test - Calculator

55 MINUTES, 38 QUESTIONS

Turn to Section 4 of your answer sheet to answer the questions in this section.

DIRECTIONS

For questions 1-30, solve each problem, choose the best answer from the choices provided, and fill in the corresponding circle on your answer sheet. **For questions 31-38,** solve the problem and enter your answer in the grid on your answer sheet. Please refer to the directions before question 31 on how to enter your answers in the grid. You may use any available space in your test booklet for scratch work.

NOTE

1. The use of a calculator **is permitted**.

2. All variables and expressions used represent real numbers unless otherwise indicated.

3. Figures provided in this test are drawn to scale unless otherwise indicated.

4. All figures lie in a plane unless otherwise indicated.

5. Unless otherwise indicated, the domain of a given function f is the set of all real numbers x for which $f(x)$ is a real number.

REFERENCE

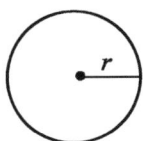 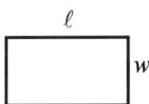 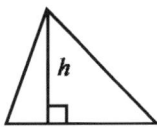 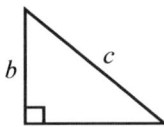 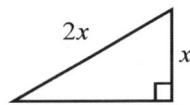 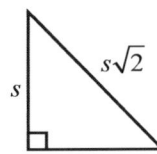

$A = \pi r^2$
$C = 2\pi r$

$A = \ell w$

$A = \frac{1}{2}bh$

$c^2 = a^2 + b^2$

Special Right Triangles

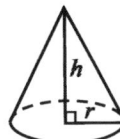 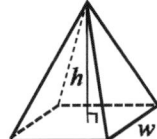

$V = \ell w h$

$V = \pi r^2 h$

$V = \frac{4}{3}\pi r^3$

$V = \frac{1}{3}\pi r^2 h$

$V = \frac{1}{3}\ell w h$

The number of degrees of arc in a circle is 360.
The number of radians of arc in a circle is 2π.
The number of the measures in degrees of the angles of a triangle is 180.

CONTINUE

1

An advertising medium charges d dollars for a basic fixed fee plus c cents for every 10 letters for an advertising campaign. If 300 letters are used for an advertising campaign, which of the following expressions represents the total amount, in dollars, of the advertisement?

A) $\dfrac{3c}{10} + d$

B) $3c + d$

C) $30c + d$

D) $300c + d$

2

$$f(x) = ax + b$$

In the function above, a and b are constants. If $f(0) = 3$ and $f(3) = -8$, what is the value of $f(6)$?

A) -22

B) -19

C) -16

D) -12

3

$$y = 2^x$$
$$y = x + 5$$

If ordered pair (x, y) is the solution to the system of equations above, what is the value of y?

A) 2

B) 8

C) 16

D) 32

4

If $f(x-5) = x^2 - 5$, which of the following is equal to $f(-2)$?

A) 4

B) 1

C) -1

D) -4

CONTINUE

5

If $a + b = 10$ and $\dfrac{1}{a} + \dfrac{1}{b} = 20$, what is the value of ab?

A) $\dfrac{1}{4}$

B) $\dfrac{1}{2}$

C) 2

D) 4

6

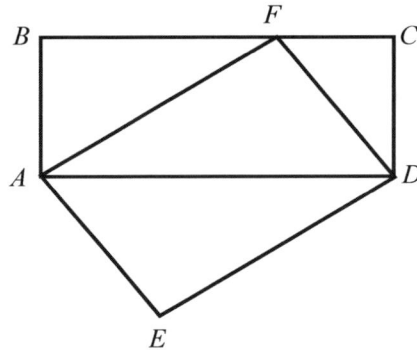

In the figure above, the area of rectangle $ABCD$ is 25. What is the area of parallelogram $AFDE$?

A) 12.5

B) 18

C) 25

D) 27.5

CONTINUE

Questions 7 and 8 refer to the following information.

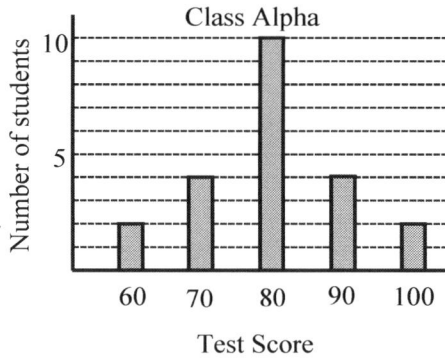

Class Alpha

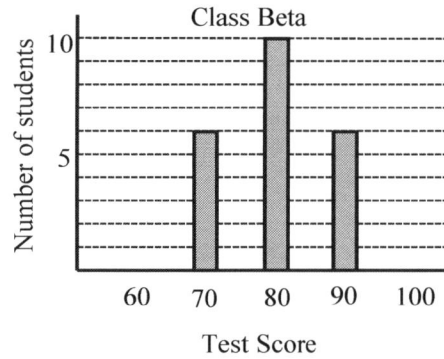

Class Beta

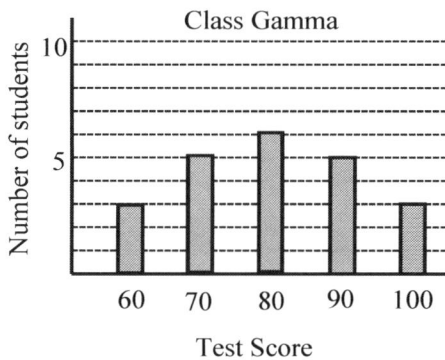

Class Gamma

The scores on a final reading test of three junior classes in a certain high school were shown on the bar graphs above.

7

Which class has the least standard deviation?

A) Class Alpha

B) Class Beta

C) Class Gamma

D) Based on the data, it cannot be determined.

8

What is the overall average score of these three combined classes?

A) 78

B) 80

C) 82

D) 84

CONTINUE

4 **4**

9

$$f(x) = x^2 - 8x + 12.$$

The function f is shown above. In the xy-plane, what are the coordinates of the vertex of the parabola defined by $g(x) = f(x-3)$?

A) $(-4, 7)$

B) $(4, 12)$

C) $(7, -4)$

D) $(7, 12)$

10

If a total of \$9,000 is invested at an annual interest rate of 2% compounded monthly, which of the following expressions shows the amount of interest after 10 years?

A) $9000\left(1 + \dfrac{2}{12}\right)^{10} - 9000$

B) $9000\left(1 + \dfrac{2}{120}\right)^{10} - 9000$

C) $9000\left(1 + \dfrac{2}{120}\right)^{120} - 9000$

D) $9000\left(1 + \dfrac{2}{1200}\right)^{120} - 9000$

11

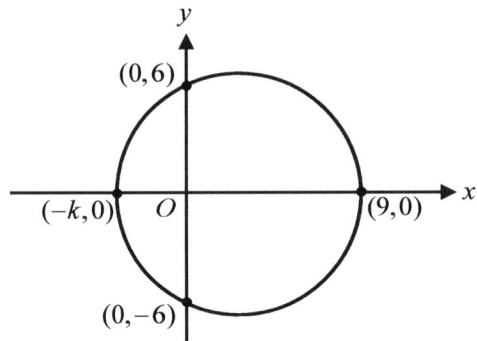

The graph of a circle in the xy-plane above intersects at four points with the x-axis and the y-axis. What is the value of k?

A) 4

B) 5

C) 6

D) 7

12

If $f(x-3) = x^2 + x + 1$, which of the following represents $f(x)$?

A) $f(x) = x^2 + x - 4$

B) $f(x) = (x-3)^2 + (x-3) + 1$

C) $f(x) = (x+3)^2 + (x+3) + 1$

D) $f(x) = (x+3)^2 + (x+3) + 3$

CONTINUE

13

If $3p + 5 \leq 15$, what is the greatest possible value of $6p - 5$?

A) 15

B) 25

C) 35

D) 85

14

Which of the following polynomials is divisible by $(x+1)$?

A) $x^3 - 1$

B) $x^3 - x^2 - x - 1$

C) $x^3 + x^2 - x + 1$

D) $x^3 + x^2 - x - 1$

15

Week	1	2	3	4	5	6	7
Height (feet)	1.5	1.7	1.8	2.2	2.9	3.7	4.8

Students in a science class observed the growth of a plant over 7 weeks. The table above shows their observations. What is the average rate of change, in feet per week, of the plant from weeks 1 to 7?

A) 0.42

B) 0.47

C) 0.55

D) 0.58

16

Ashley and Bernard work at an electronic appliance store. Ashley is paid $200 per week plus 5% of her total sales. Bernard is paid $325 per week plus 2.5% of his total sales. If their weekly pay is the same, what is the dollar amount of their sales?

A) 5,000

B) 6,200

C) 7,500

D) 8,400

CONTINUE

17

How does the graph of $f(x) = x^2 - 4x + 5$ compare with the graph of $g(x) = x^2$?

A) The graph of $g(x)$ is moved to the left 4 units and up 5 units.

B) The graph of $g(x)$ is moved to the right 4 units and up 5 units.

C) The graph of $g(x)$ is moved to the left 2 units and up 5 units.

D) The graph of $g(x)$ is moved to the right 2 units and up 1 unit.

18

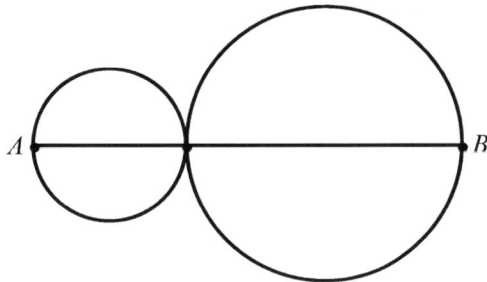

In the figure above, the circles are tangent each other and the radii are in a ratio of 1:2. If the sum of their areas is 80π, what is the length of $\overline{AB}$?

A) 12

B) 16

C) 18

D) 24

19

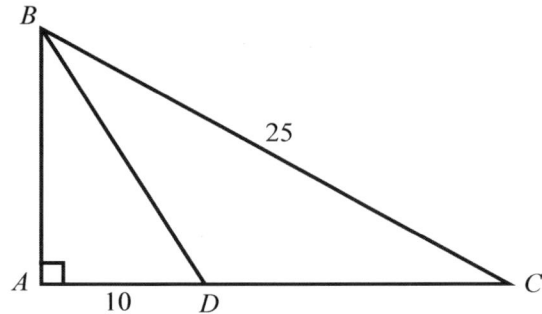

In right triangle ABC above, $AD = 10$ and $BC = 25$. If the value of $\sin \angle BCD$ is 0.6, what is the area of triangle BCD?

A) 50

B) 75

C) 100

D) 150

20

If $p = a^2 - 4a + 8$, what is the least possible value of $p + 6$?

A) 2

B) 4

C) 8

D) 10

CONTINUE

Practice Test 10

21

$$80 \le x \le 100$$
$$40 \le y \le 60$$

The intervals of x and y are shown above. If $z = x - y$, which of the following represents all possible values of z ?

A) $|z - 40| \le 20$

B) $|z - 40| \ge 20$

C) $|z - 20| \le 40$

D) $|z - 20| \ge 40$

22

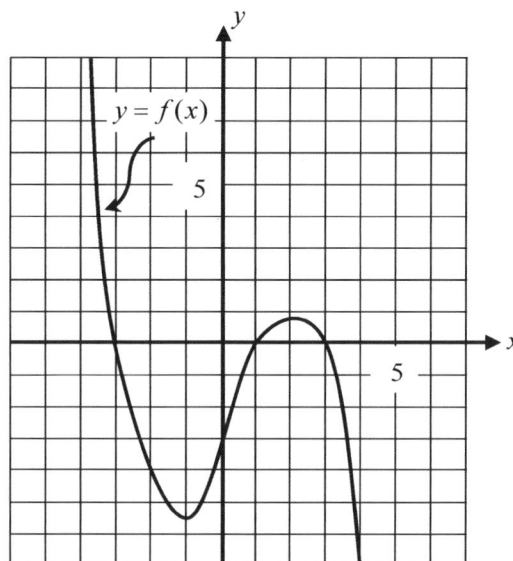

Which of the following functions could represent the graph of $f(x)$ shown in the xy-plane above?

A) $f(x) = \dfrac{1}{3}(x-3)(x^2 - 4x + 3)$

B) $f(x) = -\dfrac{1}{3}(x-3)(x^2 - 2x - 3)$

C) $f(x) = -\dfrac{1}{3}(x+3)(x^2 - 4x + 3)$

D) $f(x) = -\dfrac{1}{3}(x+3)(x^2 + 4x + 3)$

CONTINUE

23

$$f(x) = (x-4)^2 - 64$$

Which of the following is an equivalent form of the function above?

A) $f(x) = (x+3)(x-11)$

B) $f(x) = (x+6)(x-14)$

C) $f(x) = (x+4)(x-12)$

D) $f(x) = (x+6)(x-8)$

24

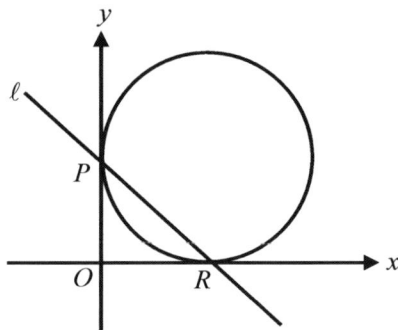

In the xy-plane above, a circle is tangent to the x-axis at R and the y-axis at P, and line ℓ passes through the points of tangency. If the area of the circle is 100π, what is the equation of line ℓ?

A) $y = -x + 5$

B) $y = -x + 10$

C) $y = -x + 50$

D) $y = -x + 100$

25

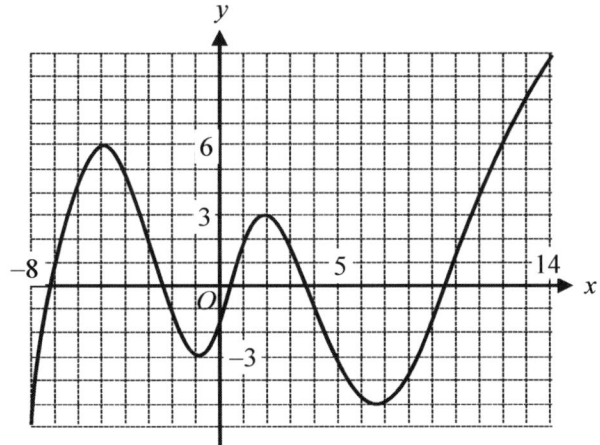

$$y = f(x)$$
$$y = k$$

In the system of equations above, k is a constant. The function $y = f(x)$ is shown in the xy-plane above for $-8 \le x \le 14$. On this closed interval, for how many values of k does the system have exactly 4 solutions?

A) 1

B) 2

C) 3

D) 4

26

Let the function f be defined by $f(x) = \sqrt{50 - 2x^2}$. What are all the values of x for which $f(x)$ is a real number?

A) $x \ge 5$

B) $x \le 5$

C) $-25 \le x \le 25$

D) $-5 \le x \le 5$

CONTINUE

27

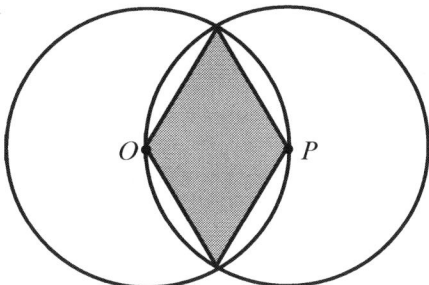

In the figure above, O and P are the centers of the circles. If the lengths of radii of the circles are each 10, what is the area of the shaded region?

A) $50\sqrt{3}$

B) $25\sqrt{3}$

C) $\dfrac{25\sqrt{3}}{2}$

D) $\dfrac{25\sqrt{3}}{4}$

28

In the xy-plane, the graph of the function is a line with a slope of 5. If $f(a) = -4$ and $f(b) = 32$, what is the value of $b - a$?

A) 6

B) 7.2

C) 8

D) 8.4

CONTINUE

Questions 29 and 30 refer to the following information.

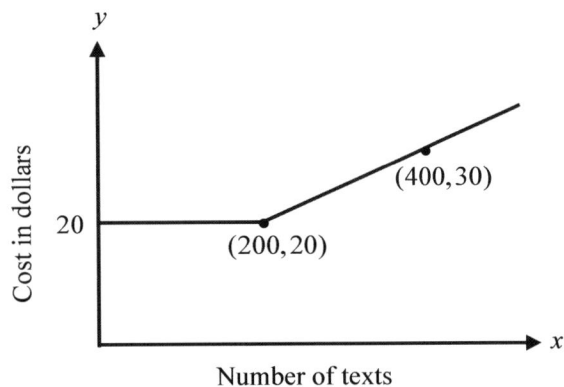

The domestic texting plan of an E-mobile telephone company is modeled by the graph in the xy-plane above.

29

Which of the following pairs of equations represents the graph of the domestic texting plan?

A) $\begin{cases} y = 20, & x \le 200 \\ y = 0.05x, & x > 200 \end{cases}$

B) $\begin{cases} y = 20, & x \le 200 \\ y = 20 + 0.05x, & x > 200 \end{cases}$

C) $\begin{cases} y = 20, & x \le 200 \\ y = 20 + 0.05(x - 200), & x > 200 \end{cases}$

D) $\begin{cases} y = 20, & x \le 200 \\ y = 30, & x > 200 \end{cases}$

30

If Jennifer uses 550 texts this month, what is her amount of money, in dollars, does she have to pay?

A) 20.00

B) 25.00

C) 32.50

D) 37.50

CONTINUE

Answer: $\frac{7}{12}$

Answer: 2.5

Write answer in boxes. → Fraction line ← Decimal point

Grid in result. →

DIRECTIONS

For questions 31-38, solve the problem and enter your answer in the grid, as described below, on the answer sheet.

1. Although not required, it is suggested that you write your answer in the boxes at the top of the columns to help you fill in the circles accurately. You will receive credit only if the circles are filled in correctly.
2. Mark no more than one circle in any column.
3. No question has a negative answer.
4. Some problems may have more than one answer.
5. **Mixed numbers** such as $3\frac{1}{2}$ must be gridded as 3.5 or 7/2. (If $3 1 / 2$ is entered into the grid, it will be interpreted as $\frac{31}{2}$, not $3\frac{1}{2}$.)
6. **Decimal answers:** If you obtain a decimal answer with more digits than the grid can accommodate, it may be either rounded or truncated, but it must fill the entire grid.

Acceptable ways to grid $\frac{2}{3}$ are:

Answer: 201
Either position is correct.

Note: You may start your answers in any column, space permitting. Columns you don't need to use should be left blank.

CONTINUE

31

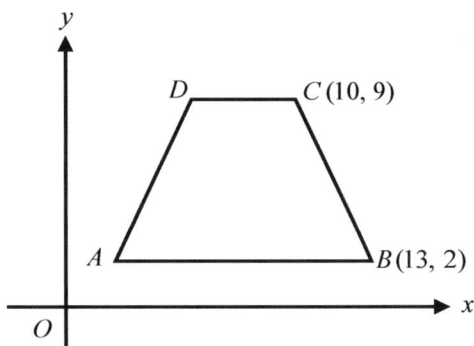

A trapezoid ADCB is in the xy-plane above. If $AD = BC$, what is the slope of $\overline{AD}$?

32

If a and b are positive integers such that $\dfrac{a}{b} = 0.48$. If $150 < b < 200$, what is the value of $a + b$?

33

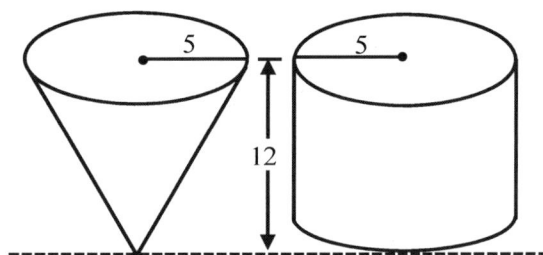

In the figure above, the cylindrical and cone-shaped containers have the same height of 12 inches and the same radius of 5 inches. If the cone-shaped container filled with water and then the water is poured into the empty cylindrical container, what will be the depth, in inches, of the water in the cylindrical container?

34

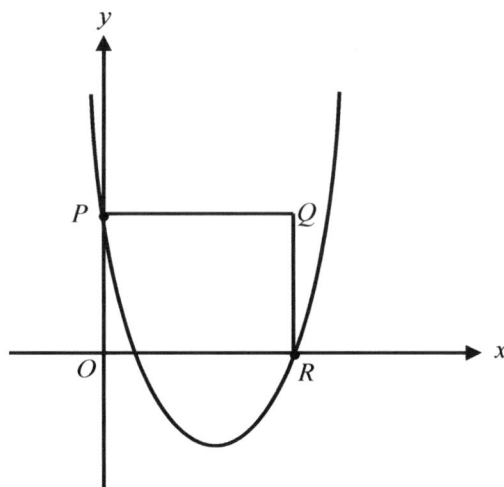

Note: Figure not drawn to scale.

In the xy-plane above, the graph of $y = 2x^2 - 19x + 9$ intersects the y-axis at P and the x-axis at R. What is the area of rectangle $OPQR$?

35

$$y \geq x^2 - 8x$$
$$y \leq 2x$$

In the xy-plane, ordered pair (a, b) is the solution of the system of inequalities above. What is the maximum possible value of b?

CONTINUE

36

$$-6 \le x \le 20$$

If the interval above is rewritten in the form $|x - a| \le k,$ what is the value of k?

37

Mr. Trump drove to work in the morning at the average speed of 60 miles per hour. He returned home in the evening along the same route and averaged 45 miles per hour. To the nearest tenth, what is his average speed, in miles per hour, for the entire trip?

38

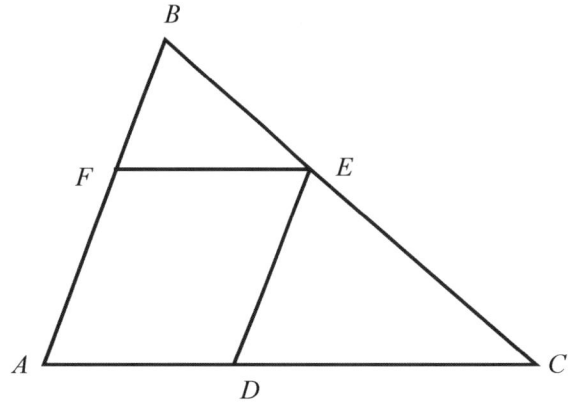

In the figure above, quadrilateral $AFED$ is a parallelogram and $\dfrac{BE}{EC} = \dfrac{1}{2}.$ If the area of the parallelogram is 10, what is the area of triangle ABC?

STOP

**If you finish before time is called, you may check your work on this section only.
Do not turn to any other section in the test.**

Math Conversion Table

Raw Score	Scaled Score	Raw Score	Scaled Score
58	800	27	500
57	800	26	490
56	800	25	480
55	800	24	470
54	790	23	460
53	780	22	460
52	770	21	450
51	750	20	440
50	740	19	430
49	730	18	430
48	720	17	420
47	710	16	420
46	700	15	410
45	690	14	400
44	670	13	390
43	680	12	380
42	670	11	370
41	660	10	360
40	650	9	450
39	640	8	340
38	630	7	330
37	620	6	310
36	610	5	290
35	600	4	280
34	590	3	270
33	580	2	260
32	560	1	240
31	550	0	200
30	540		
29	530		
28	520		

Answer Explanations

Test 10 Answers and Explanations

SECTION 3	1	2	3	4	5	6	7	8	9	10
	C	C	B	D	C	C	B	B	D	D
	11	12	13	14	15	16	17	18	19	20
	C	D	B	A	D	4.5	75	30	50	15

SECTION 4	1	2	3	4	5	6	7	8	9	10
	A	B	B	A	B	C	B	B	C	D
	11	12	13	14	15	16	17	18	19	20
	A	C	A	D	C	A	D	D	B	D
	21	22	23	24	25	26	27	28	29	30
	A	C	C	B	C	D	A	B	C	D
	31	32	33	34	35	36	37	38		
	$\frac{7}{3}$	259	4	81	20	13	51.4	22.5		

SECTION 3

1. C

$2x - 3 = x \rightarrow x = 3 \rightarrow 2x - 3 = 3$

2. C

Put the numbers in the inequalities and check. $(-2, 4) \rightarrow 2(-2) + 4 = 0 \le 3 \rightarrow (-1) - 4 = -5 < -3 \text{(OK)}$

3. B

$\frac{k}{100}(20,000 \times 10) = 2000k$

4. D

$\frac{100-k}{10} = \frac{100}{x} \rightarrow x = \frac{1000}{100-k}$.   Or $(100-k)\%$ of $x = 10 \rightarrow \frac{100-k}{100}x = 10 \rightarrow x = \frac{1000}{100-k}$

5. C

Slope

6. C

$2x^2 + 12x - 32 = 0 \rightarrow x^2 + 6x = 16 \rightarrow x^2 + 6x + (9) = 16 + (9) \rightarrow (x+3)^2 = 25$

7. B

Answer Explanations

8. B

Since the date are not exactly on the line and are correlated negatively. The best answer is -0.95.

9. D

Substitution: $2x - 3(x - 4) = 6 \rightarrow x = 6$ and $y = 6 - 4 = 2 \rightarrow (6, 2)$

10. D

11. C

First determine the value of a using $(0, 5)$. $\rightarrow 5 = a(0 + 1)(0 - 5) \rightarrow a = -1$

From the equation: Two zeros $x = -1$ and $5 \rightarrow h = \dfrac{-1 + 5}{2} = 2$ and $k = -(2 + 1)(2 - 5) = 9$

12. D

$3k = x^2 - 1 \rightarrow x^2 = 3k + 1 \rightarrow 3x^2 = 9k + 3$

13. B

$a + b + c = 45$ and $a + b - c = 5$ Addition: $2(a + b) = 50 \rightarrow a + b = 25$

14. A

$a(x + 1) + b(x - 1) = 2(x - 2) + 1 \rightarrow (a + b)x + (a - b) = 2x - 3 \rightarrow a + b = 2$ and $a - b = -3$

When you add these two equations: $2a = -1 \rightarrow a = -\dfrac{1}{2}$

15. D

$3(x - 3)^2 = 20 \rightarrow (x - 3)^2 = \dfrac{20}{3} \rightarrow x - 3 = \pm\sqrt{\dfrac{20}{3}} \rightarrow x = 3 \pm \sqrt{\dfrac{60}{9}} = 3 \pm \dfrac{\sqrt{60}}{3}$

16. 4.5

$|x - 5| \leq \dfrac{1}{2} \rightarrow -\dfrac{1}{2} \leq x - 5 \leq \dfrac{1}{2} \rightarrow 4.5 \leq x \leq 5.5 \rightarrow$ The least value is 4.5.

Or, just simply $5 - \dfrac{1}{2} = 4.5$, because midpoint is 5.

17. 75

Since the diameter is increased by 100%, the radius also is increased by 100%. Therefore,

$\pi r^2 h = \pi (2r)^2 h' \rightarrow h' = \dfrac{r^2}{4r^2} h = \dfrac{1}{4} h = 0.25h \rightarrow h' = (1 - 0.75)h$: 75% of the height will be decreased.

18. 30

Axis of symmetry: $x = \dfrac{-(-20)}{2(1/2)} = 20 \rightarrow$ 20 is the midpoint of 10 and x. $\rightarrow \dfrac{10 + x}{2} = 20 \rightarrow x = 30$

19. 50

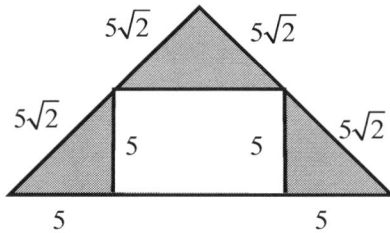

The area of the shaded region $= \dfrac{5\sqrt{2} \times 5\sqrt{2}}{2} + \dfrac{5 \times 5}{2} + \dfrac{5 \times 5}{2} = 50$

20. 15

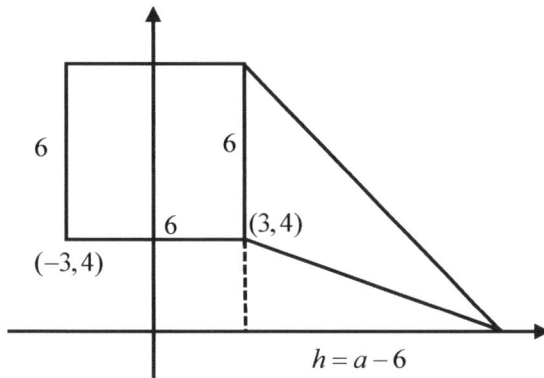

The area of the triangle $= \dfrac{6 \times (a-3)}{2} = 36 \;\rightarrow\; a-3 = 12 \;\rightarrow\; a = 15$

SECTION 4

1. A

Total amount will be $\$d + \left(\dfrac{300}{10}\right) \times \$\dfrac{c}{100} = \$\left(d + \dfrac{3c}{10}\right)$

2. B

Constant slope: Let the value of $f(6) = k$.

$(0, 3), (3, -8), (6, k) \;\rightarrow\; \text{slope} = \dfrac{-8-3}{3-0} = \dfrac{k-3}{6-0} \;\rightarrow\; 3k - 9 = -66 \;\rightarrow\; 3k = -57 \;\rightarrow\; k = -19$

3. B

When $x = 3, \; y = 8.$

Answer Explanations

4. A

When $x = 3$, $f(3-5) = 3^2 - 5 \rightarrow f(-2) = 4$.

5. B

$$\frac{1}{a} + \frac{1}{b} = \frac{a+b}{ab} = 20 \rightarrow \frac{10}{ab} = 20 \rightarrow ab = \frac{1}{2}$$

6. C

The area of $\triangle AFD$ is exactly half of the area of the rectangle. Therefore, the area of the parallelogram is $12.5 \times 2 = 25$.

7. B

Class Beta has the data closet to the mean. Class Gamma has a greatest standard deviation. You can check using a calculator.

8. B

Each class has the same average score of 80.

9. C

$$f(x) = x^2 - 8x + 12 = (x-4)^2 - 4 \rightarrow \text{vertex} (4, -4)$$

Move to the right by 3. $\rightarrow$ the vertex of $g(x)$ is $(7, -4)$.

10. D

$A = p\left(1 + \dfrac{r/100}{n}\right)^{nt}$ for annual interest of $r\%$. When $n = 12$ and $r = 2$, $\text{Interest} = 9000\left(1 + \dfrac{2}{1200}\right)^{120} - 9000$.

11. A

$9 \times k = 6 \times 6 \rightarrow k = 4$

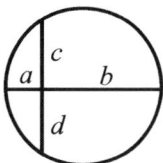

$a \times b = c \times d$

12. C

Replace x with $x + 3$. $\rightarrow$ $f((x+3)-3) = (x+3)^2 + (x+3) + 1$

13. A

$3p + 5 \leq 15 \rightarrow 3p \leq 10 \rightarrow 6p \leq 20 \rightarrow 6p - 5 \leq 20 - 5 \rightarrow 6p - 5 \leq 15$

Therefore, the greatest possible value is 15.

14. D (Factor Theorem)

When you putting $x = -1$, only choice D results in 0. $(-1)^3 + (-1)^2 - (-1) - 1 = -1 + 1 + 1 - 1 = 0$

15. C

$$\text{Average rate of change} = \text{Slope} = \frac{4.8 - 1.5}{7 - 1} = 0.55 \text{ feet/week}$$

Answer Explanations

16. A

$$200 + 0.05s = 325 + 0.025s \quad \rightarrow \quad 0.025s = 125 \quad \rightarrow \quad s = \$5,000$$

17. D

$$f(x) = x^2 - 4x + 5 \quad \rightarrow \quad f(x) = (x-2)^2 + 1 \quad \rightarrow \quad f(x) = g(x-2) + 1 \quad \text{means: move to the right by 2 and up by 1.}$$

18. D

The ratio of corresponding sides $= 1:2 \quad \rightarrow \quad$ the ratio of areas $= 1:4$. Let the areas of the circles be k and $4k$, then $k + 4k = 5k = 80\pi \quad \rightarrow \quad k = 16\pi \, (= \pi r_1^2)$ and $4k = 64\pi \, (= \pi r_2^2) \quad \rightarrow$ Therefore, $r_1 = 4$ and $r_2 = 8$.

19. B

$$AB = 25 \sin \angle BCD = 25 \times 0.6 = 15 \quad \rightarrow \quad AC = \sqrt{25^2 - 15^2} = \sqrt{400} = 20 \quad \rightarrow \quad CD = 20 - AD = 10$$

Therefore, the area of $\triangle BCD = \dfrac{10 \times 15}{2} = 75$.

20. D

$$p = a^2 - 4a + 8 = (a-2)^2 + 4 \quad \rightarrow \quad \text{The minimum of } p \text{ is } 4. \quad \rightarrow \quad \text{The minimum of } p+6 \text{ is } 4+6 = 10.$$

21. A

$$20 \le x - y \le 60 \quad \rightarrow \quad \text{mid point} = \dfrac{20 + 60}{2} = 40, \text{ the distance from mid point to the end point is } 60 - 40 = 20.$$

Therefore, $|z - \text{midpoint}| \le \text{distance} \quad \rightarrow \quad |z - 40| \le 20$

22. C

$$\frac{1}{3}(x+3)(x^2 - 4x + 3) = 0 \quad \rightarrow \quad (x+3)(x-1)(x-3) = 0 \quad \rightarrow \quad \text{zeros } x = -3, 1, 3$$

23. C

$$f(x) = (x-4)^2 - 64 = x^2 - 8x + 16 - 64 = x^2 - 8x - 48 \quad \rightarrow \quad f(x) = (x+4)(x-12)$$

24. B

$$\pi r^2 = 100\pi \quad \rightarrow \quad r = 10 \quad \rightarrow \quad OP = OR = 10 \quad \rightarrow \text{Slope} = -\dfrac{10}{10} = -1 \quad \rightarrow \quad \text{Therefore, } y = -x + 10.$$

25. C

$y = 3$ and $y = -3$ have exactly 4 points of intersection with $y = f(x)$.

26. D

$$50 - 2x^2 \ge 0 \quad \rightarrow \quad x^2 - 25 \le 0 \quad \rightarrow \quad (x+5)(x-5) \le 0 \quad \rightarrow \quad -5 \le x \le 5$$

Answer Explanations

27. A

$OP = 10$

Area of an equilateral triangle is $\dfrac{s^2\sqrt{3}}{4}$.

Therefore, the area of the shades region is $\dfrac{10^2\sqrt{3}}{4}\times 2 = 50\sqrt{3}$

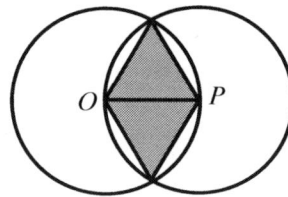

28. B

$\text{Slope} = \dfrac{f(b)-f(a)}{b-a} = 5 \;\rightarrow\; \dfrac{32-(-4)}{b-a} = 5 \;\rightarrow\; \dfrac{36}{b-a} = 5 \;\rightarrow\; b-a = 7.2$

29. C

When $x > 200$, $y = 20 + 0.05(x - 200)$

30. D

$y = 20 + 0.05(550 - 200) = 20 + 17.5 = \$37.5$

31. $\dfrac{7}{3}$

The slop of $\overline{BC} = \dfrac{9-2}{10-13} = \dfrac{7}{-3}$. Therefore, the slope of $\overline{AD}$ is $\dfrac{7}{3}$.

32. 259

$\dfrac{a}{b} = \dfrac{48}{100} = \dfrac{12}{25} \rightarrow$ Let $a = 12k$ and $b = 25k \;\rightarrow\; 150 < 25k < 200 \;\rightarrow\; 6 < k < 8 \;\rightarrow\; k = 7$

Therefore, $a = 12\times 7 = 84$ and $b = 25\times 7 = 175$. $\;\rightarrow\; a + b = 259$

33. 4

$\dfrac{\pi(5^2)(12)}{3} = \pi(5^2)h \;\rightarrow\; h = 4$

34. 81

If $x = 0, y = 9$. $\;\rightarrow\; P(0, 9)$ For x-intercept, $0 = 2x^2 - 19x + 9 = (2x-1)(x-9) \;\rightarrow\; x = \dfrac{1}{2}, 9$

Therefore, the area of the rectangle is $8\times 8 = 81$.

35. 20

$x^2 - 8x = 2x \;\rightarrow\; x^2 - 10x = 0$

$\rightarrow\; x(x-10) = 0 \;\rightarrow\; x = 0, 10$

$b = 2a \;\rightarrow\; b = 2\times 10 = 20$

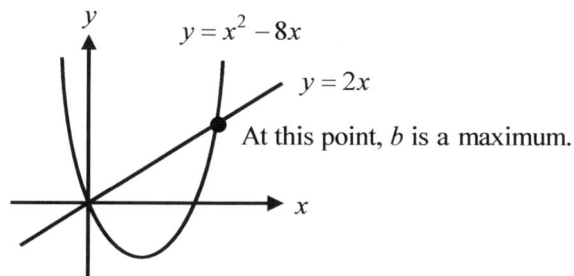

At this point, b is a maximum.

36. 13

$a = \dfrac{-6+20}{2} = 7 \;\rightarrow\; k = 20 - 7 = 13$

37. 51.4

Let distance $= 180$ miles. Average speed $\dfrac{\text{Total distance}}{\text{Total time}} = \dfrac{180 + 180}{180/60 + 180/45} = 51.42857\cdots \approx 51.4$

38. 22.5

The ratio of areas $\triangle BEF : \triangle ECD : \triangle ABC = 1^2 : 2^2 : 3^2 = 1 : 4 : 9$, because the ratio of corresponding sides is 1:2:3.

Let the areas of $\triangle BEF = k$, $\triangle ECD = 4k$, and $\triangle ABC = 9k$.

$9k - k - 4k = 4k = 10 \quad \rightarrow \quad k = 2.5 \quad \rightarrow \quad$ Area of $\triangle ABC = 9k = 9 \times 2.5 = 22.5$

No Test Material on This Page

PRACTICE TEST 11

Dr. John Chung's SAT Math

Math Test - No Calculator

25 MINUTES, 20 QUESTIONS

Turn to Section 3 of your answer sheet to answer the questions in this section.

DIRECTIONS

For questions 1–15, solve each problem, choose the best answer from the choices provided, and fill in the corresponding circle on your answer sheet. **For questions 16–20**, solve the problem and enter your answer in the grid on your answer sheet. Please refer to the directions before question 16 on how to enter your answers in the grid. You may use any available space in your test booklet for scratch work.

NOTE

1. The use of a calculator **is not permitted**.

2. All variables and expressions used represent real numbers unless otherwise indicated.

3. Figures provided in this test are drawn to scale unless otherwise indicated.

4. All figures lie in a plane unless otherwise indicated.

5. Unless otherwise indicated, the domain of a given function f is the set of all real numbers x for which $f(x)$ is a real number.

REFERENCE

$A = \pi r^2$
$C = 2\pi r$

$A = \ell w$

$A = \dfrac{1}{2} bh$

$c^2 = a^2 + b^2$

Special Right Triangles

$V = \ell wh$

$V = \pi r^2 h$

$V = \dfrac{4}{3}\pi r^3$

$V = \dfrac{1}{3}\pi r^2 h$

$V = \dfrac{1}{3}\ell wh$

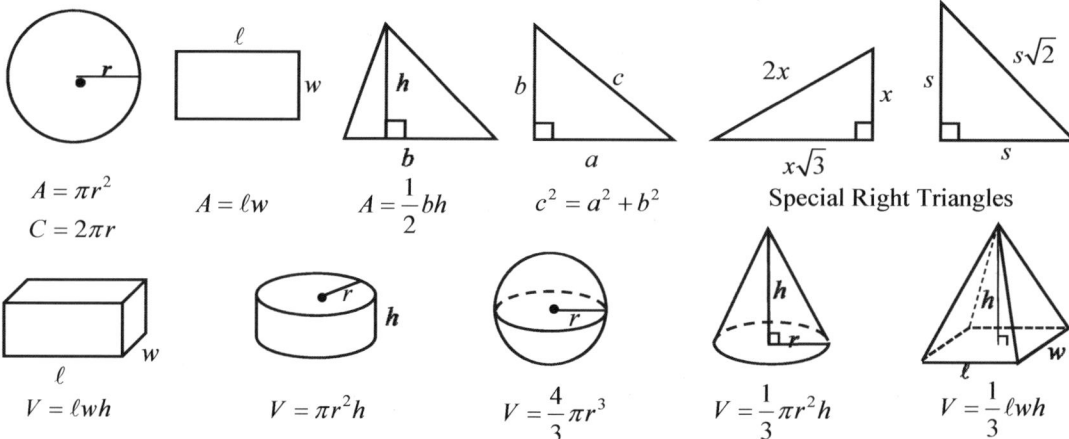

The number of degrees of arc in a circle is 360.
The number of radians of arc in a circle is 2π.
The number of the measures in degrees of the angles of a triangle is 180.

CONTINUE

1

$$\frac{3}{4}(x-2)+7=12-\frac{1}{2}(x-2)$$

What is the value of x in the equation above?

A) 4

B) 5

C) 6

D) 8

2

$$5(x^2-6)=x^2+34$$

What are the solutions to the equation above?

A) $-\sqrt{34}$ and $\sqrt{34}$

B) $-2\sqrt{6}$ and $2\sqrt{6}$

C) -6 and 6

D) -4 and 4

3

Albert and Bunny are 100 feet apart. Each person walks at a steady pace toward the other. If Albert walks at a constant rate of 3 feet per second and Bunny walks at a constant rate of 4 feet per second, how far will Albert walk, in feet, when they meet?

A) $\frac{100}{7}$

B) $\frac{300}{7}$

C) $\frac{400}{7}$

D) 70

4

Which of the following is equivalent to the expression x^3+x^2y-x-y ?

A) $(x+y)(x^2+1)$

B) $(x+y)(x^2-2y)$

C) $(xy+1)(x^2+1)$

D) $(x+y)(x+1)(x-1)$

CONTINUE

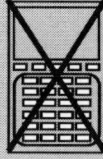

5

Which of the following values is a solution of

$$\frac{1}{5}(x-5)(x-3) < 0 ?$$

A) −4

B) 3

C) 4

D) 5

6

$$(x-5)(x+3)+10$$

Which of the following is equivalent to the expression above?

A) $(x+1)^2 + 6$

B) $(x+1)^2 - 6$

C) $(x-1)^2 + 6$

D) $(x-1)^2 - 6$

7

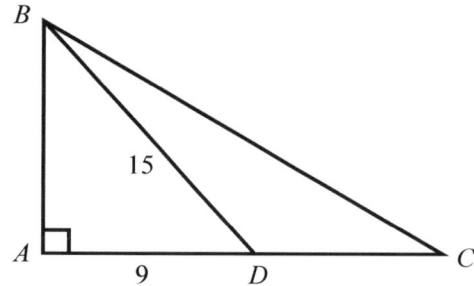

In the figure above, the area of triangle ABC is 96. If $AD = 9$ and $BD = 15,$ what is the length of $\overline{CD}$?

A) 7

B) 8

C) 9

D) 14

8

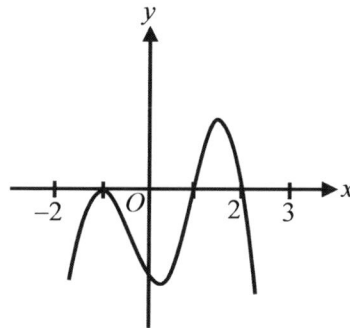

Which of the following could be the equation for the graph shown in the xy-plane above?

A) $y = (x+1)(x-1)^2(x-2)$

B) $y = (x-1)^2(x+1)(x+2)$

C) $y = (x+1)^2(x-1)(x-2)$

D) $y = -(x+1)^2(x-1)(x-2)$

CONTINUE

9

$$3x + 4y = 12$$

The graph of the equation above in the xy-plane is a line, ℓ. Line m is perpendicular to line ℓ at point (a, b) and passes through the origin. Which of the following is the value of a?

A) 1.44

B) 4.8

C) 5

D) 5.4

10

The function f is defined by $f(x) = |x + 5| - 4$. If $f(0) = f(k)$, what is the value of k?

A) −10

B) −5

C) 4

D) 10

11

If the graph of function $y = f(x)$ has a vertex at point $(-2, 5)$, which of the following is the vertex of the graph of function $y = f(x + 3) - 8$?

A) $(-5, -8)$

B) $(-5, -3)$

C) $(1, -8)$

D) $(1, -3)$

12

For $x > 0$, which of the following is equivalent to

$$\dfrac{\dfrac{x}{x+3} - \dfrac{4}{x}}{1 - \dfrac{1}{x+3}} \ ?$$

A) $\dfrac{x^2 - 4x - 3}{x^2 + 2x}$

B) $\dfrac{x^2 - 6x - 12}{x^2 + 2x}$

C) $\dfrac{x + 6}{x}$

D) $\dfrac{x - 6}{x}$

CONTINUE

13

If $\dfrac{\sqrt[5]{a^6} \cdot \sqrt[5]{a^4}}{\sqrt[3]{a^2}} = a^n$ for all values of a, what is the value of n?

A) $\dfrac{3}{4}$

B) $\dfrac{4}{3}$

C) $\dfrac{12}{7}$

D) $\dfrac{12}{5}$

14

Olivia got $2,000 from her parents to start her college fund. She is opening a new savings account and the bank offers a 2.4 percent annual interest rate, compounded quarterly. Which of the following functions best represents the amount of money, in dollars, in Olivia's account in 15 years?

A) $2,000(1.024)^{15}$

B) $2,000(1.024)^{60}$

C) $2,000(1.006)^{15}$

D) $2,000(1.006)^{60}$

15

If a function is defined by $f(x+1) = x^2 - 1$, which of the following represents $f(x-3)$?

A) $f(x-3) = (x-3)^2 - 1$

B) $f(x-3) = (x-3)^2 - (x-3)$

C) $f(x-3) = (x+4)^2 - 1$

D) $f(x-3) = (x-4)^2 - 1$

CONTINUE

DIRECTIONS

For questions 16–20, solve the problem and enter your answer in the grid, as described below, on the answer sheet.

1. Although not required, it is suggested that you write your answer in the boxes at the top of the columns to help you fill in the circles accurately. You will receive credit only if the circles are filled in correctly.

2. Mark no more than one circle in any column.

3. No question has a negative answer.

4. Some problems may have more than one answer.

5. **Mixed numbers** such as $3\frac{1}{2}$ must be gridded as 3.5 or 7/2. (If $\boxed{3\,|\,1\,|\,/\,|\,2}$ is entered into the grid, it will be interpreted as $\frac{31}{2}$, not $3\frac{1}{2}$.)

6. **Decimal answers:** If you obtain a decimal answer with more digits than the grid can accommodate, it may be either rounded or truncated, but it must fill the entire grid.

Answer: $\frac{7}{12}$

Write answer in boxes.
Fraction line
Grid in result.

Answer: 2.5

Decimal point

Acceptable ways to grid $\frac{2}{3}$ are:

Answer: 201
Either position is correct.

Note: You may start your answers in any column, space permitting. Columns you don't need to use should be left blank.

CONTINUE

16

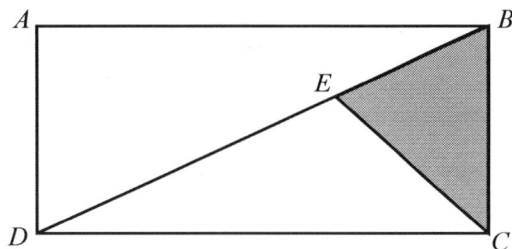

In the figure above, point E is on diagonal $\overline{BD}$ and $DE = 3BE$. If the area of the shaded region is 18, what is the area of the rectangle $ABCD$?

17

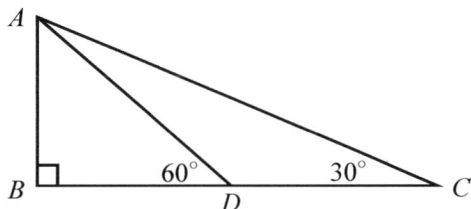

In the figure above, measure of $\angle ADB = 60^\circ$, measure of $\angle ACD = 30^\circ$, and $CD = 10\sqrt{3}$. What is the length of $\overline{AB}$?

18

$$\frac{x^2}{2} - \left(k^2 - 1\right)x + k = 0$$

In the equation above, k is a constant. If the product of the roots is 10, what is the sum of the roots?

19

$$2x + y = 5$$
$$5x - 2y = 8$$

If (a, b) is a solution to the system of equations above, what is the value of $\dfrac{a+b}{a}$?

20

$$f(x) = x^5 + k^2 x^3 + 3kx - 11$$

In the function f above, k is a positive constant. If $x - 1$ is a factor of $f(x)$, what is the value of k ?

STOP

If you finish before time is called, you may check your work on this section only.
Do not turn to any other section in the test.

No Test Material on This Page

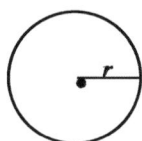
Math Test - Calculator

55 MINUTES, 38 QUESTIONS

Turn to Section 4 of your answer sheet to answer the questions in this section.

DIRECTIONS

For questions 1-30, solve each problem, choose the best answer from the choices provided, and fill in the corresponding circle on your answer sheet. **For questions 31-38,** solve the problem and enter your answer in the grid on your answer sheet. Please refer to the directions before question 31 on how to enter your answers in the grid. You may use any available space in your test booklet for scratch work.

NOTES

1. The use of a calculator **is permitted**.

2. All variables and expressions used represent real numbers unless otherwise indicated.

3. Figures provided in this test are drawn to scale unless otherwise indicated.

4. All figures lie in a plane unless otherwise indicated.

5. Unless otherwise indicated, the domain of a given function f is the set of all real numbers x for which $f(x)$ is a real number.

REFERENCE

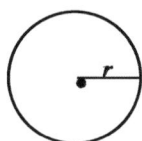

$A = \pi r^2$
$C = 2\pi r$

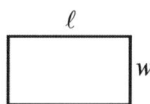

$A = \ell w$

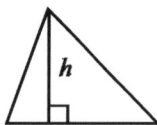

$A = \dfrac{1}{2}bh$

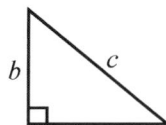

$c^2 = a^2 + b^2$

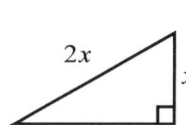

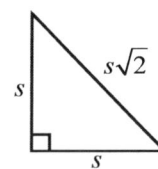

Special Right Triangles

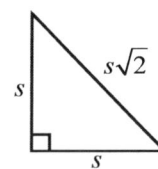

$V = \ell w h$

$V = \pi r^2 h$

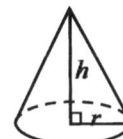

$V = \dfrac{4}{3}\pi r^3$

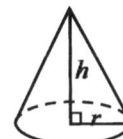

$V = \dfrac{1}{3}\pi r^2 h$

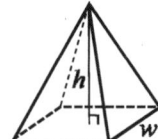

$V = \dfrac{1}{3}\ell w h$

The number of degrees of arc in a circle is 360.
The number of radians of arc in a circle is 2π.
The number of the measures in degrees of the angles of a triangle is 180.

CONTINUE

1

Ashley and Beth drove from city A to city B in exactly 5 hours, not taking rest stops. Ashley drove up to the area located halfway between the two cities at the average speed of 40 miles per hour and then Beth drove the other half at an average speed of 60 miles per hour. How many hours did Beth drive on their trip?

A) 1

B) 2

C) 3

D) 4

2

Which of the following is equivalent to
$(x+y)^2 - (x-y)^2$?

A) 0

B) $4xy$

C) $x^2 - y^2$

D) $2x^2 + 2y^2$

3

The combined weight of Adam and Bradley is 165 pounds. The combined weight of Bradley and Chris is 185 pounds. The combined weight of Chris and Adam is 175 pounds. How many pounds does Bradley weigh?

A) 80

B) 85

C) 87.5

D) 92.5

4

$$\frac{1}{x-5} = \frac{x}{x^2 - 25}$$

What is the solution of the equation shown above?

A) −10

B) 5

C) 10

D) No solution

CONTINUE

5

$$5(x+2)^2 - 15 = 45$$

Which of the following is a value of x that satisfies the equation above?

A) 4

B) $-2 + 2\sqrt{3}$

C) $-2 - 2\sqrt{2}$

D) $2 + 2\sqrt{2}$

6

In the xy-plane above, lines ℓ and m are perpendicular to each other at point P. If the graph of line m intersects the x-axis at point $(-4, 0)$, what is the area of $\triangle PQR$?

A) 84

B) 72

C) 42

D) 36

7

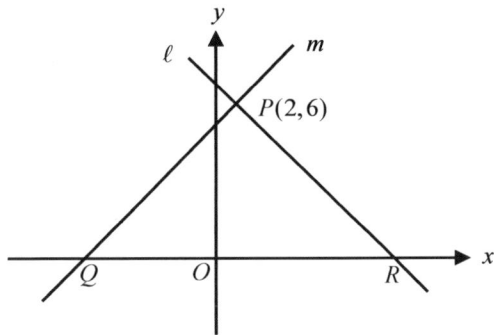

In the dot plots above, the four data sets are fund-raising totals in dollars for five clubs in Mr. Lee's class. Which data set appears to have the smallest standard deviation?

A) Club 1

B) Club 2

C) Club 3

D) Club 4

CONTINUE

8

If $x^2 + 5x = 3$, what is the value of $(x-1)(x+6)+4$?

A) 1
B) 2
C) 3
D) 4

9

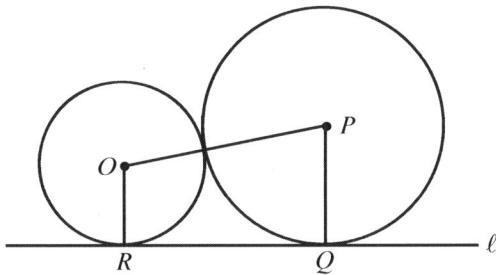

In the figure above, circles O and P are tangent to line ℓ and are tangent each other. If the radii of circles O and P are 4 and 9 respectively, what is the area of quadrilateral $OPQR$?

A) 39

B) 78

C) 108

D) 156

Questions 10 and 11 refer to the following information.

Ordered Details

20 Sandwiches
20 Soups
5 Salads
4 Sodas
Delivery Fee: 5%
Total amount: $226.17

Megan ordered lunch for her birthday party and her bill is shown above. The delivery fee was calculated as a percent of the cost of Megan's order. The total amount is the sum of the cost of Megan's order and the delivery fee.

10

What is the cost of Megan's order?

A) $200

B) $215.40

C) $225.50

D) $250

11

The total amount Megan paid for sandwiches and salads was $134. If the combined cost of one sandwich and one salad is $8.50, what is the cost of one sandwich?

A) $5.05

B) $6.10

C) $7.25

D) $7.50

CONTINUE

12

If a and b are positive integers and $8(2^{4a}) = \dfrac{2^{2b}}{2^3}$,

what is b in terms of a?

A) $a+1$

B) $a+2$

C) $2a+1$

D) $2a+3$

13

There are two different means, the arithmetic,

$A = \dfrac{a+b}{2}$ and the harmonic, $H = \dfrac{2ab}{a+b}$. If the

arithmetic mean is equal to the harmonic mean, which of the following must be true?

A) $a = 0$ and $b = 0$

B) $ab = 0$

C) $a = 1$ and $b = 1$

D) $a = b$

14

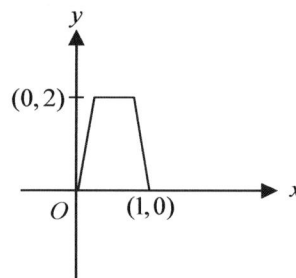

The figure above shows the graph of $y = f(x)$. Which

of the following could be the graph of $y = \dfrac{1}{2}f(x+1)$?

A)

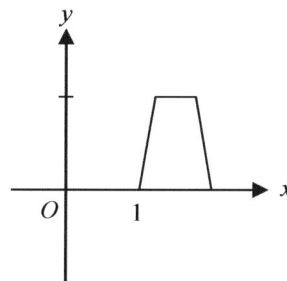

B)

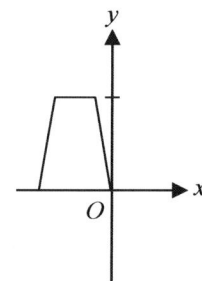

C)

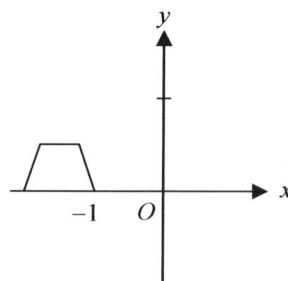

D)

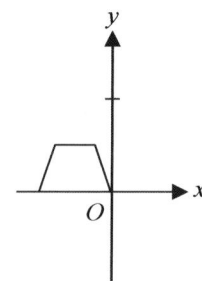

CONTINUE

15

$$y = ax + b$$
$$5x - 3y = a$$

In the system of equations above, a and b are constants. If the system has infinitely many solutions, what is the value of b?

A) -3

B) $-\dfrac{9}{5}$

C) $-\dfrac{5}{9}$

D) $\dfrac{9}{5}$

16

$$K = \frac{1}{2}mv^2$$

The kinetic energy of an object in motion is given by the equation above, where K is the kinetic energy of the object in joules, m is the mass of the object in kilograms, and v the speed of the object in meters per second. When the object is moving at a meters per second with a mass of b kilograms, the kinetic energy is 98 joules. What is the kinetic energy in joules, when the object is moving at $3a$ meters per second with a mass of $2b$ kilograms?

A) 1764

B) 1560

C) 882

D) 441

17

If $rs^2t^3u^3 > 0$ and $u < 0$, which of the following must be true?

A) $rt < 0$

B) $urt < 0$

C) $r > 0$

D) $t > 0$

18

The Maxim Telephone Company charges k cents for the first t minutes of a call and charges for any additional time at the rate of r cents per minute. If a certain customer pays \$10, which of the following could be the length of that phone call in minutes?

A) $\dfrac{1000 + rk}{r}$

B) $\dfrac{1000 + rtk}{r}$

C) $\dfrac{1000 - k - t}{r}$

D) $\dfrac{1000 - k + rt}{r}$

CONTINUE

19

CLASSES STUDENTS ARE TAKING

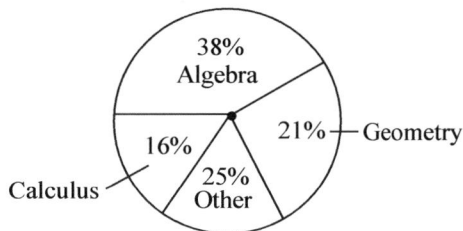

The circle graph above shows the distribution of the subjects in a certain high school. If there are 136 more students taking algebra than geometry, how many more students are taking geometry than calculus?

A) 36

B) 40

C) 42

D) 84

20

A farmer can plow a field in k days. How many days will it take two farmers working at the same rate to plow a field two times larger?

A) $2k$

B) k

C) $\dfrac{k}{2}$

D) $\dfrac{k}{4}$

21

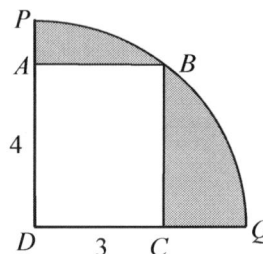

In the figure above, a rectangle is inscribed in a part of a circle. If the length of $\overline{AD}$ is 4 and the length of $\overline{DC}$ is 3, what is the area of the shaded region?

A) $25\pi - 12$

B) $\dfrac{25\pi - 12}{2}$

C) $\dfrac{25\pi - 24}{2}$

D) $\dfrac{25\pi - 48}{4}$

22

How does the graph of $f(x) = (x+2)(x-4)$ compare with the graph of $g(x) = x^2$?

A) The graph of $g(x)$ is moved to the left 1 unit and up 9 units.

B) The graph of $g(x)$ is moved to the right 1 unit and down 8 units.

C) The graph of $g(x)$ is moved to the left 1 unit and down 9 units.

D) The graph of $g(x)$ is moved to the right 1 unit and down 9 units.

CONTINUE ➤

23

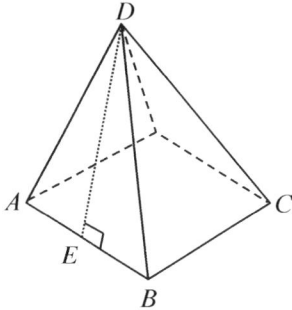

Note: Figure not drawn to scale.

In the figure above, the right regular pyramid has a square base and all four triangular faces are congruent. If $AB = 4$ and the total surface area of the pyramid is 56, what is the length of $\overline{DE}$?

A) 4

B) 5

C) $5\sqrt{2}$

D) 6

24

If $2a+b=4$ and $ab=-3$, what is the value of $|2a-b|$?

A) 4

B) $2\sqrt{10}$

C) $6\sqrt{3}$

D) 8

25

HEIGHT OF A TREE

Age (in years)	5	6	7	8	9
Height (in meters)	6	7	7.6	8.1	8.2

Which of the following graphs best represents the information in the table above?

A)

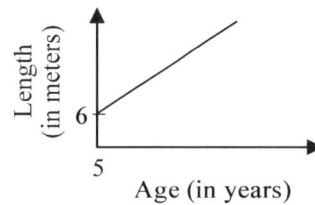

B)

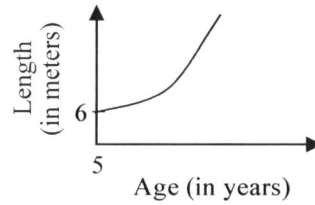

C)

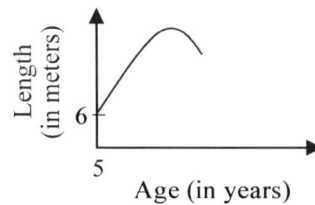

D)

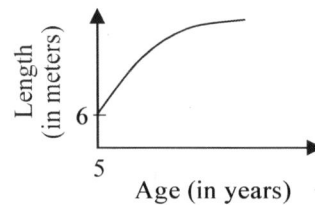

CONTINUE

26

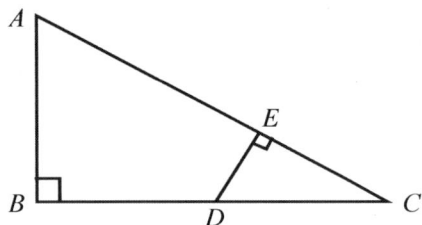

Note: Figure not drawn to scale.

In the right triangle ABC above, $AB = 9$, and

$BC = 12$. If D is the midpoint of $\overline{BC}$, what is the area of triangle CDE?

A) 27
B) 13.5
C) 9
D) 8.64

27

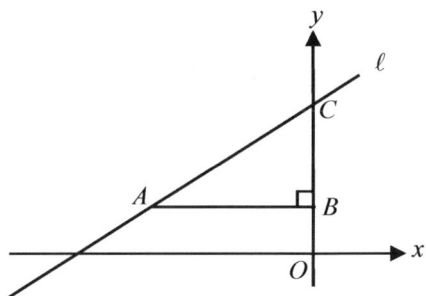

Note: Figure not drawn to scale.

In the xy-plane above, the equation of line ℓ is

$y = \dfrac{1}{2}x + 10$ and the area of $\triangle ABC$ is 36. What is the

length of $\overline{AB}$?

A) 6
B) 12
C) 15
D) 16

28

Which of the following is true about the graph of

$f(x) = 3\left(\sqrt{8x - 4} + 2\right)$ compared with the graph of

$g(x) = 3\sqrt{8x}$?

A) The graph of $g(x)$ is moved to the left by 4 units and up 2 units.

B) The graph of $g(x)$ is moved to the right by 4 units and up 2 units.

C) The graph of $g(x)$ is moved to the right by 0.5 unit and up 2 units.

D) The graph of $g(x)$ is moved to the right by 0.5 unit and up 6 units.

29

$$xy = \sqrt[3]{32}$$

$$\frac{x}{y} = \left(\frac{1}{2}\right)^{\frac{1}{3}}$$

In the system of equations, which of the following could be the value of y?

A) 1
B) 1.5
C) 2
D) 4

CONTINUE

30

For the quadratic function $f(x)$, $f(4) = 8$. If $0 \le x \le 3$ is the solution to the inequality $f(x) \le 0$, what is the value of $f(5)$?

A) 10

B) 12

C) 16

D) 20

CONTINUE

DIRECTIONS

For questions 31-38, solve the problem and enter your answer in the grid, as described below, on the answer sheet.

1. Although not required, it is suggested that you write your answer in the boxes at the top of the columns to help you fill in the circles accurately. You will receive credit only if the circles are filled in correctly.
2. Mark no more than one circle in any column.
3. No question has a negative answer.
4. Some problems may have more than one answer.
5. **Mixed numbers** such as $3\frac{1}{2}$ must be gridded as 3.5 or 7/2. (If $\boxed{3\,1\,/\,2}$ is entered into the grid, it will be interpreted as $\frac{31}{2}$, not $3\frac{1}{2}$.)
6. **Decimal answers:** If you obtain a decimal answer with more digits than the grid can accommodate, it may be either rounded or truncated, but it must fill the entire grid.

Answer: $\frac{7}{12}$ — Write answer in boxes. Fraction line. Grid in result.

Answer: 2.5 — Decimal point

Acceptable ways to grid $\frac{2}{3}$ are:

Answer: 201 — Either position is correct.

Note: You may start your answers in any column, space permitting. Columns you don't need to use should be left blank.

CONTINUE

31

$$x^3 - 3x^2 + 8x - a = (x-2)(x^2 - x + b)$$

In the equation above, a and b are constants. If the equation is true for all real values of x, what is the value of b?

32

$$x^2 - kx + k - 6 = 0$$

In the equation above, k is a constant and a and b are the solutions to the equation. If $a + b = 12$, what is the value of ab?

33

What is the distance between the points of intersection of the curves $y = x(x-4)$ and $y = x + 6$? (Round your answer to the nearest tenth.)

34

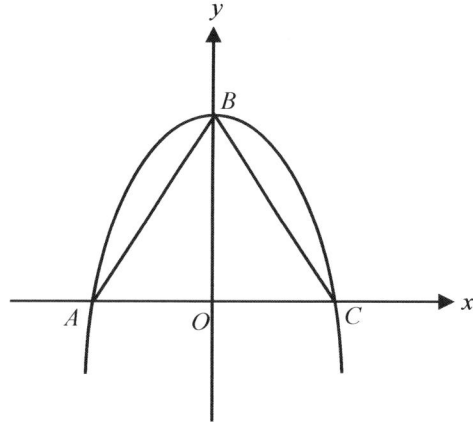

Note: Figure not drawn to scale.

In the xy-plane above, the parabola $y = -kx^2 + 1$ intersects the x- and y-axes at the points A, B, and C. If $\triangle ABC$ is an equilateral triangle, what is the value of k?

35

$$y \le -x^2 + 4x + 5$$
$$y \ge -x + 5$$

In the xy-plane, if a point with coordinates (a, b) lies in the solution set of the system of inequalities above, what is the maximum possible value of b?

CONTINUE

36

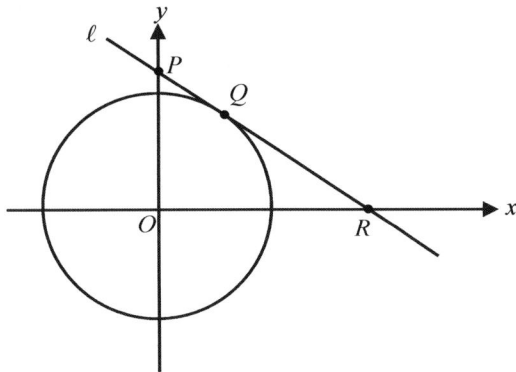

In the xy-plane above, line ℓ is tangent to the graph of circle O and the value of $\sin \angle ORP$ is 0.6. If the area of $\triangle OPR$ is 54, what is the radius of the circle?

37

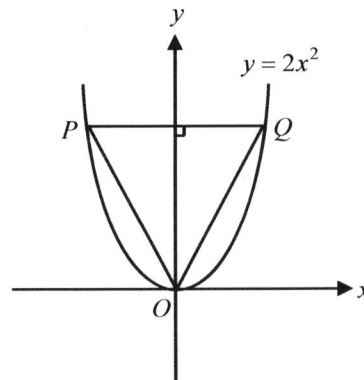

In the xy-plane above, points O, P, and Q lie on the graph of $y = 2x^2$. If the area of triangle OPQ is 54, what is the length of $\overline{PQ}$?

38

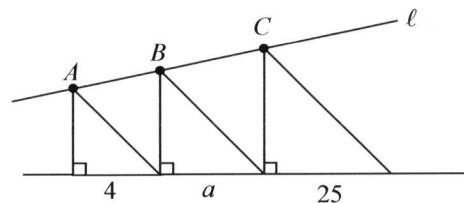

Note: Figure not drawn to scale.

The figure above shows three isosceles right triangles with sides of length 4, a, and 25, respectively. If points A, B, and C lie on line ℓ, what is the value of a?

STOP

If you finish before time is called, you may check your work on this section only.
Do not turn to any other section in the test.

No Test Material on This Page

Math Conversion Table

Raw Score	Scaled Score	Raw Score	Scaled Score
58	800	27	500
57	800	26	490
56	800	25	480
55	800	24	470
54	790	23	460
53	780	22	460
52	770	21	450
51	750	20	440
50	740	19	430
49	730	18	430
48	720	17	420
47	710	16	420
46	700	15	410
45	690	14	400
44	670	13	390
43	680	12	380
42	670	11	370
41	660	10	360
40	650	9	450
39	640	8	340
38	630	7	330
37	620	6	310
36	610	5	290
35	600	4	280
34	590	3	270
33	580	2	260
32	560	1	240
31	550	0	200
30	540		
29	530		
28	520		

Answer Explanations

Test 11   **Answers and Explanations**

	1	2	3	4	5	6	7	8	9	10
SECTION 3	C	D	B	D	C	D	A	D	A	A
	11	12	13	14	15	16	17	18	19	20
	B	D	B	D	D	144	15	48	3/2	2

	1	2	3	4	5	6	7	8	9	10
	B	B	C	D	B	D	C	A	B	B
	11	12	13	14	15	16	17	18	19	20
SECTION 4	B	D	D	D	C	A	A	D	B	B
	21	22	23	24	25	26	27	28	29	30
	D	D	B	B	D	D	B	D	C	D
	31	32	33	34	35	36	37	38		
	6	6	9.9	3	9	7.2	6	10		

SECTION 3

1. C

$$\frac{3}{4}(x-2)+7=12-\frac{1}{2}(x-2) \quad : \text{Multiply by 4}$$

$$3(x-2)+28=48-2(x-2)$$

$$5(x-2)=20 \;\rightarrow\; (x-2)=4 \;\rightarrow\; x=6$$

2. D

$$5\left(x^2-6\right)=x^2+34 \;\rightarrow\; 5x^2-30=x^2+34$$

$$4x^2=64 \;\rightarrow\; x^2=16 \;\rightarrow\; x=\pm4$$

3. B

$$3t+4t=100 \;\rightarrow\; 7t=100 \;\rightarrow\; t=\frac{100}{7}$$

For Albert, $d=3t=3\left(\dfrac{100}{7}\right)=\dfrac{300}{7}$ feet

4. D

$$x^3+x^2y-x-y \;\rightarrow\; x^2(x+y)-(x+y)$$

$$\left(x+y\right)\left(x^2-1\right) \;\rightarrow\; (x+y)(x+1)(x-1)$$

5. C

$$\frac{1}{5}(x-5)(x-3)<0 \;\rightarrow\; 3<x<5$$

6. D

$$(x-5)(x+3)+10 \;\rightarrow\; x^2-2x-15+10$$

$$\rightarrow x^2-2x-5 \;\rightarrow\; (x-1)^2-6$$

7. A

$$AB=12 \;\rightarrow\; \frac{AC\cdot12}{2}=96 \;\rightarrow\; AC=16$$

$$CD=16-9=7$$

8. D

x-intercept: $x=-1,-1,\,1,\,2$

Since the graph falls down to the right and left,

The best equation is $y=-\left(x+1\right)^2(x-1)(x-2)$.

9. **A**

Slope of $3x + 4y = 12$ → $y = -\frac{3}{4}x + 3$ is $-\frac{3}{4}$.

The slope of the perpendicular line m is $\frac{4}{3}$.

Therefore, the equation of line m is $y = \frac{4}{3}x$.

Solve the system of the equations.

$\frac{4}{3}x = -\frac{3}{4}x + 3$ → $\frac{25}{12}x = 3$ →

$x = 3 \times \frac{12}{25} = 1.44$

10. **A**

The graph of the absolute function is symmetric with respect to $x = -5$.

Therefore, $\frac{0+k}{2} = -5$ → $k = -10$.

11. **B**

$(-2, 5) \xrightarrow{\text{T}_{-3,\,-8}} (-2-3, 5-8) \to (-5, -3)$

12. **D**

$$\frac{\left(\dfrac{x}{x+3} - \dfrac{4}{x}\right) \times (x)(x+3)}{\left(1 - \dfrac{1}{x+3}\right) \times (x)(x+3)} = \frac{x^2 - 4(x+3)}{x(x+3) - x}$$

$$= \frac{x^2 - 4x - 12}{x^2 + 3x - x} = \frac{(x-6)(x+2)}{x(x+2)} = \frac{x-6}{x}$$

13. **B**

$$\frac{\sqrt[5]{a^6} \cdot \sqrt[5]{a^4}}{\sqrt[3]{a^2}} = \frac{a^{\frac{6}{5}} - a^{\frac{4}{5}}}{a^{\frac{2}{3}}} = \frac{a^2}{a^{\frac{2}{3}}} = a^{\frac{4}{3}}$$

$n = \frac{4}{3}$

14. **D**

$2{,}000\left(1 + \dfrac{0.024}{4}\right)^{4t}$ → $2{,}000(1 + 0.006)^{4 \times 15}$

→ $2{,}000(1.006)^{60}$

15. **D**

Change the variable of the first function.

$f(x+1) = x^2 - 1$ → $f(k+1) = k^2 - 1$

Now interchange $k + 1 = x - 3$ → $k = x - 4$.

Therefore, $f(x-3) = (x-4)^2 - 1 = x^2 - 8x + 15$.

16. **144**

Since the ratio of $BE : DE = 1 : 3$, the ratio of the areas of $\triangle BCE : \triangle CDE = 1 : 3$. Therefore, the area of $\triangle CDE = 18 \times 3 = 54$. The area of $ABCD = 2(54 + 18) = 144$.

17. **15**

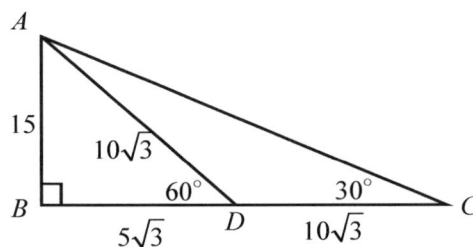

$\triangle ACD$ is isosceles. $AD = 10\sqrt{3}$ and $BD = \dfrac{AD}{2}$

18. **48**

Product $= \dfrac{k}{1/2} = 2k = 10$ → $k = 5$

Sum $= \dfrac{k^2 - 1}{1/2} = 2(k^2 - 1) = 48$

19. **3/2**

When you solve the system of equations, $a = 2$ and $b = 1$.

Therefore, $\dfrac{a+b}{a} = \dfrac{3}{2}$.

20. **2**

Factor theorem

$f(1) = 1 + k^2 + 3k - 11 = 0$

$k^2 + 3k - 10 = 0$ → $(k+5)(k-2) = 0$

$k = -5, 2$ → $k = 2 \text{(positive)}$

Answer Explanations

SECTION 4

1. B

Let $2d$ = distance between two cities.

$\dfrac{d}{40} + \dfrac{d}{60} = 5$ →(multiply by 120)→ $2d + 3d = 600$

$5d = 600$ → $d = 120$

Therefore, $t(\text{Beth}) = \dfrac{120}{60} = 2\,\text{hours}$

2. B

$(x+y)^2 - (x-y)^2 = x^2 + 2xy + y^2 - (x^2 - 2xy + y^2)$

$= 4xy$

3. C

$A + B = 165,\ B + C = 185,\ C + A = 175$

When you add all three equations,

$2(A+B+C) = 525$ → $A+B+C = 262.5$.

$175 + B = 262.5$ → $B = 87.5$

4. D

Multiply by $(x+y)(x-y)$.

$x + 5 = x$ → No solution.

You can use cross-multiplication.

5. B

$5(x+2)^2 - 15 = 45$ → $(x+2)^2 - 3 = 9$

$(x+2)^2 = 12$ → $(x+2) = \pm 2\sqrt{3}$ →

$x = -2 \pm 2\sqrt{3}$

6. D

Slope of line $m = \dfrac{6-0}{2-(-4)} = 1$

Slope of line $\ell = -1$

The equation of line ℓ is $y - 6 = -1(x-2)$.

When $y = 0$ → $-6 = -(x-2)$ → $x = 8$

$QR = 8 - (-4) = 12$ and the height $= 6$.

Therefore, area of $\triangle PQR$ is $\dfrac{12 \times 6}{2} = 36$.

7. C

The data are spread close from the mean.

8. A

$(x-1)(x+6) + 4 = x^2 + 5x - 6 + 4 = 3 - 6 + 4 = 1$

9. B

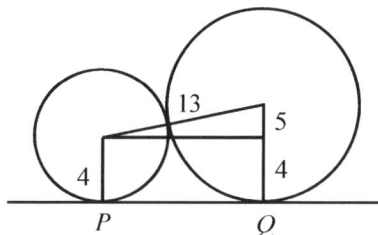

$PQ = \sqrt{13^2 - 5^2} = 12$

Therefore, the area is $\dfrac{(4+9)\cdot 12}{2} = 78$.

10. B

If the cost of Megan's order is x, then

$x + 0.05x = 226.17$ → $1.05x = 226.17$

→ $x = 215.4$

11. B

Cost of one sandwich $= a$, and cost of one salad $= b$,

then $20a + 5b = 134$ and $a + b = 8.5$.

When you solve the system of equations, $a = \$6.10$.

12. D

$8(2^{4a}) = \dfrac{2^{2b}}{2^3}$ → $2^3 \cdot 2^{4a} = 2^{2b-3}$ → $2^{3+4a} = 2^{2b-3}$

Therefore,

$3 + 4a = 2b - 3$ → $2b = 4a + 6$ → $b = 2a + 3$

13. D

$\dfrac{a+b}{2} = \dfrac{2ab}{a+b}$ → $(a+b)^2 = 4ab$ →

$a^2 + 2ab + b^2 = 4ab$ → $a^2 - 2ab + b^2 = 0$ →

$(a-b)^2 = 0$

If $a = b$, it is always true. (From the original equation, a and b cannot be 0.)

14. D

The graph of $f(x)$ is moved to the left by 1 and the height is reduced to half.

15. C

Answer Explanations

$$ax - y = -b$$
$$5x - 3y = a$$

In order to have infinitely many solutions, $\dfrac{a}{5} = \dfrac{-1}{-3} = \dfrac{-b}{a}$. $\rightarrow$ $a = \dfrac{5}{3}$ and $b = -\dfrac{5}{9}$

16. A

$$98 = \frac{1}{2}(b)(a)^2 \text{ and } K = \frac{1}{2}(2b)(3a)^2 = 18\left(\frac{1}{2}ba^2\right)$$

Therefore, $K = 18 \times 98 = 1764$ Joules.

17. A

$$rs^2t^3u^3 > 0 \rightarrow \frac{rs^2t^3u^3}{s^2t^2u^2} > \frac{0}{s^2t^2u^2} \rightarrow rtu > 0$$

Since $u < 0$, $\dfrac{rtu}{u} < 0 \rightarrow rt < 0$.

18. D

After t minutes, the additional time is $\dfrac{1000-c}{r}$.

Therefore, the total time is

$$\frac{1000-k}{r} + t = \frac{1000-k-rt}{t}.$$

19. B

$38 - 21 = 17\%$ of the students is 136 and $21 - 16 = 5\%$ of the students is x.

Therefore, $\dfrac{17}{136} = \dfrac{5}{x} \rightarrow x = 40$.

20. B

Since a farmer takes k days two farmers will takes $\dfrac{k}{2}$ days for the same job. Therefore, it takes k days for the two jobs.

$$1 \times k = 2 \times \left(\frac{k}{2}\right) \rightarrow \left(\frac{k}{2}\right) \times 2 = k \text{ days}$$

21. D

The radius of the circle is $\sqrt{3^2 + 4^2} = 5$.

The area of the sector is $\dfrac{25\pi}{4}$ and the area of the rectangle is $4 \times 3 = 12$. Therefore, the area of the shaded region is $\dfrac{25\pi}{4} - 12 = \dfrac{25\pi - 48}{4}$.

22. D

$$f(x) = x^2 - 2x - 8 = \left(x^2 - 2x + 1\right) - 9 = \left(x-1\right)^2 - 9$$

Therefore, the graph of $g(x)$ is moved to the right 1 unit and down 9 units.

23. B

The surface area is $\left(4 \cdot 4\right) + 4\left(\dfrac{AB \cdot DE}{2}\right) = 56$.

$$16 + 8 \cdot DE = 56 \rightarrow DE = \frac{56-16}{8} = 5$$

24. B

$$(2a-b)^2 = (2a+b)^2 - 8ab \rightarrow$$
$$(2a+b)^2 = 4^2 - 8 \cdot (-3) = 40 \rightarrow$$
$$|2a-b| = \sqrt{40} = 2\sqrt{10}$$

25. D

The rate of increase in height is getting slower.

26. D

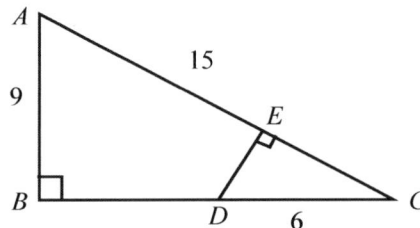

Since $AC = 15$ and $CD = 6$, the ratio of corresponding sides is 15:6 or 5:2. Because $\triangle ABC$ and $\triangle CDE$ are similar, The ratio of their area is $25:4$. Define their area as $25k$ and $4k$. $25k = \dfrac{9 \times 12}{2} = 54. \rightarrow k = 2.16$

Therefore, the area of $\triangle CDE$ is $4k = 4 \times 2.16 = 8.64$.

Or, find DE and CE using similar.

$$\text{Area} = \frac{DE \times CE}{2} = 8.64.$$

27. B

From the equation $OC = 10$ and x-intercept is -20.

Since $\triangle ABC$ and $\triangle CDO$ are similar,

$$\frac{CB}{AB} = \frac{1}{2} \rightarrow CB = k \text{ and } AB = 2k.$$

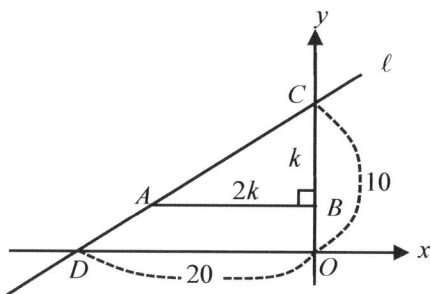

The area of $\triangle ABC$ is

$$\frac{k \times 2k}{2} = 36. \quad \rightarrow \quad k^2 = 36 \quad \rightarrow \quad k = 6$$

Therefore, $AB = 2k = 2(6) = 12$.

28. D

$$f(x) = 3\left(\sqrt{8x-4}+2\right) = 3\sqrt{8(x-0.5)} + 6$$

The graph of $g(x)$ is moved to the right by 0.5 and up 6.

29. C

Since $\dfrac{x}{y} = \left(\dfrac{1}{2}\right)^{\frac{1}{3}}$, $\dfrac{y}{x} = \dfrac{1}{\left(\dfrac{1}{2}\right)^{\frac{1}{3}}} = \left(\dfrac{1}{2}\right)^{-\frac{1}{3}} = 2^{\frac{1}{3}} = \sqrt[3]{2}$

$$(xy)\left(\dfrac{y}{x}\right) = \sqrt[3]{32} \times \sqrt[3]{2} \quad \rightarrow \quad y^2 = \sqrt[3]{64} = 4 \quad \rightarrow$$

$$y = \pm\, 2$$

30. D

We know that $0 \le x \le 3$ is the solution of the inequality $ax(x-3) \le 0$. Therefore, we can define

$f(x) = ax(x-3)$.

$f(4) = a(4)(4-3) = 8 \quad \rightarrow \quad 4a = 8 \quad \rightarrow a = 2$

Therefore, $f(x) = 2x(x-3)$.

$f(5) = 2(5)(5-3) = 20$

31. 6

Putting $x = 2$ in the equation,

$8 - 12 + 16 - a = 0 \quad \rightarrow \quad a = 12$.

$-a = -2b \quad \rightarrow \quad -12 = -2b \quad \rightarrow \quad b = 6$

Or you can put any number except 2 in the equation. For example, $x = 1$.

$1 - 3 + 8 - 12 = (1-2)(1-1+b) \quad \rightarrow \quad -6 = -b$

32. 6

From the formula:

$a + b = k = 12$

$ab = k - 6 = 12 - 6 = 6$

33. 9.9

$x(x-4) = x + 6 \quad \rightarrow \quad x^2 - 5x - 6 = 0$

$(x-6)(x+1) = 0 \rightarrow \quad x = -1, 6$

When $x = -1$, $y = 5$. When $x = 6$, $y = 12$.

Distance between $(-1, 5)$ and $(6, 12)$ is

$$d = \sqrt{\left(6 - {}^-1\right)^2 + \left(12 - 5\right)^2} = \sqrt{98} \approx 9.9 \,.$$

34. 3

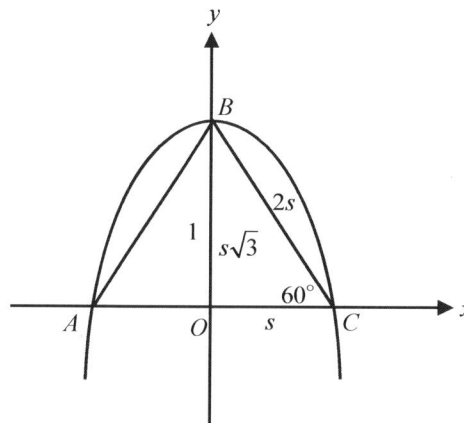

$\triangle OBC$ is a special right triangle: 30-60-90

Since y-intercept is $f(0) = 1$, $s\sqrt{3} = 1$.

$$s = \dfrac{1}{\sqrt{3}} \quad \rightarrow \quad \text{Coordinates of point } C = \left(\dfrac{1}{\sqrt{3}}, 0\right)$$

Putting this coordinates into the equation,

$$0 = -k\left(\dfrac{1}{\sqrt{3}}\right)^2 + 1 \quad \rightarrow \quad k\left(\dfrac{1}{3}\right) = 1 \quad \rightarrow \quad k = 3$$

35. 9

Find the points of intersection.

$-x^2 + 4x + 5 = -x + 5 \quad \rightarrow \quad x^2 - 5x = 0$

$x(x-5) = 0 \quad \rightarrow \quad x = 0, 5$

We know the solution set is the area between These two graphs. But the quadratic function has a

maximum at $x = \dfrac{-b}{2a} = \dfrac{-4}{2(-1)} = 2$.

Therefore, the maximum of b is

$f(2) = -2^2 + 4(2) + 5 = 9$.

The graphs of $y = -x^2 + 4x + 5$ and $y = -x + 5$
Are as follows.

$$y = -x^2 + 4x + 5 = -(x-2)^2 + 9$$

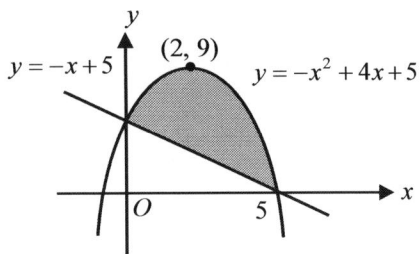

Therefore the maximum value of b is 9.

36. 7.2

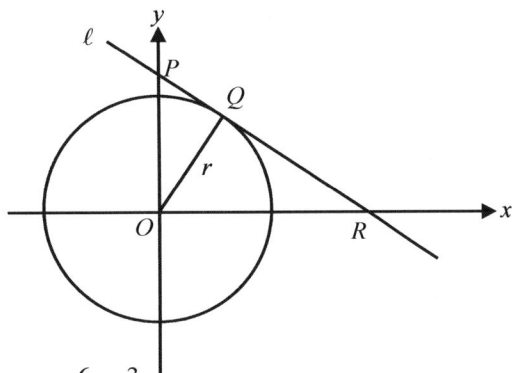

$$0.6 = \frac{6}{10} = \frac{3}{5} \quad \rightarrow \quad OP = 3k \text{ and } PR = 5k$$

Then $OR = 4k$. The area of $\triangle OPR$ is

$$\frac{(3k)(4k)}{2} = 54 \quad \rightarrow \quad 6k^2 = 54 \quad \rightarrow \quad k^2 = 9 \quad \rightarrow \quad k = 3.$$

$$PR = 5k = 5(3) = 15$$

The area of $\triangle OPR = \frac{15 \times r}{2} = 54 \quad \rightarrow \quad r = 7.2$

37. 6

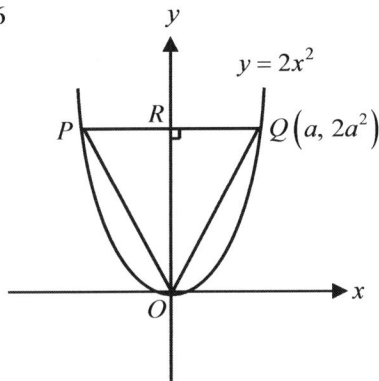

If x-coordinate is a, the coordinates of point Q
is $(a, 2a^2)$. Now we know that $PQ = 2a$ and
$OR = 2a^2$. Therefore, the area of $\triangle OPQ$ is

$$\frac{(2a)(2a^2)}{2} = 2a^3 = 54. \quad a^3 = 27 \quad \rightarrow \quad a = 3$$

$$PQ = 2(3) = 6$$

38. 10

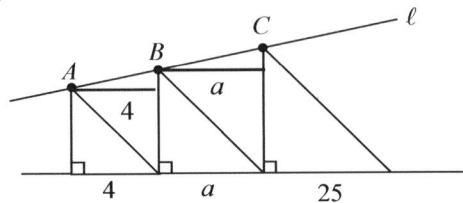

Slope of $\overline{AB} = \frac{a-4}{4}$ and slope of $\overline{BC} = \frac{25-a}{a}$

The slopes on the same line are equal.

$$\frac{a-4}{4} = \frac{25-a}{a} \quad \rightarrow \quad a^2 - 4a = 100 - 4a \quad \rightarrow$$

$$a^2 = 100 \quad \rightarrow \quad a = 10$$

PRACTICE TEST 12

Dr. John Chung's SAT Math

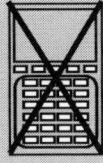

Math Test - No Calculator

25 MINUTES, 20 QUESTIONS

Turn to Section 3 of your answer sheet to answer the questions in this section.

DIRECTIONS

For questions 1–15, solve each problem, choose the best answer from the choices provided, and fill in the corresponding circle on your answer sheet. **For questions 16–20,** solve the problem and enter your answer in the grid on your answer sheet. Please refer to the directions before question 16 on how to enter your answers in the grid. You may use any available space in your test booklet for scratch work.

NOTE

1. The use of a calculator **is not permitted**.

2. All variables and expressions used represent real numbers unless otherwise indicated.

3. Figures provided in this test are drawn to scale unless otherwise indicated.

4. All figures lie in a plane unless otherwise indicated.

5. Unless otherwise indicated, the domain of a given function f is the set of all real numbers x for which $f(x)$ is a real number.

REFERENCE

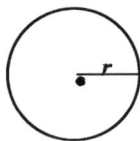

$A = \pi r^2$
$C = 2\pi r$

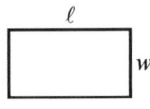

$A = \ell w$

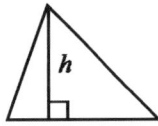

$A = \frac{1}{2}bh$

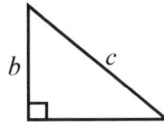

$c^2 = a^2 + b^2$

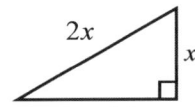

Special Right Triangles

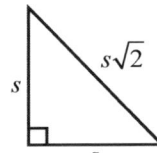

$V = \ell w h$

$V = \pi r^2 h$

$V = \frac{4}{3}\pi r^3$

$V = \frac{1}{3}\pi r^2 h$

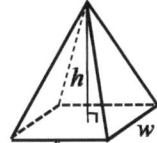

$V = \frac{1}{3}\ell w h$

The number of degrees of arc in a circle is 360.
The number of radians of arc in a circle is 2π.
The number of the measures in degrees of the angles of a triangle is 180.

CONTINUE

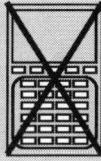

1

If $a^2b - 4ab + 42 = 0$ and $2a + 8 = 14$, what is the value of b?

A) 14

B) 12

C) 10

D) 8

2

If $a + 2b = \dfrac{1}{2}$ and $a - 2b = \dfrac{1}{3}$, what is the value of $6a^2 - 24b^2$?

A) $\dfrac{1}{3}$

B) $\dfrac{1}{2}$

C) 1

D) 2

3

If $a = 3 + 2i$ and $b = 3 - 2i$, what is the value of $a^2 - 2ab + b^2$? $\left(\text{Note: } i = \sqrt{-1}\right)$

A) -16

B) -4

C) 9

D) 36

4

$$\frac{\sqrt{x}}{2} + \sqrt{y} = 5$$

$$\sqrt{x} - \frac{\sqrt{y}}{2} = -\frac{1}{2}$$

If (a, b) is a solution to the system of equations above, what is the value of a?

A) $\dfrac{8}{5}$

B) $\dfrac{9}{5}$

C) $\dfrac{64}{25}$

D) $\dfrac{81}{25}$

CONTINUE

5

Total Cost of Renting a Car by Days

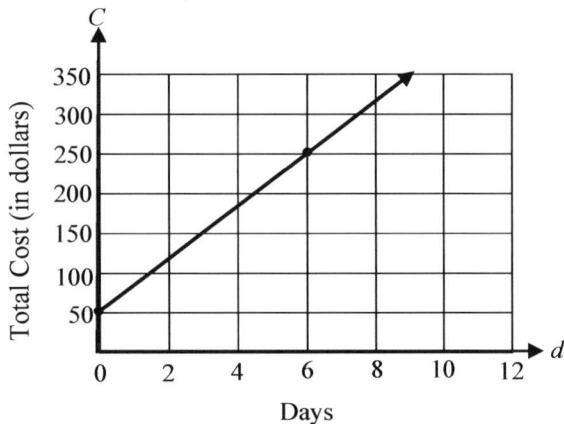

The graph above displays the total cost C, in dollars, of renting a car for d days. If Nigel rents a car for 21 days for his summer trip, what is the total rental cost?

A) $600
B) $650
C) $700
D) $750

6

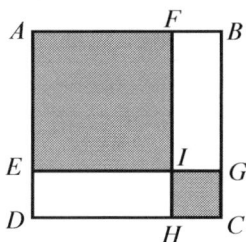

In the figure above, the area of square $AFIE$ is 9 times the area of square $HIGC$. What percent of square $ABCD$ is shaded?

A) 56.25%
B) 62.5%
C) 67.5%
D) 70.5%

Questions 7-8 refer to the following information.

A survey of 700 randomly selected people aged 15 through 20 in a certain town was conducted to gather data on how they prefer to see movies. The data are shown in the table below.

	DVD	TV provider	Smart phone downloads	Total
Ages 15 - 16	60	80	100	240
Ages 17 - 18	40	50	120	210
Ages 19 - 20	10	40	200	250
Total	110	170	420	700

7

If a person is chosen at random from those under 18 years old, which of the following is closest to the probability that the person preferred smartphone downloads?

A) 0.49
B) 0.52
C) 0.54
D) 0.56

8

How many people prefer smartphone-downloads or are aged between 15 and 16 years old?

A) 100
B) 560
C) 660
D) 665

CONTINUE

9

$$f(x) = 2x^2 + kx - 10$$

For the function f defined above, k is a constant.
If $f(-2) = f(8)$, what is the value of k?

A) -12
B) -6
C) 6
D) 12

10

If the graph of $y = x^2$ is translated up 3 units and right 5 units, what is the equation of the new graph?

A) $y = x^2 - 3x + 5$
B) $y = x^2 - 6x + 14$
C) $y = x^2 + 10x + 5$
D) $y = x^2 - 10x + 28$

11

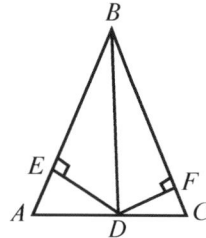

Note: Figure not drawn to scale.

Triangle ABC above is isosceles with $AB = BC$, and the ratio of DE to DF is 3:2. If the area of triangle ABC is 78, what is the area of triangle ABD?

A) 46.8
B) 54
C) 60
D) 60.25

12

$$P(x) = \left(x^2 + ax + 2\right) - 2(x - b)$$

In the polynomial $P(x)$ defined above, a and b are constants. If $P(x)$ is divisible by $(x-1)$ and $(x-3)$, what is the value of a?

A) -4
B) -2
C) 2
D) 4

CONTINUE

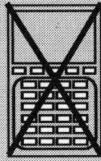

13

$$y = x^2 - 2x$$
$$y = x + k$$

In the system of equations defined above, k is a constant. If the system has no solution, which of the following could be the value of k?

A) 1

B) 0

C) −2

D) −3

14

The distance between two parallel lines is defined as the length of the perpendicular segment between them. What is the distance between $y = x$ and $y = x - 4$?

A) $\sqrt{2}$

B) 2

C) $2\sqrt{2}$

D) 4

15

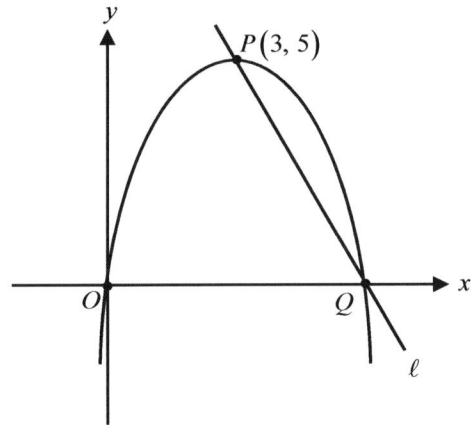

Note: Figure not drawn to scale.

In the xy-plane above, the graph of a quadratic function $y = f(x)$ intersects the x-axis at points O and Q, and has a vertex at point P. If line ℓ passes through points P and Q, what is the y-intercept of line ℓ?

A) 8

B) 10

C) 12

D) 15

CONTINUE

DIRECTIONS

For questions 16–20, solve the problem and enter your answer in the grid, as described below, on the answer sheet.

1. Although not required, it is suggested that you write your answer in the boxes at the top of the columns to help you fill in the circles accurately. You will receive credit only if the circles are filled in correctly.

2. Mark no more than one circle in any column.

3. No question has a negative answer.

4. Some problems may have more than one answer.

5. **Mixed numbers** such as $3\frac{1}{2}$ must be gridded as 3.5 or 7/2. (If $\boxed{3\,|\,1\,|\,/\,|\,2}$ is entered into the grid, it will be interpreted as $\frac{31}{2}$, not $3\frac{1}{2}$.)

6. **Decimal answers:** If you obtain a decimal answer with more digits than the grid can accommodate, it may be either rounded or truncated, but it must fill the entire grid.

Answer: $\frac{7}{12}$

Write answer in boxes.

Fraction line

Grid in result.

Answer: 2.5

Decimal point

Acceptable ways to grid $\frac{2}{3}$ are:

Answer: 201
Either position is correct.

Note: You may start your answers in any column, space permitting. Columns you don't need to use should be left blank.

CONTINUE

16

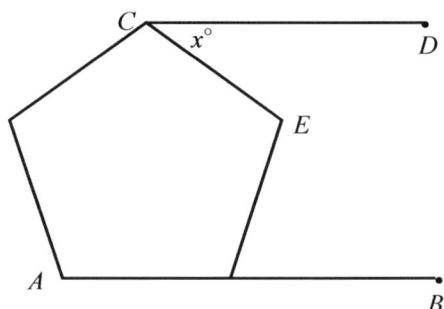

In the figure of a regular pentagon above, $\overline{CD}$ is parallel to $\overline{AB}$. If $x°$ is the measure of $\angle DCE$, what is the value of x?

17

$$\frac{\sqrt{x^2}+2\sqrt{x}}{6}=8$$

If $x>0$, what is the solution to the equation above?

18

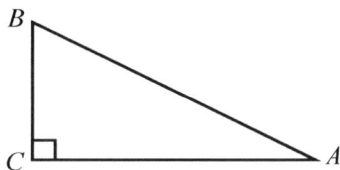

In the triangle above, the sine of $\angle BAC$ is 0.6 and the area of triangle ABC is 216. What is the length of $\overline{BC}$?

19

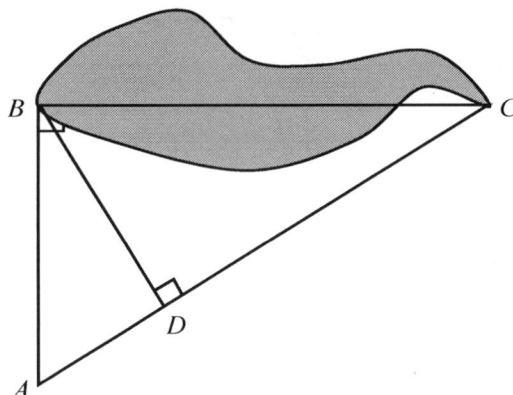

A surveyor drew the diagram above to find the distance across a pond. If $AB=150$ meters and $AD=90$ meters, what is the distance, in meters, from point B to point C?

20

$$x^2+y^2\le100$$
$$y\le x+2$$

In the xy-plane, if a point with coordinates (a,b) lies in the solution set of the system of inequalities above, what is the maximum possible valuable of b?

STOP

If you finish before time is called, you may check your work on this section only.
Do not turn to any other section in the test.

No Test Material on This Page

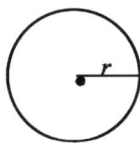
Math Test - Calculator

55 MINUTES, 38 QUESTIONS

Turn to Section 4 of your answer sheet to answer the questions in this section.

DIRECTIONS

For questions 1-30, solve each problem, choose the best answer from the choices provided, and fill in the corresponding circle on your answer sheet. **For questions 31-38,** solve the problem and enter your answer in the grid on your answer sheet. Please refer to the directions before question 31 on how to enter your answers in the grid. You may use any available space in your test booklet for scratch work.

NOTES

1. The use of a calculator **is permitted**.

2. All variables and expressions used represent real numbers unless otherwise indicated.

3. Figures provided in this test are drawn to scale unless otherwise indicated.

4. All figures lie in a plane unless otherwise indicated.

5. Unless otherwise indicated, the domain of a given function f is the set of all real numbers x for which $f(x)$ is a real number.

REFERENCE

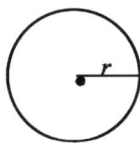 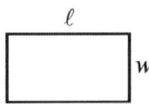 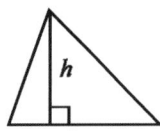 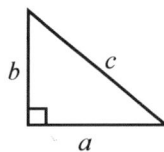 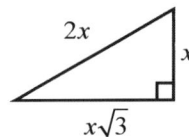 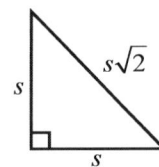

$A = \pi r^2$
$C = 2\pi r$

$A = \ell w$

$A = \dfrac{1}{2}bh$

$c^2 = a^2 + b^2$

Special Right Triangles

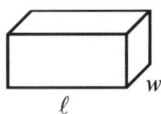 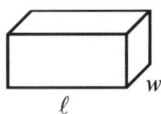   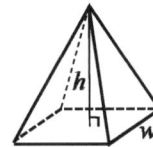

$V = \ell w h$

$V = \pi r^2 h$

$V = \dfrac{4}{3}\pi r^3$

$V = \dfrac{1}{3}\pi r^2 h$

$V = \dfrac{1}{3}\ell w h$

The number of degrees of arc in a circle is 360.

The number of radians of arc in a circle is 2π.

The number of the measures in degrees of the angles of a triangle is 180.

CONTINUE

4

1

Janet subscribes to an online movie service that charges a monthly fee of $5.00 and $0.75 per movie. Which of the following functions gives Janet's cost, in dollars, for a month in which she saw x movies?

A) $C(x) = 5.75x$

B) $C(x) = 5x + 0.75$

C) $C(x) = 5 + 0.75x$

D) $C(x) = 5(x + 0.75)$

Questions 2-3 refer to the following information.

	Boys	Girls	Total
Junior	20		
Senior		17	
Total	40		64

The incomplete table shows the distribution of grades and gender for 64 students who entered an essay contest.

2

If the contest winner will be selected at random, what is the probability that the winner will be a senior?

A) $\dfrac{29}{64}$

B) $\dfrac{32}{64}$

C) $\dfrac{37}{64}$

D) $\dfrac{47}{64}$

3

How many girls or juniors are entered in the essay contest?

A) 7

B) 17

C) 44

D) 53

4

Monthly Rainfall in City K

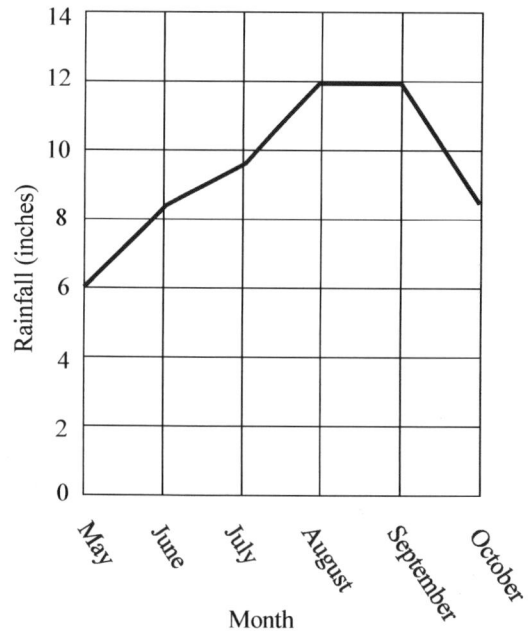

The line graph above shows the monthly rainfall from May to October in City K. What is the average rate of change of rainfall, in inches per month, from May to September?

A) 1.5

B) 2

C) 2.5

D) 3

5

$$k = \sqrt{\frac{ab}{a+b}}$$

In the equation above, which of the following represents b in terms of a and k?

A) $b = \dfrac{ak}{a-k}$

B) $b = \dfrac{ak}{a+k}$

C) $b = \dfrac{ak^2}{a-k}$

D) $b = \dfrac{ak^2}{a-k^2}$

6

$$C(x) = 200x + 150$$
$$R(x) = -0.5x^2 + 600x + 100$$

In the equations above, a company produces x units of a product per month, where $C(x)$ represents the total cost and $R(x)$ represents the total revenue for the month. The profit is the difference between revenue and the cost where $P(x) = R(x) - C(x)$. For what value of x will the company create the maximum profit for the month?

A) 200

B) 400

C) 600

D) 1200

7

$$x^3 - 3x + k = (x-1)(ax^2 + bx + c)$$

In the equation above, a, b, c, and k are constants. If the equation is true for all values of x, what is the value of c?

A) 4

B) 2

C) –2

D) –4

8

A car traveled at an average speed of p miles per hour for h hours and consumed fuel at a rate of m miles per gallon. How many gallons of fuel did the car use for the trip?

A) $\dfrac{p}{hm}$

B) phm

C) $\dfrac{ph}{m}$

D) $\dfrac{m}{ph}$

CONTINUE

9

$$f(x) = (x+a)^2 - 36$$

In the function f above, a is a constant. If sum of the zeros of the function is -10, which of the following is the product of the zeros of the function?

A) -36

B) -26

C) -11

D) 26

10

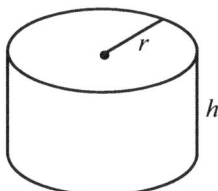

In the cylinder above, r is the radius of the base and h is the height of the cylinder. If the volume of the cylinder is equal to the surface area of the cylinder. Which of the following is true about the cylinder?

A) $r = \dfrac{h}{2}$

B) $r = \dfrac{2h}{h-2}$

C) $r = \dfrac{h+2}{h}$

D) $r = \dfrac{h-2}{h}$

11

In triangle ABC, measure of $m\angle A + m\angle B = 90°$. Which of the following is always true?

A) $\sin \angle A = \sin \angle B$

B) $\cos \angle A = \cos \angle B$

C) $\sin \angle A = \cos \angle B$

D) $\tan \angle A = \tan \angle B$

12

Omega-3 in five products

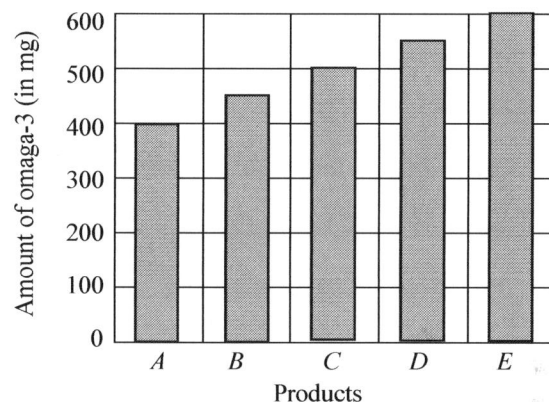

The graph above shows the amount of omega-3 supplied by five different products A, B, C, D, and E. The cost of the products A, B, C, D, and E are $40.00, $50.00, $55.00, $58.00, and $62.00, respectively. Which of the five products supplies the most omega-3 per dollars?

A) A
B) C
C) D
E) E

CONTINUE

Questions 13-14 refer to the following information.

Near the surface of the earth, the speed (v) of an object falling freely from rest and the distance (s) it falls during time t, where g is 9.8 meters/s^2, can be found as follows.

$$v(t) = gt \quad \text{and} \quad s(t) = \frac{1}{2}gt^2$$

13

When an object falls freely near the earth, what is the distance traveled by the object during the time interval 1 second to 2 seconds?

A) 19.6 meters
B) 18.2 meters
C) 14.7 meters
D) 12.6 meters

14

When an object falls freely near the earth, what is the average speed, in meters per second, during the time interval 2 seconds to 4 seconds?

A) 9.8
B) 19.6
C) 29.4
D) 39.2

15

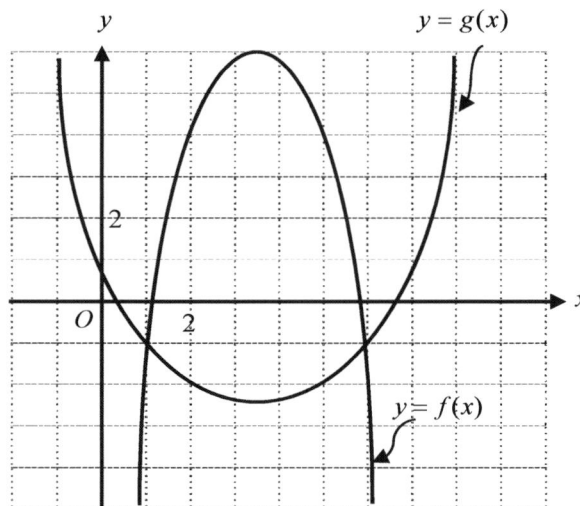

Graphs of the quadratic functions f and g are shown in the xy-plane above. For which of the following values of x does $f(x) + 2g(x) = 0$?

A) 1

B) 3

C) 3.5

D) 5

CONTINUE

16

Yellow Cab of Charleston charges $5.00 for the first two miles and 50 cents for each additional $\frac{1}{5}$ of a mile. If Peter paid a yellow Cab $25.00 for his trip, how many miles did he travel for the trip?

A) 6 miles
B) 8 miles
C) 10 miles
D) 12 miles

17

$$y = 3^{\left(-x^2 + 8x - 12\right)}$$

In the equation above, what is the maximum value of y?

A) 27
B) 81
C) 243
D) 729

18

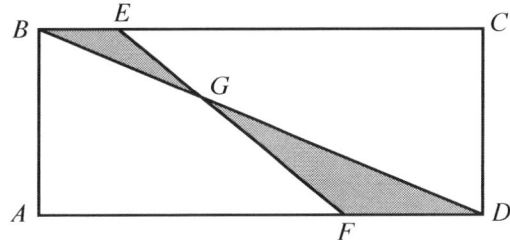

In rectangle $ABCD$ above, $AB = 8$, $BE = 4$, $FD = 6$, and $BC = 20$. If $\overline{BD}$ and $\overline{EF}$ intersect at point G, what is the area of the shaded region?

A) 18.4

B) 20.8

C) 36.8

D) 41.6

19

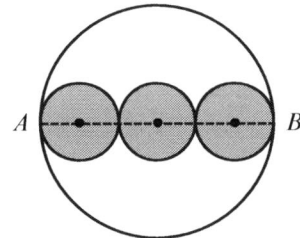

In the figure above, three identical circles are tangent to each other and are inscribed in a larger circle. $\overline{AB}$ is the diameter of the larger circle. If the area of the shaded region is 15, what is the area of the larger circle?

A) 27
B) 39.2
C) 42
D) 45

CONTINUE

20

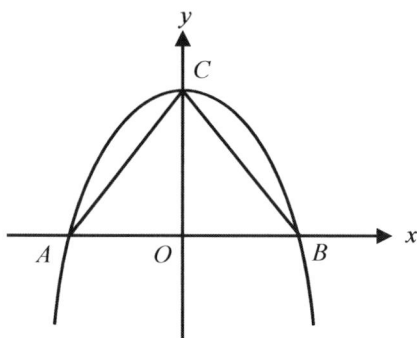

Note: Figure not drawn to scale.

In the xy-plane above, vertices of triangle ABC are lying on the graph of $y = -x^2 + k$, where k is a constant. If the area of triangle ABC is 27, what is the value of k?

A) 3

B) 9

C) 18

D) 27

21

The sum of four numbers is 2,835. One of the numbers, a, is 25% more than the sum of the other three numbers. What is the average of the other three numbers?

A) 365
B) 420
C) 475
D) 600

22

When the graph of $y = x^2$ is moved to the right 3 units and down 4 units, which of the following is the equation of the resulting graph?

A) $y = (x-1)(x-5)$

B) $y = (x+1)(x-4)$

C) $y = (x-2)(x-4)$

D) $y = (x+2)(x-2)$

23

For a quadratic function $f(x) = x^2 + bx + c$ in the xy-plane, $f(2) = f(6) = 4$. What is the value of c?

A) 4
B) 8
C) 12
D) 16

CONTINUE

24

$$y = (x-5)^2$$
$$y = k$$

In the system of equations above, k is a constant. If the system has two points of intersection and the distance between these two points is 10, what is the value of k?

A) 20
B) 25
C) 30
D) 35

25

$$x^2 - 6x + 2 = 0$$

If a and b are the roots of the equation above, what is the value of $(a+1)(b+1)$?

A) 9
B) 10.5
C) 11
D) 12

26

For a polynomial $p(x)$, the remainder when $p(x)$ is divided by $(x-1)$ is 2, and the remainder when $p(x)$ is divided by $(x+1)$ is -4. If $p(x)$ is divided by $(x^2 - 1)$, what is the reminder?

A) $2x+1$
B) $3x-1$
C) $4x+1$
D) -8

27

If $\cos(3x+20)^\circ = \sin(x+10)^\circ$, what is the value of $\sin(2x)^\circ$?

A) $\dfrac{1}{\sqrt{2}}$

B) $\dfrac{1}{2}$

C) $\dfrac{\sqrt{3}}{2}$

D) $\sqrt{3}$

CONTINUE

28

If the function f is defined by $f(x-1) = x^2 - 1$, which of the following represents $f(x)$?

A) $f(x) = x^2$

B) $f(x) = x^2 + 1$

C) $f(x) = x^2 + 2x$

D) $f(x) = x^2 + 2x - 1$

29

$$y > -2x + a$$
$$y < 2x + b$$

In the xy-plane, if $(-1, 2)$ is a solution to the system of inequalities above, which of the following must be true about relationship between a and b?

A) $a + b > 0$

B) $a + b < 0$

C) $a - b > 0$

D) $a - b < 0$

30

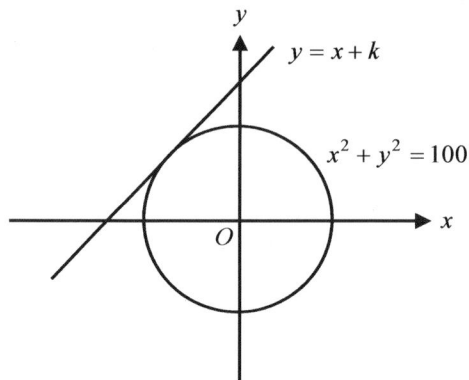

The graph of a circle, whose equation is $x^2 + y^2 = 100$, is shown in the xy-plane above. If the graph of $y = x + k$, where k is a constant, is tangent to the graph of the circle, which of the following is closest to the value of k?

A) 12.03

B) 14.14

C) 15.05

D) 17.31

CONTINUE

DIRECTIONS

For questions 31-38, solve the problem and enter your answer in the grid, as described below, on the answer sheet.

1. Although not required, it is suggested that you write your answer in the boxes at the top of the columns to help you fill in the circles accurately. You will receive credit only if the circles are filled in correctly.
2. Mark no more than one circle in any column.
3. No question has a negative answer.
4. Some problems may have more than one answer.
5. **Mixed numbers** such as $3\frac{1}{2}$ must be gridded as 3.5 or 7/2. (If $3\;1\;/\;2$ is entered into the grid, it will be interpreted as $\frac{31}{2}$, not $3\frac{1}{2}$.)
6. **Decimal answers:** If you obtain a decimal answer with more digits than the grid can accommodate, it may be either rounded or truncated, but it must fill the entire grid.

Answer: $\frac{7}{12}$

Write answer in boxes. → Fraction line

Grid in result. →

Answer: 2.5

← Decimal point

Acceptable ways to grid $\frac{2}{3}$ are:

Answer: 201
Either position is correct.

Note: You may start your answers in any column, space permitting. Columns you don't need to use should be left blank.

CONTINUE ➡

31

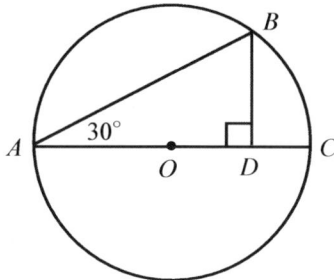

In the figure above, the radius of circle O is 10. If the measure of $\angle BAC$ is $30°$, what is the length of $\overline{AD}$?

32

If the graph of $f(x) = x^2 - 2x + k$ intersects the line $f(x) = 3$ at exactly one point, what is the value of k ?

33

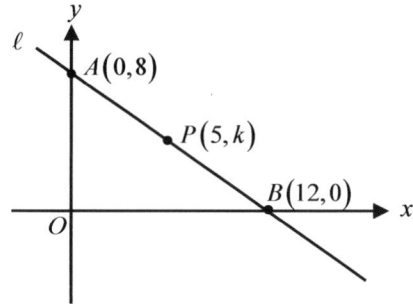

In the xy-plane above, if line ℓ passes through point P, what is the value of k ?

34

For $i = \sqrt{-1}$, $a + b + (ab)i = 4 + 2i$, where a and b are real numbers. What is the value of $a^2 + b^2$?

35

The mean score of 10 students of an algebra class was 85. When two new students enrolled, the mean increased to 86. What was the average of the new students?

CONTINUE

36

If $f\left(\dfrac{x}{5}\right) = x^2 - 3x + 1$, what is the value of $f(-1)$?

38

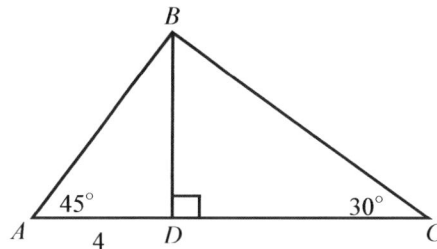

In the figure above, $AD = 4$, measure of $\angle BAD$ is $45°$, and measure of $\angle BCD$ is $30°$. What is the area of triangle ABC? (Round your answer to the nearest tenth)

37

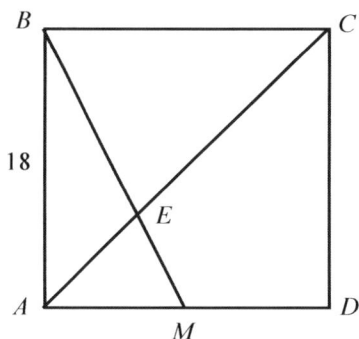

In the figure above, $ABCD$ is a square and $AB = 18$. If M is the midpoint of $\overline{AD}$, what is the area of $\triangle ABE$?

STOP

If you finish before time is called, you may check your work on this section only.
Do not turn to any other section in the test.

Math Conversion Table

Raw Score	Scaled Score	Raw Score	Scaled Score
58	800	27	500
57	800	26	490
56	800	25	480
55	800	24	470
54	790	23	460
53	780	22	460
52	770	21	450
51	750	20	440
50	740	19	430
49	730	18	430
48	720	17	420
47	710	16	420
46	700	15	410
45	690	14	400
44	670	13	390
43	680	12	380
42	670	11	370
41	660	10	360
40	650	9	450
39	640	8	340
38	630	7	330
37	620	6	310
36	610	5	290
35	600	4	280
34	590	3	270
33	580	2	260
32	560	1	240
31	550	0	200
30	540		
29	530		
28	520		

Answer Explanations

Test 12 — Answers and Explanations

SECTION 3	1	2	3	4	5	6	7	8	9	10
	A	C	A	C	D	B	A	B	A	D
	11	12	13	14	15	16	17	18	19	20
	A	B	D	C	B	36	36	18	200	8

SECTION 4	1	2	3	4	5	6	7	8	9	10
	C	C	C	A	D	B	C	C	C	B
	11	12	13	14	15	16	17	18	19	20
	C	A	C	C	D	C	B	B	D	B
	21	22	23	24	25	26	27	28	29	30
	B	A	D	B	A	B	B	C	D	B
	31	32	33	34	35	36	37	38		
	15	4	$\frac{14}{3}$	12	91	41	54	21.9		

SECTION 3

1. **A**

   $2a + 8 = 14 \rightarrow a = 3$ Putting $a = 4$ in the equation

   $9b - 12b + 42 = 0 \rightarrow 3b = 42 \rightarrow b = 14$

2. **C**

   $6a^2 - 24b^2 = 6(a^2 - 4b^2) = 6(a+2b)(a-2b) =$

   $6\left(\frac{1}{2}\right)\left(\frac{1}{3}\right) = 1$

3. **A**

   $a - b = (3+2i) - (3-2i) = 4i$

   $a^2 - 2ab + b^2 = (a-b)^2 \rightarrow (4i)^2 = -16$

4. **C**

   $\begin{cases} \dfrac{\sqrt{x}}{2} + \sqrt{y} = 5 \\ \sqrt{x} - \dfrac{\sqrt{y}}{2} = -\dfrac{1}{2} \end{cases} \rightarrow \begin{cases} \dfrac{\sqrt{x}}{2} + \sqrt{y} = 5 \\ 2\sqrt{x} - \sqrt{y} = -1 \end{cases}$

   Using addition,

   $\dfrac{5\sqrt{x}}{2} = 4 \rightarrow \sqrt{x} = \dfrac{8}{5} \rightarrow x = \dfrac{64}{25}$

   When $x = \dfrac{64}{25}$, $y = \dfrac{441}{25}$.

   Therefore, $a = \dfrac{64}{25}$

5. **D**

   The equation of the line is $C = \left(\dfrac{250 - 50}{6}\right)d + 50$.

   Therefore, $C(21) = \left(\dfrac{200}{6}\right)(21) + 50 = 750$.

6. **B**

   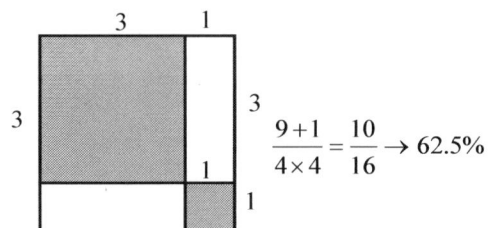

   $\dfrac{9+1}{4 \times 4} = \dfrac{10}{16} \rightarrow 62.5\%$

7. A

There are 450 under 18 years and 220 prefer smartphone to see movies. Therefore,

$$\frac{220}{450} = 0.48\overline{8} \approx 0.49$$

8. B

$$n(A \vee B) = n(A) + n(B) - n(A \wedge B)$$
$$\rightarrow 420 + 240 - 100 = 560$$

9. A

Axis of symmetry: $x = \dfrac{-k}{4}$

Midpoint of -2 and 8 is $\dfrac{-2+8}{2} = 3$

They must be equal. $\rightarrow \dfrac{-k}{4} = 3 \rightarrow k = -12$

10. D

The resulting equation is

$$y = (x-5)^2 + 3 \rightarrow y = x^2 - 10x + 28.$$

11. A

Since $\triangle AED \sim \triangle CDF$, $\dfrac{AD}{DC} = \dfrac{3}{2}$ and

$\dfrac{\text{area of } \triangle ABD}{\text{area of } \triangle CBD} = \dfrac{3}{2}$. Therefore, the area of

$\triangle ABD$ is $78 \times \dfrac{3}{5} = 46.8$.

12. B

$$p(3) = (9 + 3a + 2) - 2(3 - b) = 0 \rightarrow 3a + 2b = -5$$
$$p(1) = (1 + a + 2) - 2(1 - b) = 0 \rightarrow a + 2b = -1$$

Using subtracting, $2a = -4 \rightarrow a = -2$.

13. D

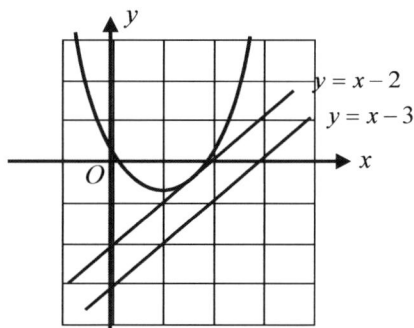

14. C

$y = x(x-2) = x^2 - 2x \rightarrow y = (x-1)^2 - 1$

From the equation, x – intercept $\rightarrow$ 0 and 2

And vertex: $(1, -1)$

In the xy-plane above, when $k = -3$, no solution.

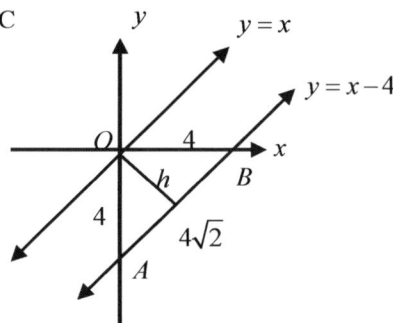

From $y = x - 4$, $AO = 4$, $BO = 4$, and $AB = 4\sqrt{2}$.

Area of $\triangle AOB = \dfrac{4 \times 4}{2} = \dfrac{4\sqrt{2} \times h}{2} \rightarrow h = 2\sqrt{2}$

15. B

Coordinates of point $B \rightarrow (6, 0)$ and slope of

$\overline{PQ}$ is $-\dfrac{5}{3}$. Equation of the line: $y = -\dfrac{5}{3}x + b$

Putting $(3, 5)$ in the equation, $5 = -\dfrac{5}{3}(3) + b$

$\rightarrow b = 10$.

16. 36

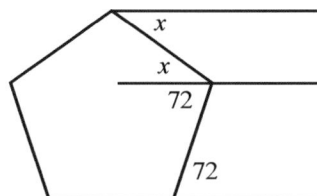

An exterior angle of the regular pentagon is

$\dfrac{360}{5} = 72$ and an interior angle is 108.

Therefore, $x + 72 = 108 \rightarrow x = 36$.

17. 36

$$\frac{\sqrt{x^2} + 2\sqrt{x}}{6} = 8 \rightarrow (\sqrt{x})^2 + 2\sqrt{x} - 48 = 0$$
$$(\sqrt{x} + 8)(\sqrt{x} - 6) = 0 \rightarrow \sqrt{x} = 6, \cancel{-8} \, (\sqrt{x} > 0)$$

Answer Explanations

Therefore, $x = 36$.

18. 18

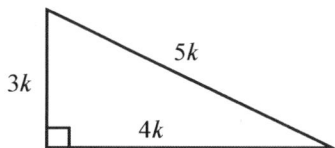

Since $\sin \angle BAC = 0.6 = \dfrac{3}{5}$, let $AB = 5k$ and $AC = 4k$. Area of the triangle is

$$\dfrac{(3k)(4k)}{2} = 216. \;\rightarrow\; 6k^2 = 216 \;\rightarrow\; k = 6$$

Therefore, $BC = 3k = 3(6) = 18$.

19. 200

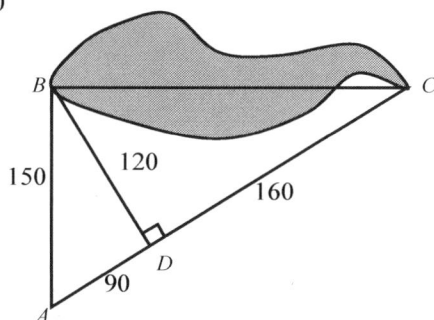

$$BD^2 = AD \cdot DC \;\rightarrow\; 120^2 = 90 \cdot DC \;\rightarrow\; DC = 160$$

Therefore, $BC = 200$. (Pythagorean Theorem)

20. 8

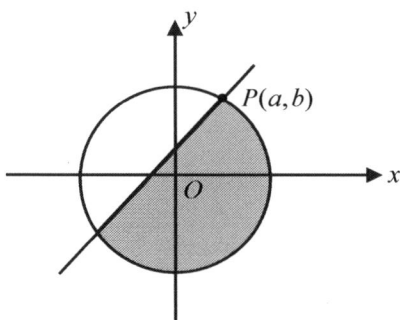

In the shaded region of the solution set, b has a maximum at point P. P is a point of intersection of the system of equations.

$$\begin{cases} x^2 + y^2 = 100 \\ x = y - 2 \end{cases}$$

$$(y-2)^2 + y^2 = 100 \;\rightarrow\; 2y^2 - 4y + 4 = 100$$

$$y^2 - 2y - 96 = 0 \;\rightarrow\; (y-8)(y+6) = 0$$

$$y = 8 \text{ or } -6.$$

The maximum of $b = 8$.

1. C

2. C
There are 37 seniors out of 64.

3. C
$24(\text{girls}) + 27(\text{juniors}) - 7(\text{girls} \wedge \text{juniors}) = 44$

4. A
Average rate of change
$$= \dfrac{12 - 6}{4} = 1.5 \text{ inches per month}$$

5. D
$$k = \sqrt{\dfrac{ab}{a+b}} \;\rightarrow\; k^2 = \dfrac{ab}{a+b} \;\rightarrow\; ak^2 + bk^2 = ab$$

$$ak^2 = ab - bk^2 \;\rightarrow\; ak^2 = b(a - k^2) \;\rightarrow\; b = \dfrac{ak^2}{a - k^2}$$

6. B
$$P(x) = -0.5x^2 + 600x + 100 - 200x - 150$$
$$P(x) = -0.5x^2 + 400x - 50$$
P has a maximum at $x = \dfrac{-(400)}{2(-0.5)} = 400$.

7. C
For $x = 1 \;\rightarrow\; 1 - 3 + k = 0 \;\rightarrow\; k = 2$.
$c = k \times (-1) = 2 \times (-1) = -2$

8. C
Total travel distance is ph miles.
$$\# \text{ of gallons} = \dfrac{ph}{m}$$

9. C
$$f(x) = (x + a)^2 - 36$$
Zeros: $x^2 + 2ax + a^2 - 36 = 0$
$$\text{Sum} = \dfrac{-2a}{1} = -10 \;\rightarrow\; a = 5$$
$$\text{Product} = \dfrac{a^2 - 36}{1} = \dfrac{5^2 - 36}{1} = -11$$

10. B

$$\pi r^2 h = 2\pi r^2 + 2\pi rh \;\rightarrow\; rh = 2r + 2h$$

$$r(h-2) = 2h \;\rightarrow\; r = \frac{2h}{h-2}$$

11. C

Cofunction: If $A + B = 90$, then $\sin A = \cos B$.

12. A

$$A = \frac{400}{40} = 10 \qquad B = \frac{450}{50} = 9 \qquad C = \frac{500}{55} = 9.09$$

$$D = \frac{550}{58} = 9.48 \qquad E = \frac{600}{62} = 9.68$$

13. C

$$d = s(2) - s(1) = \frac{1}{2}(9.8)(4) - \frac{1}{2}(9.8)(1) = 14.7$$

14. C

Average speed

$$= \frac{s(4) - s(2)}{4 - 2} = \frac{\frac{1}{2}(9.8)(16) - \frac{1}{2}(9.8)(4)}{2} = 29.4$$

15. D

For $x = 5$, $f(5) = 4$ and $g(5) = -2$.

$$f(5) + 2g(5) = 4 + 2(-2) = 0$$

16. C

50 cents per 0.2 mile $\rightarrow$ \$2.50 per mile

$$C = 5 + 2.5(x - 2) \;\rightarrow\; 25 = 5 + 2.5(x - 2)$$

$$(x - 2) = 8 \;\rightarrow\; x = 10 \text{ miles}$$

17. B

$y = 3^{k(x)}$ has a maximum at a maximum value of

k. $\quad k(x) = -x^2 + 8x - 12 \;\rightarrow\; k(x) = -(x - 4)^2 + 4$

The maximum of k is 4.

Therefore, the maximum of y is $3^4 = 81$.

18. B

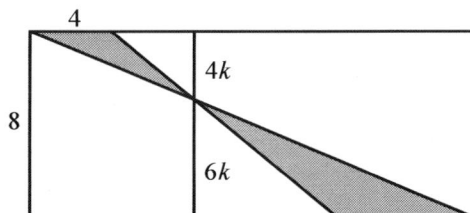

The two triangles are similar and ratio of corresponding sides are 4:6. Therefore, the ratio of their heights are also 4:6. Let define their heights as $4k$ and $6k$. $\quad 4k + 6k = 8 \;\rightarrow\; k = 0.8$

The heights are $4k = 3.2$ and $6k = 6.8$.

Therefore, sum of the areas of the triangles is

$$\frac{4 \times 3.2}{2} + \frac{6 \times 4.8}{2} = 20.8 .$$

19. D

The ratio of the radius of the three identical circles and the largest circle is 1:1:1:3.

The ratio of their areas is 1:1:1:9.

Define their areas as k, k, k, $9k$.

Since $3k = 15 \;\rightarrow\; k = 5$. Therefore the area of the largest circle is $9k = 9(5) = 45$.

20. B

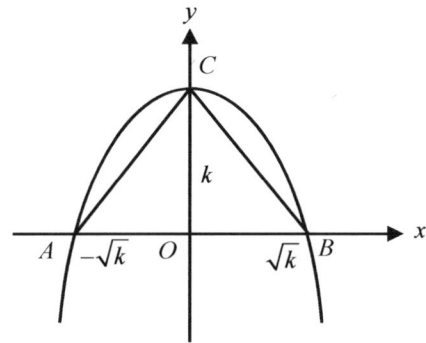

Since $OC = k$ and $AB = 2\sqrt{x}$,

$$\frac{2\sqrt{k} \times k}{2} = 27 \;\rightarrow\; k^{\frac{3}{2}} = 27 \;\rightarrow\; k = 9$$

21. B

Let $S = $ sum of the three numbers.

$$a + s = 1.25S + S \;\rightarrow\; 2.25S = 2835 \;\rightarrow\; S = 1260$$

Therefore, average is $\dfrac{1260}{3} = 420$.

22. A

The resulting equation is $y = (x - 3)^2 - 4$.

$$y = x^2 - 6x + 9 - 4 \;\rightarrow\; y = x^2 - 6x + 5 \;\rightarrow$$

$$y = (x - 1)(x - 5)$$

23. D

Answer Explanations

Axis of symmetry: $x = \dfrac{2+6}{2} = 4$

From the equation, axis of symmetry is

$x = -\dfrac{b}{2}. \rightarrow -\dfrac{b}{2} = 4 \rightarrow b = -8$

Now, $f(x) = x^2 - 8x + c$.

Using the point $(2, 4)$ or $(6, 4)$,

$4 = 2^2 - 8(2) + c \rightarrow c = 16$.

24. B

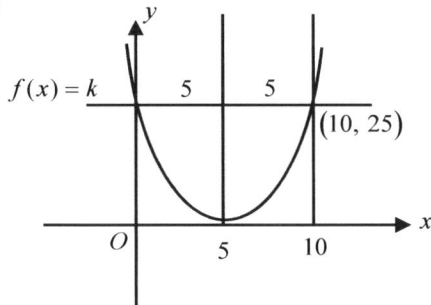

$y = (x-5)^2 \rightarrow$ At $x = 10$, $y = (10-5)^2 = 25$

Therefore, $k = 25$.

25. A

From the equation: $a + b = 6$ and $ab = 2$.

$(a+1)(b+1) = ab + a + b + 1 = 2 + 6 + 1 = 9$

26. B

The remainder is $ax + b$ when $p(x)$ is divided by

$(x+1)(x-1)$. Since divisor is degree 2

polynomial, the remainder is degree one

polynomial.

$p(x) = (x-1)(x+1)q_3(x) + ax + b$ (1)

$p(x) = (x-1)q_1(x) + 2$ (2)

$p(x) = (x+1)q_2(x) - 4$ (3)

$p(1) = 2 = a + b \rightarrow$ from (1) and (2)

$p(-1) = -4 = -a + b \rightarrow$ from (1) and (3)

Solve the system of equations.

$a = 3$ and $b = -1$. Therefore, the remainder is

$3x - 1$.

27. B

$\cos(3x+20)^\circ = \sin(x+10)^\circ \rightarrow$ cofunction $\rightarrow$

$3x + 20 + x + 10 = 90$

$4x = 60 \rightarrow 2x = 30$

Therefore, $\sin 2x = \sin 30 = \dfrac{1}{2}$.

28. C

1) Change variable.

$f(x-1) = x^2 - 1 \rightarrow f(k-1) = k^2 - 1$

$k - 1 = x \rightarrow k = x + 1$

2) Replace k with x.

$f(x) = (x+1)^2 - 1 \rightarrow f(x) = x^2 + 2x$

Or, simply substitute $x+1$ in the equation.

29. D

Since $(-1, 2)$ is a solution,

$2 > -2(-1) + a \rightarrow a < 0$ and

$2 < 2(-1) + b \rightarrow b > 4$

Since $a - b < 0$, choice D is always true.

30. B

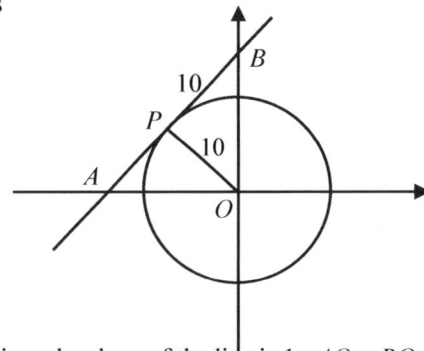

Since the slope of the line is 1, $AO = BO$.

Triangle BOP is isosceles with $PO = PB$.

Therefore, $OB = 10\sqrt{2} \approx 14.14$.

31. 15

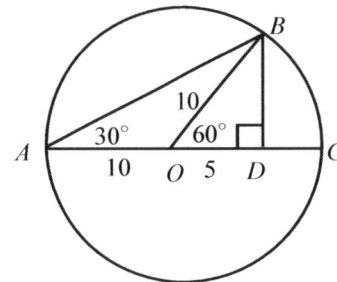

$\triangle ABO$ is isosceles with $AO = BO = 10$.

$OD = \dfrac{1}{2} \cdot OB = 5 \rightarrow AD = 10 + 5 = 15$

32. 4

$f(x) = 3$ intersects the vertex of

$f(x) = x^2 - 2x + k \rightarrow f(x) = (x-1)^2 + k - 1$

Therefore, $k - 1 = 3 \rightarrow k = 4$.

33. $\dfrac{14}{3}$

Collinear: Slopes between any two points are equal.

$$\dfrac{8-0}{0-12} = \dfrac{k-0}{5-12} \rightarrow k = \dfrac{14}{3}$$

34. 12

From $a + b + (ab)i = 4 + 2i$,

$a + b = 4$ and $ab = 2$.

$(a+b)^2 = 4^2 \rightarrow a^2 + b^2 + 2ab = 16$

$a^2 + b^2 = 16 - 2ab \rightarrow a^2 + b^2 = 16 - 4 = 12$

35. 91

$10 \times 85 = 850$ and $12 \times 86 = 1032$

The average of the new students is

$\dfrac{1032 - 850}{2} = 91$.

36. 41

For $x = -5$, $f(-1) = (-5)^2 - 3(-5) + 1 = 41$

37. 54

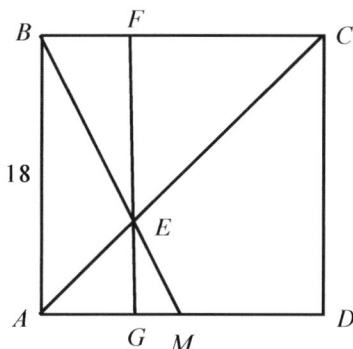

$\triangle AEM$ and $\triangle BCE$ are similar.

$\dfrac{BC}{AM} = \dfrac{FE}{GE} = \dfrac{2}{1} \rightarrow FE = 12$ and $GE = 6$

Area of $\triangle ABE$ = area of $\triangle ABC$ − area of $\triangle BEC$

$= \dfrac{18 \times 18}{2} - \dfrac{18 \times 12}{2} = 162 - 108 = 54$.

38. 21.9

$BD = 4$ and $CD = 4\sqrt{3}$

Area of $\triangle ABC = \dfrac{(4 + 4\sqrt{3})4}{2} \approx 21.9$

PRACTICE TEST 13

Dr. John Chung's SAT Math

Math Test - No Calculator

25 MINUTES, 20 QUESTIONS

Turn to Section 3 of your answer sheet to answer the questions in this section.

DIRECTIONS

For questions 1–15, solve each problem, choose the best answer from the choices provided, and fill in the corresponding circle on your answer sheet. **For questions 16–20,** solve the problem and enter your answer in the grid on your answer sheet. Please refer to the directions before question 16 on how to enter your answers in the grid. You may use any available space in your test booklet for scratch work.

NOTE

1. The use of a calculator **is not permitted**.

2. All variables and expressions used represent real numbers unless otherwise indicated.

3. Figures provided in this test are drawn to scale unless otherwise indicated.

4. All figures lie in a plane unless otherwise indicated.

5. Unless otherwise indicated, the domain of a given function f is the set of all real numbers x for which $f(x)$ is a real number.

REFERENCE

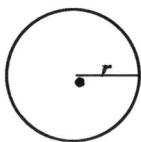 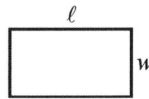 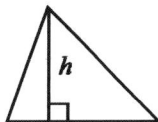 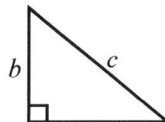 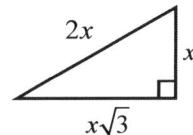 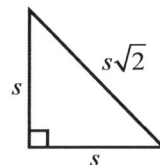

$A = \pi r^2$
$C = 2\pi r$

$A = \ell w$

$A = \frac{1}{2} bh$

$c^2 = a^2 + b^2$

Special Right Triangles

 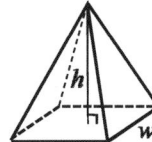

$V = \ell wh$

$V = \pi r^2 h$

$V = \frac{4}{3}\pi r^3$

$V = \frac{1}{3}\pi r^2 h$

$V = \frac{1}{3}\ell wh$

The number of degrees of arc in a circle is 360.
The number of radians of arc in a circle is 2π.
The number of the measures in degrees of the angles of a triangle is 180.

CONTINUE

1

Which of the following expressions is equal to 0 for some value of x?

A) $\dfrac{|x-2|}{2}+1$

B) $\dfrac{|x-1|}{2}+2$

C) $\dfrac{|x+2|}{2}+3$

D) $\dfrac{|x-1|}{2}-4$

2

$$f(x) = x^2 + 2x + k$$

In the function above, k is a constant. If the minimum value of $f(x)$ is 8, what is the value of k?

A) 4

B) 6

C) 8

D) 9

3

$$y = (x-2)(x+2)$$
$$y = -3$$

How many ordered pairs (x, y) satisfy the system of equations shown above?

A) 0

B) 1

C) 2

D) Infinitely many

4

If $f(x-1) = -5x + 2$, what is $f(x+2)$ equal to?

A) $-5x - 2$

B) $-5x - 13$

C) $-5x - 14$

D) $-5x^2 - 10x + 2$

CONTINUE

5

$$2x^2 + 8x + 10$$

Which of the following is equivalent to the expression above?

A) $(2x+2)(x+5)$

B) $(2x-2)(x-5)$

C) $2(x+2)^2 + 10$

D) $2(x+2)^2 + 2$

6

If $\dfrac{2\sqrt{a} - \sqrt{b}}{2\sqrt{b}} = \dfrac{1}{4}$, what is the value of $\dfrac{a}{b}$?

A) $\dfrac{2}{3}$

B) $\dfrac{3}{4}$

C) $\dfrac{4}{9}$

D) $\dfrac{9}{16}$

7

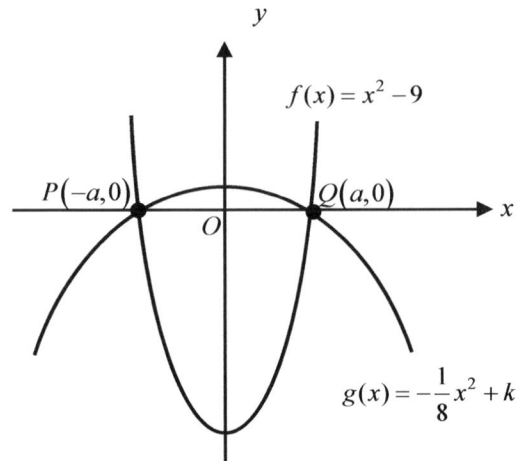

The functions f and g are defined by $f(x) = x^2 - 9$ and $g(x) = -\dfrac{1}{8}x^2 + k$, where k is a constant. In the xy-plane above, the graphs of f and g intersect at the points P and Q. What is the value of k?

A) 1

B) $\dfrac{8}{9}$

C) $\dfrac{9}{8}$

D) $\dfrac{3}{2}$

CONTINUE

8

$$\frac{x^2}{4} - x = \frac{k}{2}$$

In the quadratic equation above, k is a constant. For what following values of k does the equation have two unequal real roots?

A) -6

B) -4

C) -2

D) -1

9

$$(2-i)(3+2i) - 5i = a + bi$$

In the expression above, a and b are real numbers. What is the value of b? $\left(\text{Note}: i = \sqrt{-1}\right)$

A) -4

B) -2

C) 4

D) 8

10

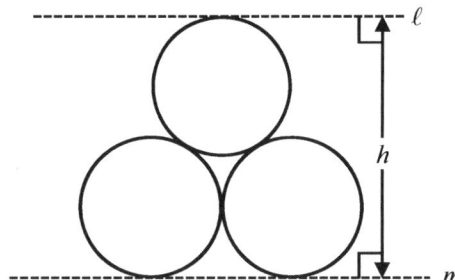

In the figure above, three identical circles with radius of 5 are tangent to one another. If lines ℓ and m are parallel and tangent to the circles, what is the value of h, the distance between the two lines?

A) $5\sqrt{2} + 10$

B) $10\sqrt{3} + 5$

C) $5\sqrt{3} + 10$

D) $10\sqrt{3} + 10$

11

The population of a nation decreases by 2.5% every 10 years. If the population was 10,000,000 people in 2000, which of the following expressions estimates the population of the nation in the year 2040?

A) $10,000,000(0.025)^{40}$

B) $10,000,000(0.025)^{4}$

C) $10,000,000(0.975)^{40}$

D) $10,000,000(0.975)^{4}$

CONTINUE

12

$$x^2 - 5x - 11 = a\left(x^2 - 1\right) + b\left(x + 1\right) + c$$

In the equation above, a, b, and c are constants. If the equation has infinitely many solutions, what is the value of c?

A) -7

B) -5

C) 2

D) 7

13

Friday, Peter drives from home to his office in 24 minutes. Saturday, there is no traffic, so he can drive to his office 15 miles per hour faster and gets to the office in 15 minutes. How far, in miles, is it from home to his office?

A) 10

B) 12

C) 20

D) 25

14

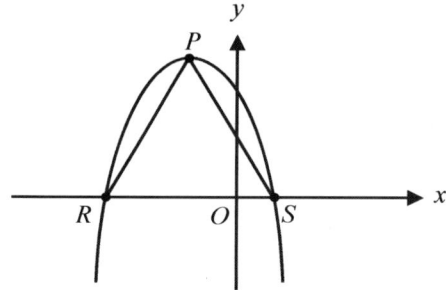

In the xy-plane above, the graph of the equation $y = a(x+4)(x-2)$ has a vertex at point P. If the graph intersects the x-axis at points R and S, what is the area of $\triangle RPS$?

A) $-54a$

B) $-27a$

C) $27a$

D) $54a$

15

When $2x^2 - 5x + 4$ is divided by $(x-1)(x+1)$, the resulting remainder is $ax + b$, where a and b are constants. What is the value of a?

A) 6

B) 5

C) -5

D) -6

CONTINUE

DIRECTIONS

For questions 16–20, solve the problem and enter your answer in the grid, as described below, on the answer sheet.

1. Although not required, it is suggested that you write your answer in the boxes at the top of the columns to help you fill in the circles accurately. You will receive credit only if the circles are filled in correctly.
2. Mark no more than one circle in any column.
3. No question has a negative answer.
4. Some problems may have more than one answer.
5. **Mixed numbers** such as $3\frac{1}{2}$ must be gridded as 3.5 or 7/2. (If $3\ 1\ /\ 2$ is entered into the grid, it will be interpreted as $\frac{31}{2}$, not $3\frac{1}{2}$.)
6. **Decimal answers:** If you obtain a decimal answer with more digits than the grid can accommodate, it may be either rounded or truncated, but it must fill the entire grid.

Answer: $\frac{7}{12}$

Write answer in boxes. → Fraction line

Grid in result.

Answer: 2.5

← Decimal point

Acceptable ways to grid $\frac{2}{3}$ are:

Answer: 201
Either position is correct.

Note: You may start your answers in any column, space permitting. Columns you don't need to use should be left blank.

CONTINUE

16

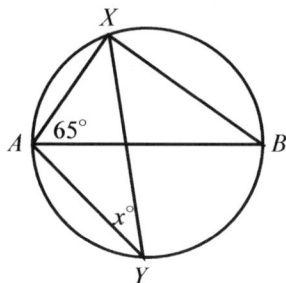

In the figure above, $\angle XAB$ and $\angle XYA$ are inscribed in the circle. If $\overline{AB}$ is the diameter of the circle, and the measure of $\angle XAB$ is $65°$, what is the value of x?

17

$$x^2 - 8x - 3 + y^2 - 6y = k^2$$

The graph of the equation above in the xy-plane is a circle, where k is a positive real number. If the radius of the circle is 8, what is the value of k?

18

$$(2x - 2)(x - 7) = 2(x - a)^2 - b$$

In the equation above, a and b are constants. If the equation is true for all real values of x, what is the value of b?

19

The function f is defined by $f(x) = x^3 - ax^2 + bx - 4$ where a and b are constants. In the xy-plane, the graph of f intersects the x-axis at the three points $(-1, 0)$, $(-2, 0)$, and $(k, 0)$. What is the value of k?

20

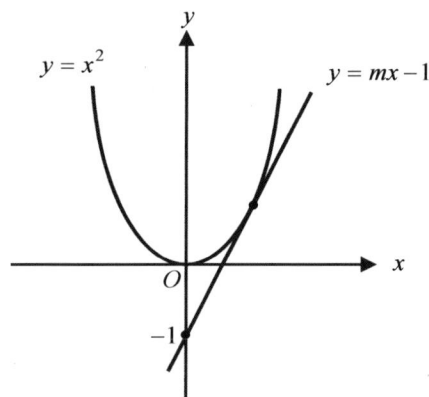

The graphs of $y = x^2$ and $y = mx - 1$ are shown in the xy-plane above, where m is the slope of the line. If the graph of $y = mx - 1$ is tangent to the graph of $y = x^2$, what is the value of m?

STOP

If you finish before time is called, you may check your work on this section only.
Do not turn to any other section in the test.

No Test Material on This Page

Math Test - Calculator

55 MINUTES, 38 QUESTIONS

Turn to Section 4 of your answer sheet to answer the questions in this section.

DIRECTIONS

For questions 1-30, solve each problem, choose the best answer from the choices provided, and fill in the corresponding circle on your answer sheet. **For questions 31-38,** solve the problem and enter your answer in the grid on your answer sheet. Please refer to the directions before question 31 on how to enter your answers in the grid. You may use any available space in your test booklet for scratch work.

NOTES

1. The use of a calculator **is permitted**.

2. All variables and expressions used represent real numbers unless otherwise indicated.

3. Figures provided in this test are drawn to scale unless otherwise indicated.

4. All figures lie in a plane unless otherwise indicated.

5. Unless otherwise indicated, the domain of a given function f is the set of all real numbers x for which $f(x)$ is a real number.

REFERENCE

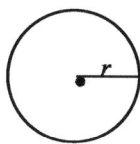 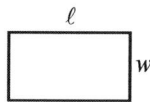 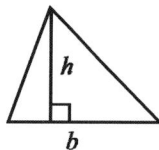 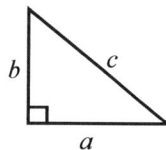 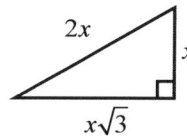 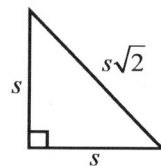

$A = \pi r^2$
$C = 2\pi r$

$A = \ell w$

$A = \dfrac{1}{2}bh$

$c^2 = a^2 + b^2$

Special Right Triangles

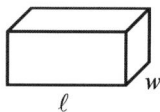  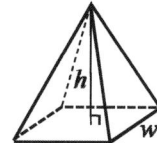

$V = \ell wh$

$V = \pi r^2 h$

$V = \dfrac{4}{3}\pi r^3$

$V = \dfrac{1}{3}\pi r^2 h$

$V = \dfrac{1}{3}\ell wh$

The number of degrees of arc in a circle is 360.
The number of radians of arc in a circle is 2π.
The number of the measures in degrees of the angles of a triangle is 180.

CONTINUE

1

Peter did push-ups every day. Each day after the first day, he did m more push-ups than the day before. If he did k push-ups on the first day, which of the following was the number of push-ups on the 7th day?

A) $7(m+k)$

B) $k+7(m-1)$

C) $k+7m$

D) $k+6m$

2

One soccer ball and two soccer shirts together cost P dollars. Three soccer balls and four soccer shirts together cost Q dollars. What is the cost, in dollars, of one soccer ball and one soccer shirt?

A) $Q+P$

B) $Q-2P$

C) $Q-P$

D) $\dfrac{Q-P}{2}$

3

$$x+16 < 2x+1 < x+46$$

Which of the following is equivalent to the inequality above?

A) $|x-15| < 45$

B) $|x-29| \le 15$

C) $|x-30| < 15$

D) $|x+15| < 30$

4

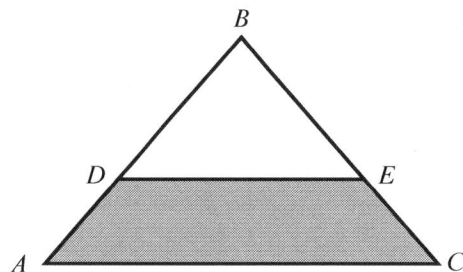

In the figure above, $\triangle ABC$ and $\triangle DBE$ are similar and $\dfrac{AD}{AB} = \dfrac{1}{3}$. If the area of $\triangle ABC$ is 45, what is the area of the shaded region?

A) 15

B) 17.5

C) 25

D) 30

CONTINUE

5

Which of the following are the zeros of the function

$$f(x) = \frac{1}{10}x^2 + \frac{1}{5}x + \frac{1}{20}?$$

A) $x = \dfrac{-5 \pm \sqrt{2}}{5}$

B) $x = \dfrac{-4 \pm 2\sqrt{2}}{5}$

C) $x = \dfrac{-4 \pm 2\sqrt{2}}{5}$

D) $x = \dfrac{-2 \pm \sqrt{2}}{2}$

6

Which of the following equations has no real solution?

A) $0 = -(x-5)^2 + \sqrt{3}$

B) $0 = -3(x-5)(x+9)$

C) $0 = 2(x-5)^2 - \dfrac{1}{9}$

D) $0 = (x-1)(x-7) + 10$

Questions 7-8 refer to the following information.

t (years)	100	200	300	400
y (radioactive grams)	400	200	100	50

Radioactive decay is an exponential function where the amount, y, of radioactive material is reduced by one half over certain period of time t. The table above represents such exponential decay over 400 years.

7

Which of the following functions represents the exponential decay shown in the table?

A) $f(x) = 400\left(\dfrac{1}{2}\right)^{t}$

B) $f(x) = 400\left(\dfrac{1}{2}\right)^{\frac{t}{100}}$

C) $f(x) = 800\left(\dfrac{1}{2}\right)^{t}$

D) $f(x) = 800\left(\dfrac{1}{2}\right)^{\frac{t}{100}}$

8

How much of this material, in grams, would remain radioactive after 1000 years?

A) 0.78125

B) 0.90254

C) 1.5625

D) 3.125

CONTINUE

9

Which of the following describes the transformation of the quadratic function $f(x) = x^2$ that results in the function $g(x) = x^2 + 12x + 20$?

A) The graph of f has been translated 12 units to the right and moved up 20 units.

B) The graph of f has been translated 6 units to the right and moved down 20 units.

C) The graph of f has been translated 6 units to the left and moved up 16 units.

D) The graph of f has been translated 6 units to the left and moved down 16 units.

10

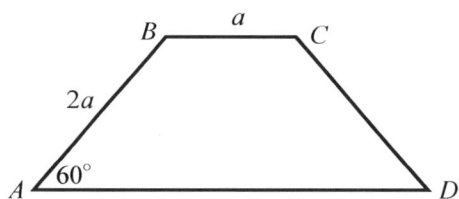

Isosceles trapezoid $ABCD$ is shown above. $AB = 2a$, $BC = a$, and $\angle BAD = 60°$. What is the area of the isosceles trapezoid?

A) $\dfrac{a^2\sqrt{2}}{2}$

B) $2a^2\sqrt{3}$

C) $\dfrac{a^2\sqrt{3}}{2}$

D) $a^2\sqrt{3}$

11

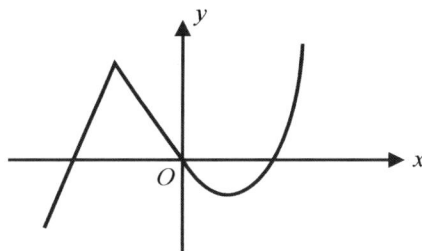

The graph of $y = f(x)$ is shown in the xy-plane above. If $g(x) = f(-x)$, which of the following graphs bet represents the graph of $y = g(x)$?

A)

B)

C)

D)

12

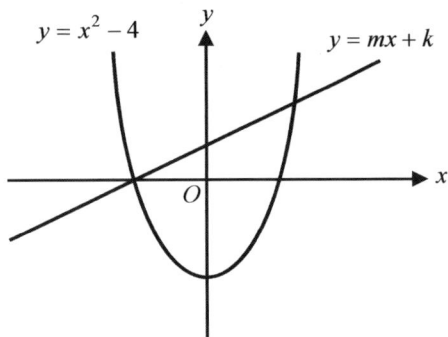

In the xy-plane above, the graphs of $y = x^2 - 4$ and $y = mx + k$ intersect at $x = -2$ and $x = 3$. What is the value of k?

A) 1
B) 2
C) 3
D) 4

13

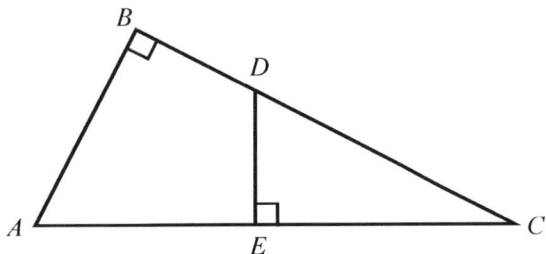

In the figure above, $AE = 18$, $DE = 9$, and $CD = 15$. What is the length of $\overline{AB}$?

A) 15
B) 18
C) 21
D) 24

14

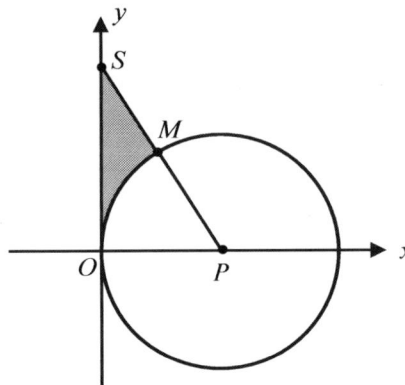

In the xy-plane above, the equation of circle P is $(x - 12)^2 + y^2 = 144$ and M is the midpoint of $\overline{PS}$. What is the area of the shaded region?

A) $72\sqrt{3}$
B) $72\sqrt{3} - 36\pi$
C) $72\sqrt{3} - 24\pi$
D) $72\sqrt{3} - 12\pi$

15

The function $f(x) = -x^2 - 2x + 3$ is defined for $-2 \le x \le 2$. Which of the following represents the range of the function?

A) $-3 \le f \le 4$
B) $-5 \le f \le 3$
C) $-5 \le f \le 4$
D) $-3 \le f \le 5$

CONTINUE

16

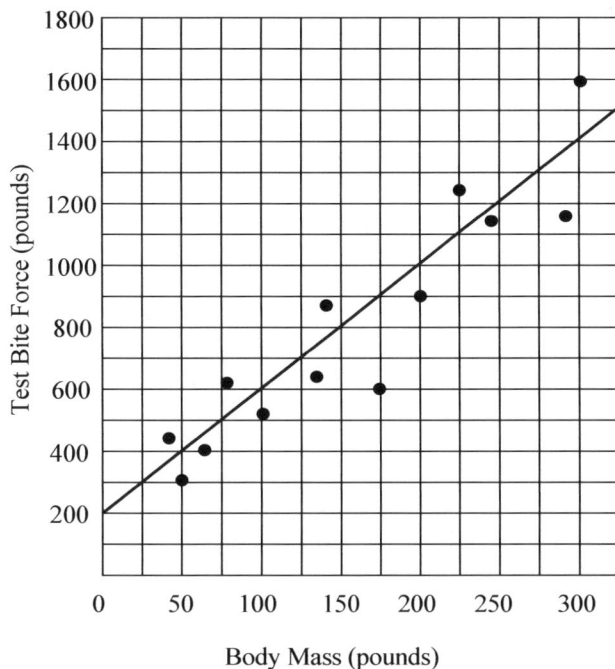

Body Mass (pounds)

The scatter plot shows the relationship between body mass and test bite force for lions. The linear model of best fit for the data is also shown. For a body mass of 175 pounds, which of the following best approximates the percent increase from the test bite force to the force that the model predicts?

A) 50%

B) 100%

C) 125%

D) 150%

17

$$\sqrt[3]{a^{2x+3}} = \sqrt{a^{x-2}}$$

In the equation above, a is a positive number. What is the value of x?

A) 12

B) 10

C) −10

D) −12

18

A truck leaves City A traveling at an average speed of 40 miles per hour. Three hours later, a car leaves City A, on the same route, traveling at an average speed of 60 miles per hour. How many hours after the car leaves City A will the car catch up to the truck?

A) 4

B) 6

C) 9

D) 12

CONTINUE

19

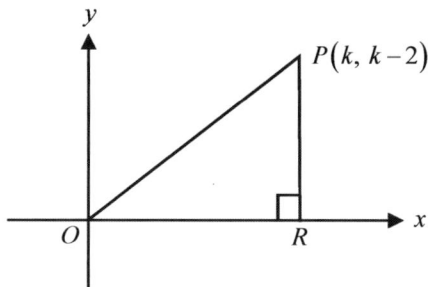

In the xy-plane above, the area of $\triangle OPR$ is 24. What is the slope of $\overline{OP}$?

A) $\dfrac{4}{3}$

B) $\dfrac{3}{4}$

C) $\dfrac{2}{3}$

D) $\dfrac{1}{2}$

20

If $z = a + bi$ and $\dfrac{z-i}{z+1} = 3$, where a and b are real numbers, which of the following is the value of a?

A) -0.5

B) -1.5

C) -2

D) -2.5

21

Gender	Junior	Senior
Female		
Male		
Total	66	18

The incomplete table above shows the number of juniors and seniors in a certain reading group. There are three times as many female-juniors as there are female-seniors, and there are five times as many male-juniors as there are male-seniors. How many male-juniors are there in the reading group?

A) 6

B) 20

C) 30

D) 40

22

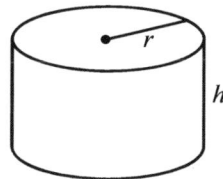

The cylinder above was altered by increasing its radius of the base by 10 percent and decreasing its height by k percent. If these alterations decreased the volume of the cylinder by 8 percent, which of the following is the closest to the value of k?

A) 10

B) 12

C) 24

D) 32

CONTINUE

23

$$h = -16t^2 + 80t + 108$$

The equation above gives the height h, in feet, of a ball t seconds after it is thrown straight up with an initial velocity 80 feet per second. Which of the following is true for the interval from 3 to 5 seconds?

A) The ball is moving upward at an average rate of 48 feet per second.

B) The ball is moving upward at an average rate of 96 feet per second.

C) The ball is moving downward at an average rate of 48 feet per second.

D) The ball is moving downward at an average rate of 96 feet per second.

Questions 24-25 refer to the following information.

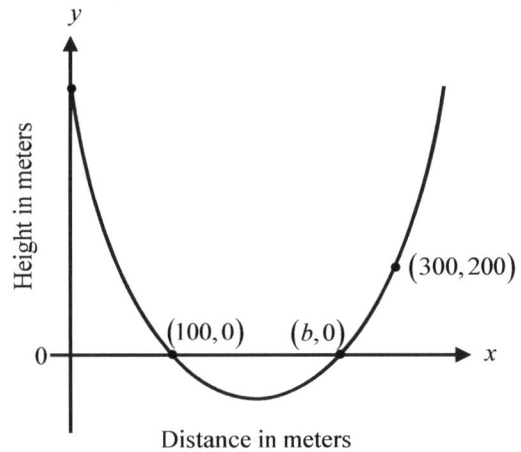

Distance in meters

In the xy-plane above, the cross-section view of a certain river is modeled by the graph of an equation

$y = \dfrac{1}{50}(x-100)(x-b)$, where b is a constant. The length

of a bridge that spans the river is the distance from the point $(100,0)$ to the other point $(b,0)$.

24

How long is the bridge in meters?

A) 100

B) 125

C) 150

D) 175

25

How high, in meters, is it from the bottom of the river to the bridge?

A) 112.5

B) 125

C) 127.5

D) 130

CONTINUE

26

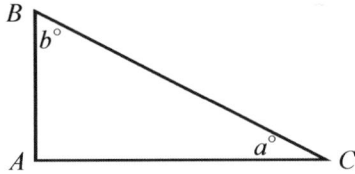

Note: Figure not drawn to scale.

In the figure above, $\sin(a^\circ) = \cos(b^\circ)$ and $BC = 10$. If $a = 2x - 15$ and $b = 5x - 21$, which of the following is closest to the area of triangle ABC?

A) 17

B) 20

C) 24

D) 48

27

$$x^2 + y^2 - 4x + 4y - 8 = 0$$

The graph of the equation above in the xy-plane is a circle. If the center of this circle is translated 2 units up and the radius increased by 2, which of the following is an equation of the resulting circle?

A) $(x-2)^2 + y^2 = 18$

B) $(x-2)^2 + (y+4)^2 = 36$

C) $(x-2)^2 + y^2 = 36$

D) $(x-2)^2 + (y-2)^2 = 36$

28

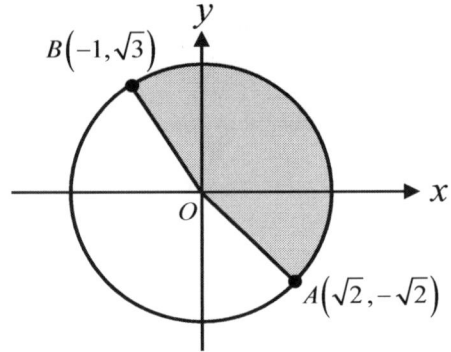

The circle above in the xy-plane is centered at the origin, O, and the coordinates of points A and B are also shown. What is the area of the shaded region?

A) $\dfrac{5\pi}{6}$

B) $\dfrac{3\pi}{2}$

C) $\dfrac{5\pi}{3}$

D) $\dfrac{11\pi}{6}$

29

If $x = 1 - i$, where $i = \sqrt{-1}$, what is the value of $(x-1)^4 + 2(x-1)^2 + x - 1$?

A) $-1 + i$

B) $-1 - i$

C) $1 + i$

D) $1 - i$

CONTINUE

30

If the function f is defined by $f\left(\dfrac{x-1}{2}\right) = x^2 + 1$,

what is the value of $f(-3)$?

A)　32

B)　26

C)　10

D)　−10

CONTINUE

DIRECTIONS

For questions 31-38, solve the problem and enter your answer in the grid, as described below, on the answer sheet.

1. Although not required, it is suggested that you write your answer in the boxes at the top of the columns to help you fill in the circles accurately. You will receive credit only if the circles are filled in correctly.
2. Mark no more than one circle in any column.
3. No question has a negative answer.
4. Some problems may have more than one answer.
5. **Mixed numbers** such as $3\frac{1}{2}$ must be gridded as 3.5 or 7/2. (If $3\ 1\ /\ 2$ is entered into the grid, it will be interpreted as $\frac{31}{2}$, not $3\frac{1}{2}$.)
6. **Decimal answers:** If you obtain a decimal answer with more digits than the grid can accommodate, it may be either rounded or truncated, but it must fill the entire grid.

Answer: $\frac{7}{12}$

Write answer in boxes.

Grid in result.

← Fraction line

Answer: 2.5

← Decimal point

Acceptable ways to grid $\frac{2}{3}$ are:

Answer: 201
Either position is correct.

Note: You may start your answers in any column, space permitting. Columns you don't need to use should be left blank.

CONTINUE

31

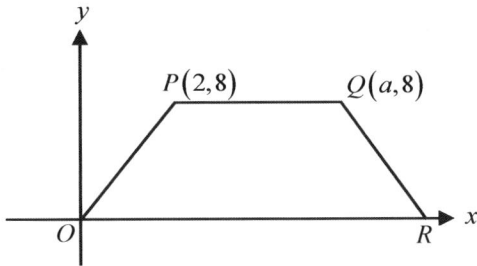

Note: Figure not drawn to scale.

In the xy-plane above, the area of isosceles trapezoid $OPQR$ is 128. What is the value of a?

32

If $\sqrt{x-1}+7=x$, what is the value of x?

33

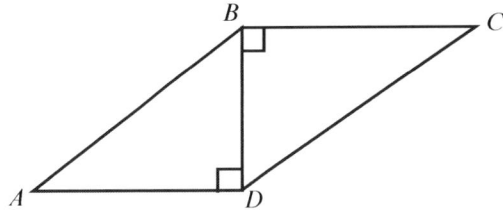

In the figure above, $\sin \angle BAD = 0.4$ and $\sin \angle BCD = 0.5$. If the length of $\overline{AB}$ is 10, what is the length of $\overline{CD}$?

34

$$\left(\sqrt{x}+2\right)^2 - 2\left(\sqrt{x}+2\right) - 3 = 0$$

In the equation above, what is the value of x?

35

$$(x-1)^2 - 3(x-1) = y$$
$$6x - y = 6$$

If (x, y) is a solution of the system of equations above and $y \neq 0$, what is the value of x?

CONTINUE

36

The graph of $y = \dfrac{x^2 - 4}{(x-3)(x-2)}$ has a vertical asymptote at $x = a$. What is the value of a?

37

$$\frac{2}{x^2 + 3x + 2} = \frac{a}{x+1} + \frac{b}{x+2}$$

The equation above is always true for all values of x except $x = -1$ and $x = -2$, where a and b are constants. What is the value of a?

38

$$\frac{x}{x-2} = \frac{x-k}{x-4}$$

In the equation above, k is a constant. If the equation has no solution, what is the value of k?

STOP

If you finish before time is called, you may check your work on this section only.
Do not turn to any other section in the test.

No Test Material on This Page

Math Conversion Table

Raw Score	Scaled Score	Raw Score	Scaled Score
58	800	27	500
57	800	26	490
56	800	25	480
55	800	24	470
54	790	23	460
53	780	22	460
52	770	21	450
51	750	20	440
50	740	19	430
49	730	18	430
48	720	17	420
47	710	16	420
46	700	15	410
45	690	14	400
44	670	13	390
43	680	12	380
42	670	11	370
41	660	10	360
40	650	9	450
39	640	8	340
38	630	7	330
37	620	6	310
36	610	5	290
35	600	4	280
34	590	3	270
33	580	2	260
32	560	1	240
31	550	0	200
30	540		
29	530		
28	520		

Answer Explanations

Test 13   Answers and Explanations

<table>
<tr><th colspan="2" rowspan="2">SECTION
3</th><th>1</th><th>2</th><th>3</th><th>4</th><th>5</th><th>6</th><th>7</th><th>8</th><th>9</th><th>10</th></tr>
<tr><td>D</td><td>D</td><td>C</td><td>B</td><td>D</td><td>D</td><td>C</td><td>D</td><td>A</td><td>C</td></tr>
<tr><td colspan="2"></td><td>11</td><td>12</td><td>13</td><td>14</td><td>15</td><td>16</td><td>17</td><td>18</td><td>19</td><td>20</td></tr>
<tr><td colspan="2"></td><td>D</td><td>B</td><td>A</td><td>B</td><td>C</td><td>25</td><td>6</td><td>18</td><td>2</td><td>2</td></tr>
<tr><th colspan="2" rowspan="2">SECTION
4</th><th>1</th><th>2</th><th>3</th><th>4</th><th>5</th><th>6</th><th>7</th><th>8</th><th>9</th><th>10</th></tr>
<tr><td>D</td><td>D</td><td>C</td><td>C</td><td>D</td><td>D</td><td>D</td><td>A</td><td>D</td><td>B</td></tr>
<tr><td colspan="2"></td><td>11</td><td>12</td><td>13</td><td>14</td><td>15</td><td>16</td><td>17</td><td>18</td><td>19</td><td>20</td></tr>
<tr><td colspan="2"></td><td>C</td><td>B</td><td>B</td><td>C</td><td>C</td><td>A</td><td>D</td><td>B</td><td>B</td><td>B</td></tr>
<tr><td colspan="2"></td><td>21</td><td>22</td><td>23</td><td>24</td><td>25</td><td>26</td><td>27</td><td>28</td><td>29</td><td>30</td></tr>
<tr><td colspan="2"></td><td>C</td><td>C</td><td>C</td><td>C</td><td>A</td><td>A</td><td>C</td><td>D</td><td>B</td><td>B</td></tr>
<tr><td colspan="2"></td><td>31</td><td>32</td><td>33</td><td>34</td><td>35</td><td>36</td><td>37</td><td>38</td><td></td><td></td></tr>
<tr><td colspan="2"></td><td>6</td><td>10</td><td>8</td><td>1</td><td>10</td><td>3</td><td>2</td><td>2</td><td></td><td></td></tr>
</table>

SECTION 3

1. **D**

$$\frac{|x-1|}{2} - 4 = 0 \;\rightarrow\; \frac{|x-1|}{2} = 4 \;\rightarrow\; x = 9, -7$$

Because absolute value is always greater than or equal to zero. A, B, and C are all positive.

2. **D**

$$f(x) = (x+1)^2 + k - 1 \;\rightarrow\; \text{at } x = -1, \text{ minimum is}$$
$$k - 1 = 8 \;\rightarrow\; k = 9$$

3. **C**

Axis of symmetry $\rightarrow x = \dfrac{2 + (-2)}{2} = 0$

Minimum of y is $f(0) = -4$. $y = -3$ has two points of intersection.

4. **B**

Replace x with $x + 3$.

$$f(x + 3 - 1) = -5(x + 3) + 2 = -5x - 13$$

Or, change the variable with k.

$$f(k - 1) = -5k + 2$$

Now exchange $k - 1 = x + 2 \;\rightarrow\; k = x + 3$

$$f(x + 2) = -5(x + 3) + 2 = -5x - 13$$

5. **D**

$$2x^2 + 8x + 10 = 2(x+2)^2 + 2$$

6. **D**

$$\frac{2\sqrt{a} - \sqrt{b}}{2\sqrt{b}} = \frac{1}{4} \;\rightarrow\; 8\sqrt{a} - 4\sqrt{b} = 2\sqrt{b}$$

$$8\sqrt{a} = 6\sqrt{b} \;\rightarrow\; \frac{\sqrt{a}}{\sqrt{b}} = \frac{6}{8} = \frac{3}{4} \;\rightarrow\; \frac{a}{b} = \frac{9}{16}$$

7. **C**

$$0 = x^2 - 9 \;\rightarrow\; x = \pm 3 \,(\text{zeros})$$

At point Q, $g(3) = -\dfrac{1}{8}(3)^2 + k = 0 \;\rightarrow\; k = \dfrac{9}{8}$

8. **D**

$$\frac{x^2}{4} - x = \frac{k}{2} \;\rightarrow\; x^2 - 4x - 2k = 0$$

In order to have two unequal real roots, discriminant $D = b^2 - 4ab > 0$. Therefore,

$$D = (-4)^2 - 4(1)(-2k) > 0 \;\rightarrow\; 8k > -16$$
$$\rightarrow\; k > -2$$

9. A

$$(2-i)(3+2i) - 5i = a + bi \;\rightarrow\; 8 - 4i = a + bi$$
$$a = 8,\; b = -4$$

10. C

$AB = 5\sqrt{3}$ (special triangle)

Therefore, $h = 5 + 5 + 5\sqrt{3} = 10 + 5\sqrt{3}$.

11. D

$$p = p_0 (1 - 0.025)^{\frac{t}{10}} = p_0 (0.975)^{\frac{40}{10}} = p_0 (0.975)^4$$

12. B

$$\begin{cases} x^2 - 5x - 11 = a(x^2 - 1) + b(x + 1) + c \\ x^2 - 5x - 11 = ax^2 + bx - a + b + c \end{cases}$$

From the equation, $a = 1$, $b = -5$, and $c = -5$.

13. A

	Speed	Time	Distance
Office	x	0.4hr	$0.4x$
Home	$x+15$	0.25hr	$0.25x+3.75$

Distance is equal. $0.4x = 0.25x + 3.75 \;\rightarrow\; x = 25$
Therefore, $d = 0.4 \times 25 = 10$ miles.

14. B

Two zeros at points $R(-4, 0)$ and $S(2, 0)$.

$$RS = 2 - (-4) = 6$$

$$\text{Axis of symmetry} = \frac{(-4) + 2}{2} = -1$$

$$f(-1) = a(-1 + 4)(-1 - 2) = -9a$$

Therefore, area of $\triangle RPS = \frac{1}{2}(6)(-9a) = -27a$

15. C

$$2x^2 - 5x + 4 = (x+1)(x-1)Q(x) + ax + b$$

When you divide degree 2 polynomial, the remainder is degree 1 $(ax + b)$.

Using Remainder Theorem,
$$P(-1) = 11 = -a + b \text{ and } P(1) = 1 = a + b$$

From the system of equations above,
$a = -5$, $b = 6$.

16. 25

$\triangle AXB$ is a right triangle with $\angle X = 90°$.

$\angle B = 25°$, and $\angle B = \angle X = 25°$ (Inscribed angles for the same arc are equal in measure.)

17. 6

$$x^2 - 8x - 3 + y^2 - 6y = k^2 \;\rightarrow\;$$
$$(x^2 - 8x + 16) + (y^2 - 6y + 9) = k^2 + 3 + 16 + 9 \;\rightarrow\;$$
$$(x - 4)^2 + (y - 3)^2 = k^2 + 28$$

Therefore,
$$r^2 = k^2 + 28 = 8^2 \;\rightarrow\; r^2 = 36 \;\rightarrow\; r = 6.$$

18. 18

$$(2x - 2)(x - 7) = 2(x - a)^2 - b \;\rightarrow\;$$
$$2x^2 - 16x + 14 = 2x^2 - 4ax + 2a^2 - b$$

Identical equation: $-4a = -16 \;\rightarrow\; a = 4$

$$2a^2 - b = 14 \;\rightarrow\; b = 2a^2 - 14 = 2(4)^2 - 14 = 18$$

19. 2

$$x^3 - ax^2 + bx - 4 = (x + 1)(x + 2)(x - k)$$

At $x = -1$, $-1 - a - b - 4 = 0 \;\rightarrow\; -a - b = 5$
At -2, $-8 - 4a - 2b - 4 = 0 \;\rightarrow\; -2a - b = 6$

From the system of equations, $a = -1, b = -4$.

$$f(x) = x^3 + x^2 - 4x - 4 = 0 \;\rightarrow\;$$
$$x^2(x + 1) - 4(x + 1) = 0 \;\rightarrow\; (x + 1)(x^2 - 4) = 0$$
$$(x + 1)(x + 2)(x - 2) = 0 \;\rightarrow\; x = -1, 1, 2$$

Therefore, $k = 2$.
Or, simply, product of the roots is

$$(-1)(-2)(k) = \frac{-4}{1}(-1) = 4 \;\rightarrow\; 2k = 4 \;\rightarrow\; k = 2.$$

20. 2

$$x^2 = mx - 1 \rightarrow x^2 - mx + 1 = 0$$

Because the line is tangent to the parabola,

$$D = b^2 - 4ac = 0 \rightarrow D = m^2 - 4 = 0$$

Therefore, $m = \pm 2 \rightarrow m = 2$ (Positive slope)

SECTION 4

1. D

Let $x =$ number of days. Slope is m (rate) and
y-intercept is 4. Therefore, $y = mx + k$.

When $x = 6$, $y = 6m + k$.

Or use arithmetic sequence.

$$a_7 = k + (7 - 1)m = k + 6m$$

2. D

$3b + 4s = Q$ and $b + 2s = P$

Subtract the system of equations above.

$$2b + 2s = Q - P \rightarrow b + s = \frac{Q - P}{2}$$

3. C

$x + 16 < 2x + 1 < x + 46 \rightarrow x + 15 < 2x < x + 45$

When subtract x, $\rightarrow 15 < x < 45$

midpoint $= \dfrac{15 + 45}{2} = 30$, and distance from MP to

the end point is $45 - 30 = 15$. Now substitute in

the formula. $|x - 30| < 15$

4. C

Since $\triangle ABC \sim \triangle DBE$, $\dfrac{AB}{DB} = \dfrac{3}{2}$ and the ratio of

their areas is $3^2 : 2^2 = 9 : 4$. Let their areas be
$9k$ and $4k$. The area of shaded region is
$9k - 4k = 5k$. Because $9k = 45$, $k = 5$ and
$5k = 5(5) = 25$.

5. D

$$\frac{1}{10}x^2 + \frac{1}{5}x + \frac{1}{20} = 0 \rightarrow 2x^2 + 4x + 1 = 0$$

Using the quadratic formula,

$$x = \frac{-4 \pm \sqrt{4^2 - 4(2)(1)}}{4} = \frac{-2 \pm 2\sqrt{2}}{2}$$

6. D

A: Graph is open downward and maximum is
$\sqrt{3}$. It has two roots.

B: Two roots $x = 5, -9$

C: Graph open upward and its minimum is
negative. It has two roots.

D: axis of symmetry is $x = \dfrac{1 + 7}{2} = 4$

$f(4) = (4 - 1)(4 - 7) + 10 = 19$ Graph open

upward and minimum is 19. There is no
x-intercept.

7. D

Initial amount is 800. Therefore,

$$f(x) = 800\left(1 - \frac{1}{2}\right)^{\frac{t}{100}} = 800\left(\frac{1}{2}\right)^{\frac{t}{100}}$$

8. A

$$800\left(\frac{1}{2}\right)^{\frac{1000}{100}} = 800\left(\frac{1}{2}\right)^{10} = 0.78125$$

9. D

$$g(x) = x^2 + 12x + 20 = (x + 6)^2 - 16$$

Compare with $f(x) = x^2$.

10. B

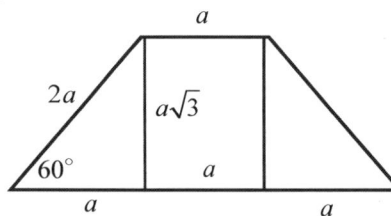

Use special triangle's ratio. The area of the

trapezoid is $\dfrac{(3a + a) \times a\sqrt{3}}{2} = 2a^2\sqrt{3}$.

11. C

$f(x) = f(-x) \rightarrow$ Reflection in the y-axis

12. B

The points of intersection of the graphs are
$(-2, 0)$ and $(3, 5)$. Slope of the line is

$\dfrac{5 - 0}{3 - (-2)} = 1$. The line is $y = x + k$.

Answer Explanations

Substitute either of the points. $\rightarrow (-2,0)$

$0 = -2 + k \rightarrow k = 2$

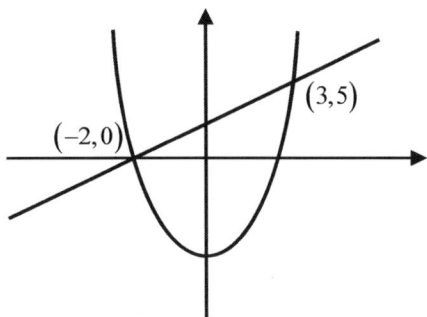

13. B

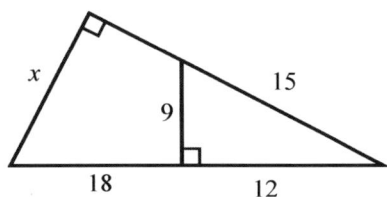

Similar triangles: $\dfrac{9}{x} = \dfrac{15}{30} \rightarrow x = 18$

14. C

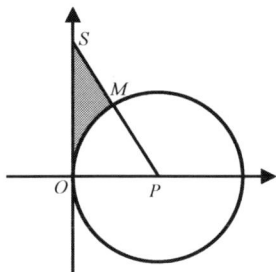

Radius is 12. $OP = 12, SP = 24$ and $\angle OPS = 60°$

Therefore, $OS = 12\sqrt{3}$. Area of $\triangle OSP$ is

$\dfrac{1}{2}(12)(12\sqrt{3}) = 72\sqrt{3}$, and area of the sector is

$\pi(12)^2 \times \dfrac{1}{6} = 24\pi$.

Area of the shaded area is $72\sqrt{3} - 24\pi$.

15. C

$f(-2) = -4 + 4 + 3 = 3$ and

$f(2) = -4 - 4 + 3 = -5$

Graph has a maximum at $x = -1$

(Axis of symmetry: $x = \dfrac{-b}{2a} = \dfrac{2}{-2} = -1$)

$f(-1) = -1 + 2 + 3 = 4 \rightarrow 4 > 3$

Therefore, the range of f is $-5 \le f \le 4$.

16. A

In the scatterplot, for 175 pounds
Actual force is 600 and predicted force is 900.

Percent increase is $\dfrac{900 - 600}{600} \times 100 = 50\%$.

17. D

Since $\sqrt[3]{a^{2x+3}} = \sqrt{a^{x-2}} \rightarrow a^{\frac{2x+3}{3}} = a^{\frac{x-2}{2}}$,

$\dfrac{2x+3}{3} = \dfrac{x-2}{2} \rightarrow 4x + 6 = 3x - 6 \rightarrow x = -12$.

18. C

	speed	time	distance
truck	40	x	$40x$
car	60	$x-3$	$60x-180$

Same distance: $40x = 60x - 180 \rightarrow x = 9$

Therefore, $x - 3 = 9 - 3 = 6$.

19. B

Area of $\triangle OPR$ is $\dfrac{k(k-2)}{2} = 24$. From the

equation, $k^2 - 2k - 48 = 0 \rightarrow (k-8)(k+6) = 0$

$k = 8 \ (\because k > 0)$, slope is $\dfrac{k-2}{k} = \dfrac{8-2}{8} = \dfrac{3}{4}$.

20. B

$\dfrac{z-i}{z+1} = \dfrac{a+bi-i}{a+bi+1} = 3 \rightarrow a+bi-i = 3a+3bi+3$

$a + (b-1)i = 3a + 3 + 3bi$

Because it is an identical equation,

$a = 3a + 3 \rightarrow a = -\dfrac{3}{2}$.

21. C

Gender	Junior	Senior
Female	$3x$	x
Male	$5y$	y
Total	66	18

From the table above,
$3x + 5y = 66$ and $x + y = 18$.

From the system of equations, $y = 6$.

Therefore, male-juniors is $5y = 5(6) = 30$.

22. **C**

$V = \pi r^2 h$ and $V' = \pi(1.1r)^2 h'$, where

$V' =$ new volume and $h' =$ new height.

Since new volume decreases by 8%, $V' = (1 - 0.08)V$.

$\pi(1.1r)^2 h' = 0.92\pi r^2 h \rightarrow 1.21h' = 0.92h$

Therefore, $h' = \dfrac{0.92}{1.21}h = 0.76h \rightarrow h' = (1 - 0.24)h$.

$0.24 = 24\% \rightarrow k = 24$

23. **C**

Axis of symmetry: $t = \dfrac{-80}{2(-16)} = 2.5$

For 3 to 5 seconds, the ball is moving down.

$h(3) = -144 + 240 + 108 = 204$

$h(5) = -400 + 400 + 180 = 108$

Average rate of change is

$\dfrac{108 - 204}{5 - 3} = -48$ feet/second. (negative means"

moving downward")

24. **C**

$y = \dfrac{1}{50}(x - 100)(x - b)$

For point $(300, 200)$

$200 = \dfrac{1}{50}(200)(300 - b) \rightarrow 50 = 300 - b$

Therefore, $b = 250$. The length of the bridge is

$250 - 100 = 150$ meters.

25. **A**

Axis of symmetry is $x = \dfrac{100 + 250}{2} = 175$

$f(175) = \dfrac{1}{50}(175 - 100)(175 - 250) =$

$\dfrac{1}{50}(75)(-75) = -112.5$ meters

Therefore, the depth of the river is 112.5 meters.

26. **A**

Since $\sin(a^\circ) = \cos(b^\circ)$, $\triangle ABC$ is a right triangle.

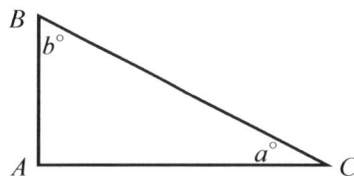

By the Cofunction formula,

$(2x - 15) + (5x - 21) = 90 \rightarrow 7x = 126 \rightarrow x = 18$.

$a = 2(18) - 15 = 21$ and $b = 5(8) - 21 = 69$

Therefore, $AB = 10\cos(69)$ and $AC = 10\cos(21)$.

Area of $\triangle ABC$ is

$\dfrac{1}{2}(10\cos 69)(10\cos 21) = 16.72\cdots \approx 17$.

27. **C**

$x^2 + y^2 - 4x + 4y - 8 = 0 \rightarrow (x - 2)^2 + (y + 2)^2 = 16$

Radius is 4. New radius is $4 + 2 = 6$.

For 2 units up,

$(x - 2)^2 + (y + 2 - 2)^2 = 6^2 \rightarrow (x - 2)^2 + y^2 = 36$

28. **D**

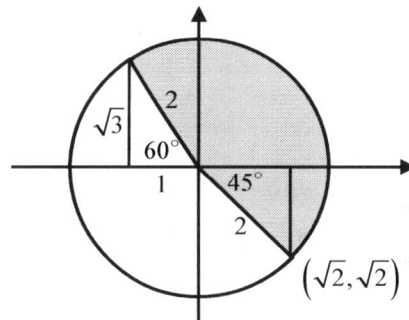

Radius of the circle is 2, and central angle is 165°.

Therefore, area of the sector is $\pi(2)^2 \times \dfrac{165}{360} = \dfrac{11}{6}\pi$.

29. **B**

$(x - 1)^4 + 2(x - 1)^2 + x - 1 \rightarrow$

$(1 - i - 1)^4 + 2(1 - i - 1)^2 + 1 - i - 1 = (-i)^4 + 2(-i)^2 - i$

$= 1 - 2 - i = -1 - i$

30. **B**

$\dfrac{x - 1}{2} = -3 \rightarrow x = -5$

Substitute into the equation: $f(-3) = (-5)^2 + 1 = 26$

31. 6

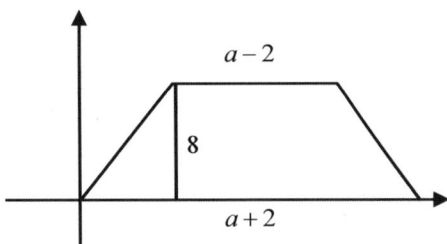

Area is $\dfrac{(a+2)(a-2)8}{2} = 4(a^2-4) = 128$.

$a^2 - 4 = 32 \rightarrow a^2 = 36 \rightarrow a = 6$

32. 10

$\sqrt{x-1} + 7 = x \rightarrow \sqrt{x-1} = x-7$

$x - 1 = x^2 - 14x + 49 \rightarrow 0 = x^2 - 15x + 50$

$(x-5)(x-10) = 0 \rightarrow x = 5$ or 10

For $x = 5$, $\sqrt{5-1} \neq 5-7$. Answer is 10.

33. 8

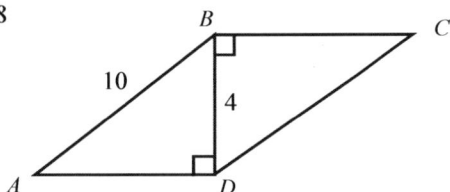

$BD = 10\sin\angle BAD = 10(0.4) = 4$

$\sin\angle BCD = \dfrac{4}{CD} \rightarrow CD = \dfrac{4}{0.5} = 8$

34. 1

Let $\sqrt{x} + 2 = k$. $\rightarrow k > 0$

$k^2 - 2k - 3 = 0 \rightarrow (k-3)(k+1) = 0 \rightarrow k = 3, -1$

$k = -1$ is not working.

For $k = 3$, $\sqrt{x} + 2 = 3 \rightarrow \sqrt{x} = 1 \rightarrow x = 1$

35. 10

$(x-1)^2 - 3(x-1) = y$

$6x - 6 = y \rightarrow y = 6(x-1)$

Substitute $y = 6(x-1)$ in the equation.

$(x-1)^2 - 3(x-1) = 6(x-1) \rightarrow (x-1)^2 - 9(x-1) = 0$

$(x-1)(x-1-9) = 0 \rightarrow (x-1)(x-10) = 0$

$x = 1$ or 10

But, when $x = 1$, $y = 0$ (not working $y \neq 0$).

Therefore, the value of x is 10.

36. 3

$y = \dfrac{x^2-4}{(x-3)(x-2)} = \dfrac{(x+2)(x-2)}{(x-3)(x-2)} = \dfrac{x+2}{x-3}$

Denominator $x - 3 = 0 \rightarrow x = 3$: vertical asymptote

37. 2

$\dfrac{2}{x^2+3x+2} = \dfrac{a}{x+1} + \dfrac{b}{x+2}$

Multiply by $(x+1)(x+2)$.

$2 = a(x+2) + b(x+1) \rightarrow 2 = (a+b)x + 2a + b$

From the equation,

$a + b = 0$ and $2a + b = 2$.

Therefore, $a = 2$ and $b = -2$.

38. 2

$\dfrac{x}{x-2} = \dfrac{x-k}{x-4} \rightarrow x(x-4) = (x-2)(x-k)$

$x^2 - 4x = x^2 - (2+k)x + 2k \rightarrow (k-2)x = 2k$

Therefore, $x = \dfrac{2k}{k-2}$. $\rightarrow$ If $k = 2$, no solution.

PRACTICE TEST 14

Dr. John Chung's SAT Math

Math Test - No Calculator

25 MINUTES, 20 QUESTIONS

Turn to Section 3 of your answer sheet to answer the questions in this section.

DIRECTIONS

For questions 1–15, solve each problem, choose the best answer from the choices provided, and fill in the corresponding circle on your answer sheet. **For questions 16–20,** solve the problem and enter your answer in the grid on your answer sheet. Please refer to the directions before question 16 on how to enter your answers in the grid. You may use any available space in your test booklet for scratch work.

NOTE

1. The use of a calculator **is not permitted**.

2. All variables and expressions used represent real numbers unless otherwise indicated.

3. Figures provided in this test are drawn to scale unless otherwise indicated.

4. All figures lie in a plane unless otherwise indicated.

5. Unless otherwise indicated, the domain of a given function f is the set of all real numbers x for which $f(x)$ is a real number.

REFERENCE

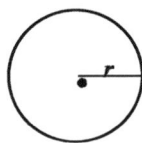 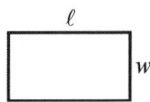 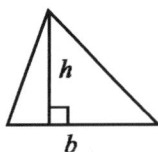 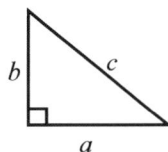 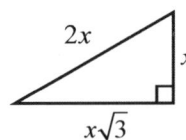 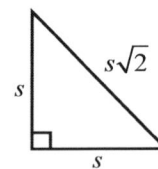

$A = \pi r^2$
$C = 2\pi r$

$A = \ell w$

$A = \frac{1}{2}bh$

$c^2 = a^2 + b^2$

Special Right Triangles

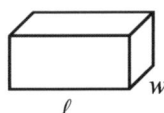

$V = \ell wh$

$V = \pi r^2 h$

$V = \frac{4}{3}\pi r^3$

$V = \frac{1}{3}\pi r^2 h$

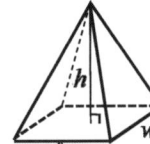

$V = \frac{1}{3}\ell wh$

The number of degrees of arc in a circle is 360.

The number of radians of arc in a circle is 2π.

The number of the measures in degrees of the angles of a triangle is 180.

CONTINUE

1

Which of the following is equal to $10^{-\frac{2}{3}}$?

A) $\sqrt{100}$

B) $\sqrt[3]{100}$

C) $\sqrt[3]{\dfrac{1}{100}}$

D) $\dfrac{1}{\sqrt[3]{100^2}}$

2

If $a + 2b = 2$ and $a - 2b = 10$, what is the value of $2a^2 - 8b^2$?

A) 40

B) 30

C) 20

D) 10

3

Which of the following is the graph of an even function?

A)

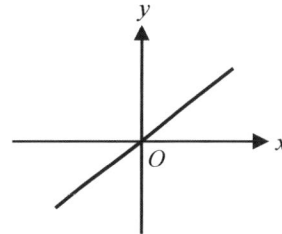

B)

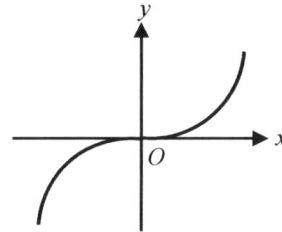

C)

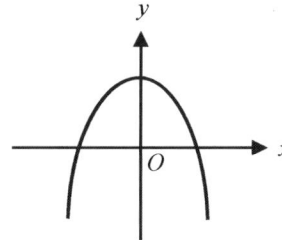

D)

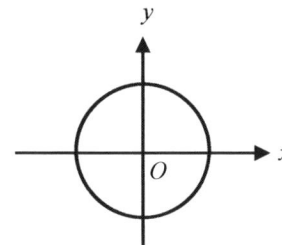

CONTINUE

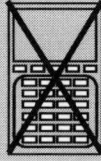

4

Which of the following numbers is contained in the domain of the function $f(x) = \dfrac{1}{x-5} - \dfrac{1}{\sqrt{x-5}}$?

A)   6

B)   5

C) −5

D) −6

5

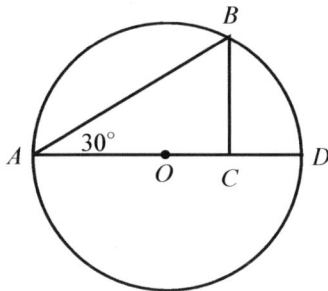

Note: Figure not drawn to scale.

In the figure above, the radius of circle O is 10 and $\overline{BC}$ is perpendicular to $\overline{AD}$. What is the length of $\overline{AC}$?

A) 12

B) 13

C) 15

D) 18

6

If the graph of $y = x^2 - 2x + k$ in the xy-plane intersects the line $y = 3$ exactly once, what is the value of k?

A) 5

B) 4

C) 3

D) 2

7

If the volume of a cube is a cubic meters, which of the following is the expression for the surface area of this cube?

A) $\sqrt{6a^3}$

B) $6\sqrt[3]{a^2}$

C) $6a\sqrt{a}$

D) $6a^2$

CONTINUE

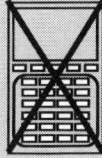

8

$$x^2 - 4x + 2 = 0$$

If a and b are the roots of the equation above, what is the value of $(a+1)(b+1)$?

A) 5

B) 6

C) 7

D) 10

9

$$3x - 5y = 5$$
$$5x - 3y = 35$$

If (x, y) is a solution to the system of equations above, what is the value of $x - y$?

A) 5

B) 10

C) 20

D) 40

10

x	$f(x)$
0	6
1	0
2	−2
3	0
4	6

Some values of x and $f(x)$ are shown in the table above. If the function f is defined by a quadratic polynomial, which of the following defines f?

A) $f(x) = (x-1)(x-3)$

B) $f(x) = 3x(x-6)$

C) $f(x) = 2(x-1)(x-3)$

D) $f(x) = 2(x+1)(x+4)$

11

If $a^5 = 10$ and $a^7 = \dfrac{2}{p}$, which of the following is an expression for p?

A) $2a^2$

B) $\dfrac{5}{a^2}$

C) $\dfrac{1}{5a^2}$

D) $\dfrac{a^2}{100}$

CONTINUE

12

A movie theater sold n tickets today. Some of these were adult tickets and the rest were child tickets. An adult ticket costs twelve dollars and a child ticket costs eight dollars. If the theater sold $2000 worth of tickets, and the number of child tickets sold is c, which of the following could be the value of n?

A) $n = \dfrac{2000}{c}$

B) $n = \dfrac{500 - 4c}{3}$

C) $n = \dfrac{500 - c}{3}$

D) $n = \dfrac{500 + c}{3}$

13

$$f(x) = x(x+2)(x-3)$$

In the function $f(x)$ above, which of the following about the graph of f in the xy-plane could not be true?

A) The graph of f rises to the right as x approaches positive infinity.

B) The function f falls to the left as x approaches negative infinity.

C) The function f has three distinct real zeros.

D) The graph of f has a negative y-intercept.

14

Each dot plot represents the number of pets owned by students in a class. Which of the following data sets appears to have the smallest standard deviation?

A)

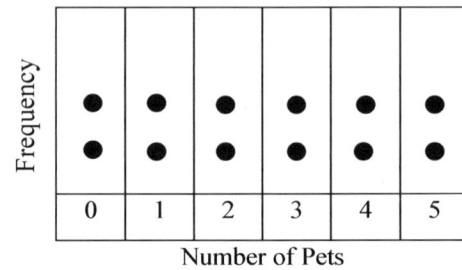

Number of Pets

B)

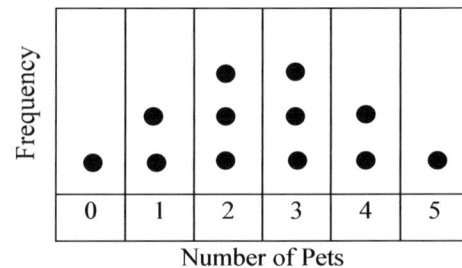

Number of Pets

C)

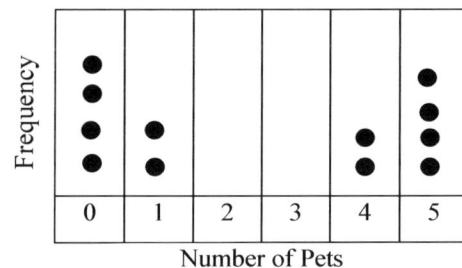

Number of Pets

D)

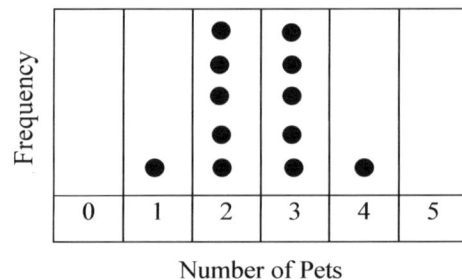

Number of Pets

CONTINUE

15

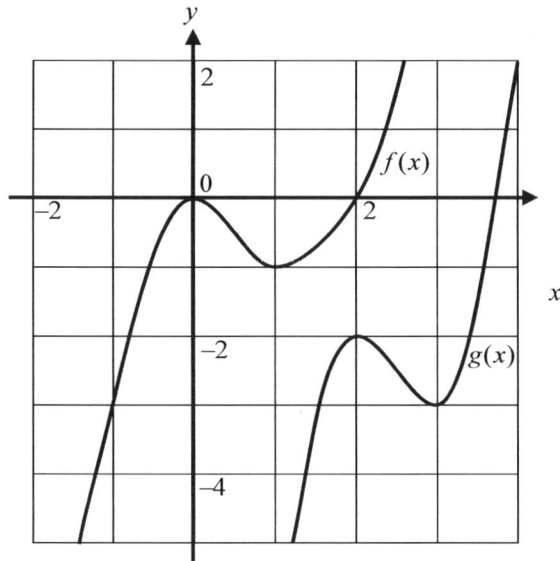

In the xy-plane above, the graphs of $f(x)$ and $g(x)$ are shown. The graph of $g(x)$ is the result of a translation of $f(x)$. If $f(x) = x^2(x-2)$, which of the following is an equation of $g(x)$?

A) $g(x) = (x-2)^3 + 2(x-2) - 2$

B) $g(x) = (x+2)^3 - 2(x+2)^2 - 2$

C) $g(x) = (x-2)^3 - 2(x-2)^2 - 2$

D) $g(x) = (x+2)^3 - 2(x+2)^2 - 2$

CONTINUE

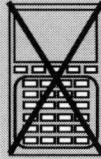

DIRECTIONS

For questions 16–20, solve the problem and enter your answer in the grid, as described below, on the answer sheet.

1. Although not required, it is suggested that you write your answer in the boxes at the top of the columns to help you fill in the circles accurately. You will receive credit only if the circles are filled in correctly.
2. Mark no more than one circle in any column.
3. No question has a negative answer.
4. Some problems may have more than one answer.
5. **Mixed numbers** such as $3\frac{1}{2}$ must be gridded as 3.5 or 7/2. (If $3\,1\,/\,2$ is entered into the grid, it will be interpreted as $\frac{31}{2}$, not $3\frac{1}{2}$.)
6. **Decimal answers:** If you obtain a decimal answer with more digits than the grid can accommodate, it may be either rounded or truncated, but it must fill the entire grid.

Answer: $\frac{7}{12}$

Write answer in boxes.

← Fraction line

Answer: 2.5

← Decimal point

Grid in result.

Acceptable ways to grid $\frac{2}{3}$ are:

Answer: 201
Either position is correct.

Note: You may start your answers in any column, space permitting. Columns you don't need to use should be left blank.

CONTINUE →

16

If $(x+y)^2 = 10$ and $(x-y)^2 = 4$, what is the value of xy?

17

$$f\left(\frac{x}{3}\right) = x^2 - 2x + k$$

In the equation above, k is a constant. If $f(2) = 20$, what is the value of $f(-2)$?

18

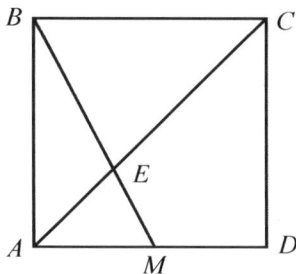

In the figure above, the area of square $ABCD$ is 36 and M is the midpoint of $\overline{AD}$. What is the area of $\triangle ABE$?

19

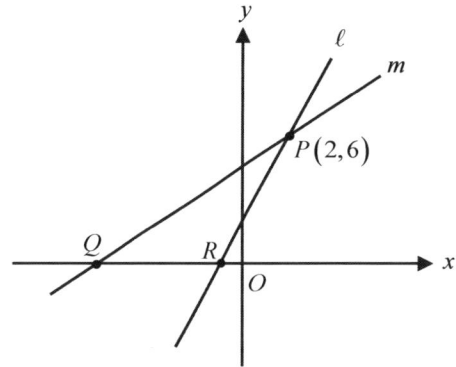

In the xy-plane above, the slope of line ℓ is 2 and the slope of line m is 1. Lines ℓ and m intersect at point P. What is the area of $\triangle PQR$?

20

$$y = |x - 3|$$
$$y = x$$

If (a,b) is a solution to the system of equations above, what is the value of a?

STOP

**If you finish before time is called, you may check your work on this section only.
Do not turn to any other section in the test.**

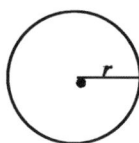
Math Test - Calculator

55 MINUTES, 38 QUESTIONS

Turn to Section 4 of your answer sheet to answer the questions in this section.

DIRECTIONS

For questions 1-30, solve each problem, choose the best answer from the choices provided, and fill in the corresponding circle on your answer sheet. **For questions 31-38**, solve the problem and enter your answer in the grid on your answer sheet. Please refer to the directions before question 31 on how to enter your answers in the grid. You may use any available space in your test booklet for scratch work.

NOTES

1. The use of a calculator **is permitted**.

2. All variables and expressions used represent real numbers unless otherwise indicated.

3. Figures provided in this test are drawn to scale unless otherwise indicated.

4. All figures lie in a plane unless otherwise indicated.

5. Unless otherwise indicated, the domain of a given function f is the set of all real numbers x for which $f(x)$ is a real number.

REFERENCE

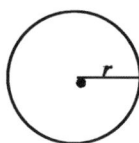 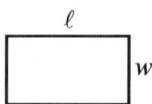 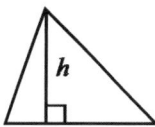 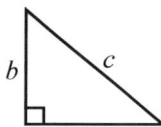 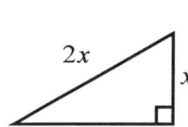 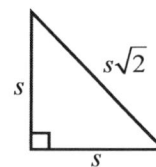

$A = \pi r^2$
$C = 2\pi r$

$A = \ell w$

$A = \dfrac{1}{2}bh$

$c^2 = a^2 + b^2$

Special Right Triangles

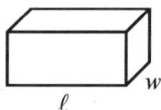 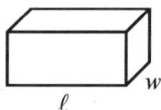   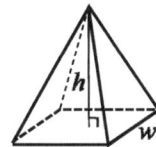

$V = \ell wh$

$V = \pi r^2 h$

$V = \dfrac{4}{3}\pi r^3$

$V = \dfrac{1}{3}\pi r^2 h$

$V = \dfrac{1}{3}\ell wh$

The number of degrees of arc in a circle is 360.
The number of radians of arc in a circle is 2π.
The number of the measures in degrees of the angles of a triangle is 180.

CONTINUE

1

Hillary's new car costs m dollars per month for car payment and insurance. She estimates that gas and maintenance cost k cents per mile. What is her total monthly cost, in dollars, as a function of the miles, x, driven during the month?

A) $(m+k)x$

B) $mx + \dfrac{kx}{100}$

C) $m + \dfrac{kx}{100}$

D) $100m + kx$

2

If x is greater than y, x is how much, in percent, greater than y?

A) $\dfrac{x}{y}$

B) $\dfrac{x-y}{y}$

C) $\dfrac{100x}{y}$

D) $\dfrac{100(x-y)}{y}$

3

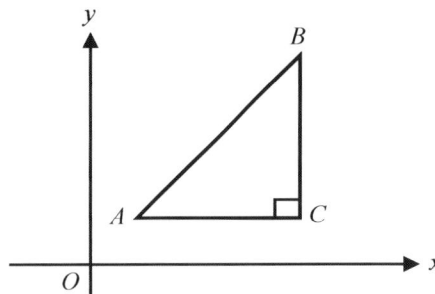

Note: Figure not drawn to scale.

In the xy-plane above, $BC = AC$ and $AB = 3\sqrt{2}$. If the coordinates of point A are $(1,2)$, what are the coordinates point B?

A) $(3,3)$

B) $(3,4)$

C) $(4,5)$

D) $(5,6)$

4

If $\sec(3x+10)^\circ = \csc(x-20)^\circ$, which of the following could be the value of x?

A) 20

B) 25

C) 30

D) 35

CONTINUE

5

Which of the following could be the graph of a fifth degree equation with a leading coefficient of five and a constant of three?

A)

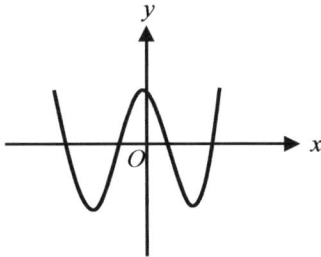

B)

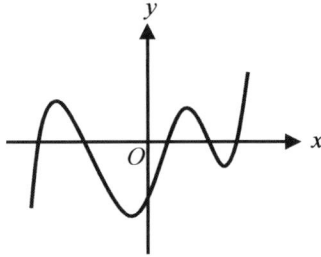

C)

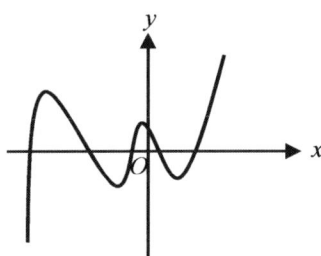

D)

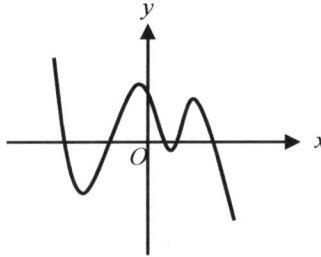

6

What is the range of the function $y = 2x^2 - 8x - 3$?

A) All real numbers greater than or equal to -13.

B) All real numbers less than or equal to -13.

C) All real numbers greater than or equal to -11.

D) All real numbers less than or equal to -11.

7

The equation $P = 20,000(0.93)^{10}$ is being used to calculate the cost of money of an automobile. What does 0.93 represent in this equation?

A) 0.93% decay

B) 0.93% growth

C) 7% decay

D) 7% growth

CONTINUE

8

Variation of Average Temperature

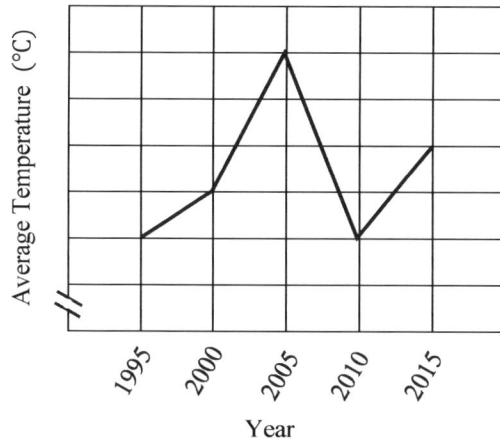

The graph above shows the variation in the average temperature of a certain area from 1995 to 2015. During which year did the temperature variation change the most per year?

A) 1995-2000

B) 2000-2005

C) 2005-2010

D) 2010-2015

9

If $x \leq -10$ or $x \geq 16$, which of the following represents the values of x?

A) $|x+3| \leq 13$

B) $|x+3| \geq 13$

C) $|x-3| \geq 13$

D) $|x-3| \leq 13$

10

If the vertex of $y = x^2 - ax + b$ has coordinates $(2,-5)$, what is the y-intercept of this graph in the xy-plane?

A) $(0,2)$

B) $(0,1)$

C) $(0,-1)$

D) $(0,-2)$

11

$$y = -|x-2|+3$$
$$y = (x-2)^2 + k$$

In the system of equations above, for which of the following values of k does the system have no solution?

A) -2

B) 0

C) 3

D) 4

CONTINUE

12

$$f(x) = a(x-b)^2 + 10$$

In the xy-plane, the graph of the function f above has x-intercepts at -2 and 6. What is the value of a?

A) -5

B) $-\dfrac{5}{8}$

C) $\dfrac{5}{8}$

D) 5

13

$$2x - 3y = 1$$
$$ax + by = c$$

In the system of equations above, a, b, and c are constants. For which of the following values of a, b, and c does the system have only one solution?

A) $a = 2, b = -3, c = 1$

B) $a = 4, b = -6, c = 5$

C) $a = 6, b = -9, c = 5$

D) $a = 10, b = 10, c = 1$

14

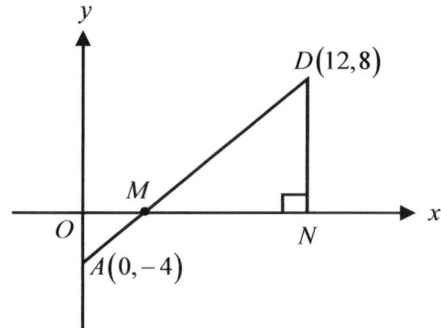

In the xy-plane above, what are the coordinates of point M?

A) $(2, 0)$

B) $(3, 0)$

C) $(4, 0)$

C) $(5, 0)$

15

$$F = \frac{9}{5}C + 32$$

Based on the equation above, if there is an increase of 27 in F, how much is the increase in C?

A) 9

B) 15

C) 27

D) 59

CONTINUE

16

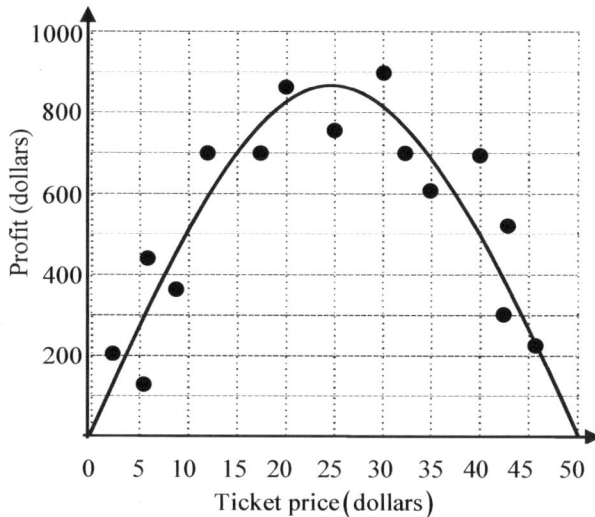

The scatterplot above shows the ticket price for a school music concert and the profit made when the ticket was sold at different prices. A quadratic model that best fits the data was drawn on the scatterplot. For a ticket price of $40, which of the following is closest to the percent decrease from the actual profit to the profit predicted by the curve of best fit?

A) 25%

B) 29%

C) 40%

D) 50%

17

Gender	Chocolate	Strawberry	Total
Male	20		
Female		15	25
Total	30		70

The incomplete table above summarizes a survey of students' favorite ice cream flavors in a math club. What percent of students prefer strawberry ice cream are male?

A) 34%

B) 40%

C) 60%

D) 62.5%

18

$$\frac{x+1}{x^2-3x+2} = \frac{a}{x-1} + \frac{b}{x-2}$$

The equation above is true for all values of x except 1 and 2, where a and b are constants. What is the value of a?

A) -4

B) -2

C) 3

D) 4

CONTINUE

19

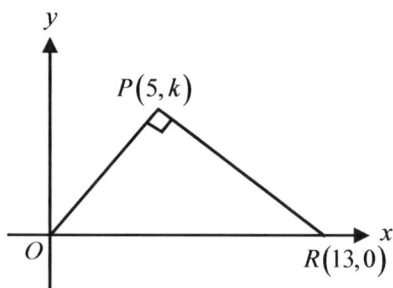

Note: Figure not drawn to scale.

In the xy-plane above, $\overline{OP}$ is perpendicular to $\overline{PR}$. What is the value of k?

A) 5

B) $2\sqrt{10}$

C) 8

D) $3\sqrt{5}$

20

$$y \leq 4 - x^2$$
$$y \geq x + 2$$

If (a,b) is a solution to the system of inequalities above, what is the maximum value of b?

A) 3

B) 4

C) 5

D) 6

21

Machine	10 inches	20 inches	Hours available
A	1 hour	2 hours	20 hours
B	2 hours	3 hours	15 hours

A small computer monitor manufacturing company produces 10-inch and 20-inch sized monitors. The table above shows how many hours are required on each machine per day in order to produce the monitors. If x represents the number of 10-inch monitors and y represents the number of 20-inch monitors, which of the following systems represents all the constraints that x and y must satisfy?

A) $\begin{cases} x \geq 0 \\ y \geq 0 \\ x + 2y \geq 20 \\ 2x + 3y \geq 15 \end{cases}$

B) $\begin{cases} 0 \leq x \leq 3 \\ 0 \leq y \leq 5 \\ x + 2y \leq 20 \\ 2x + 3y \leq 15 \end{cases}$

C) $\begin{cases} x \geq 0 \\ y \geq 0 \\ x + 2y \leq 20 \\ 2x + 3y \leq 15 \end{cases}$

D) $\begin{cases} x \geq 0 \\ y \geq 0 \\ x + 2y = 20 \\ 2x + 3y = 15 \end{cases}$

CONTINUE

22

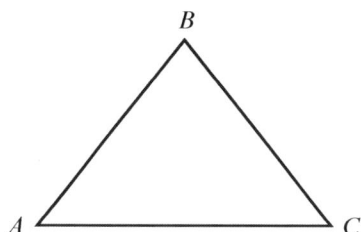

The triangle above was altered by increasing its base by 20 percent and decreasing its height by k percent. If these alterations decreased the area of the triangle by 16 percent, which of the following is the value of k?

A) 20

B) 25

C) 30

D) 40

23

$$P(x) = x^2 - 5x + k$$

In the function above, k is a constant. When a polynomial $P(x)$ is divided by $x-1$, the remainder is 3. If $P(x)$ is divided by $x+1$, what is the remainder?

A) 3

B) 4

C) 7

D) 13

24

$$(2-3i)a + (3-2i)b = 5i$$

In the equation above, a and b are real numbers. What is the value of b? (Note: $i = \sqrt{-1}$)

A) −3

B) −2

C) 2

D) 3

25

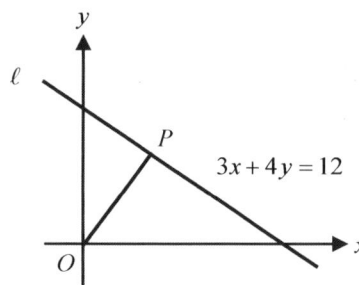

In the xy-plane above, line ℓ is defined by the equation $3x + 4y = 12$ and $\overline{OP}$ is perpendicular to line ℓ. What is the length of $\overline{OP}$?

A) 1.6

B) 2.4

C) 3.2

D) 3.6

CONTINUE

26

If $f(x) = 2^x$ and $f(x-1) = f(x) - 8$, what is the value of x?

A) 2
B) 4
C) 6
D) 8

27

If $f\left(\dfrac{x-2}{2}\right) = 5x - 10$, which of the following represents $f(x)$?

A) $f(x) = 10x - 5$
B) $f(x) = 10x + 5$
C) $f(x) = 10x$
D) $f(x) = 10x + 10$

28

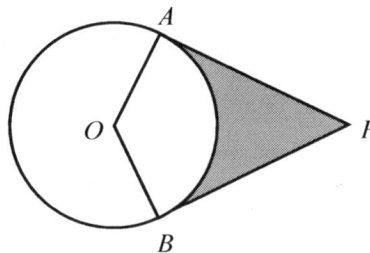

In the figure above, $\overline{PA}$ and $\overline{PB}$ are tangent to circle O with radius $AO = 6$. If the measure of $\angle APB$ is $60°$, which of the following is closest to the area of the shaded region?

A) 20
B) 25
C) 28
D) 30

CONTINUE

546

29

$$f(x) = x^2 + mx + 4$$

In the function above, $f(x) > 0$ for all values of x. Which of the following could be the value of m?

A) 8

B) 6

C) 4

D) 2

30

Which of the following functions is an odd function?

A) $f(x) = x + 5$

B) $f(x) = x^2$

C) $f(x) = x^3$

D) $f(x) = x^3 + 5$

CONTINUE

DIRECTIONS

For questions 31-38, solve the problem and enter your answer in the grid, as described below, on the answer sheet.

1. Although not required, it is suggested that you write your answer in the boxes at the top of the columns to help you fill in the circles accurately. You will receive credit only if the circles are filled in correctly.
2. Mark no more than one circle in any column.
3. No question has a negative answer.
4. Some problems may have more than one answer.
5. **Mixed numbers** such as $3\frac{1}{2}$ must be gridded as 3.5 or 7/2. (If $\boxed{3\,1\,/\,2}$ is entered into the grid, it will be interpreted as $\frac{31}{2}$, not $3\frac{1}{2}$.)
6. **Decimal answers:** If you obtain a decimal answer with more digits than the grid can accommodate, it may be either rounded or truncated, but it must fill the entire grid.

Answer: $\frac{7}{12}$

Write answer in boxes. → ← Fraction line

Grid in result. →

Answer: 2.5

← Decimal point

Acceptable ways to grid $\frac{2}{3}$ are:

Answer: 201
Either position is correct.

Note: You may start your answers in any column, space permitting. Columns you don't need to use should be left blank.

CONTINUE

31

If $\dfrac{1}{10}x + \dfrac{1}{20}y = 5$, what is the value of $4x + 2y$?

32

$$2\sqrt{x} + 8 = x$$

What is the solution to the equation above?

33

If $x^2 - y^2 + (x+y)i = 12 + 5i$, what is the value of $x - y$? $\left(\text{Note: } i = \sqrt{-1}\right)$

34

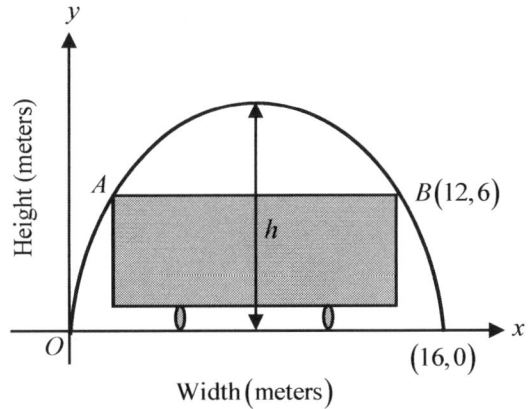

Note: Figure not drawn to scale.

In the xy-plane above, the cross-section view of a tunnel under a bridge with an opening in the shape of a parabolic arch is shown. A large rectangular house trailer can be moved along a highway that passes through the tunnel. If the trailer is 8 meters wide and 6 meters tall, what is the height, h in meters, of the tunnel in the center?

35

In the xy-plane, a parabola with equation $y = (x-5)^2 - 10$ intersects a line with equation $y = 6$ at two points, A and B. What is the length of $\overline{AB}$?

CONTINUE

36

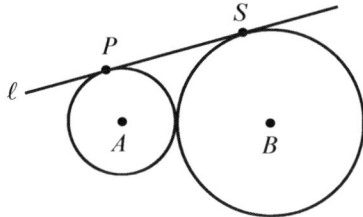

In the figure above, line ℓ is tangent to both circles at points A and B respectively. The radius of circle A is 4 and the radius of circle B is 9. What is the length of $\overline{PS}$?

37

$$C(n) = 3,400 - n$$

A manufacturing company sells a certain product for $6 per unit. The cost , C, of producing n units is estimated by the formula above. How many units must the company produce and sell to earn a profit of $2,200?

38

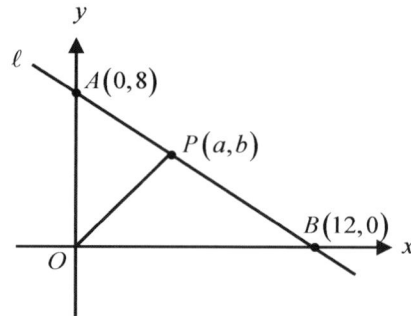

Note: Figure not drawn to scale.

In the xy-plane above, the area of $\triangle POB$ is three times the area of $\triangle POA$, and points A, B, and P lie on line ℓ. What is the value of a?

STOP

If you finish before time is called, you may check your work on this section only.
Do not turn to any other section in the test.

No Test Material on This Page

Math Conversion Table

Raw Score	Scaled Score	Raw Score	Scaled Score
58	800	27	500
57	800	26	490
56	800	25	480
55	800	24	470
54	790	23	460
53	780	22	460
52	770	21	450
51	750	20	440
50	740	19	430
49	730	18	430
48	720	17	420
47	710	16	420
46	700	15	410
45	690	14	400
44	670	13	390
43	680	12	380
42	670	11	370
41	660	10	360
40	650	9	450
39	640	8	340
38	630	7	330
37	620	6	310
36	610	5	290
35	600	4	280
34	590	3	270
33	580	2	260
32	560	1	240
31	550	0	200
30	540		
29	530		
28	520		

Answer Explanations

Test 14 — Answers and Explanations

	1	2	3	4	5	6	7	8	9	10
SECTION **3**	C	A	C	A	C	B	B	C	A	C
	11	12	13	14	15	16	17	18	19	20
	C	D	D	D	C	3/2	44	6	9	1.5

	1	2	3	4	5	6	7	8	9	10
	C	D	C	B	C	C	C	C	C	C
	11	12	13	14	15	16	17	18	19	20
SECTION **4**	D	B	D	C	B	B	D	B	B	B
	21	22	23	24	25	26	27	28	29	30
	C	C	D	C	B	B	C	B	D	C
	31	32	33	34	35	36	37	38		
	200	16	12/5 2.4	8	8	12	800	3		

SECTION 3

1. C
$$10^{-\frac{2}{3}} = \sqrt[3]{10^{-2}} = \sqrt[3]{\frac{1}{100}}$$

2. A
$$2a^2 - 8b^2 = 2(a+2b)(a-2b) = 2(2)(10) = 40$$

3. C
The graph of an even function is symmetric with respect to the y-axis. D: Circle is not a function.

4. A
Domain of the graph is $x > 5$.

5. D

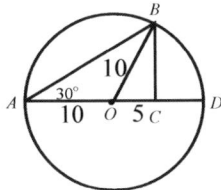

Triangle ABO is isosceles.
$BO = 10$ and $OC = 5$. Therefore,
$AC = 10 + 5 = 15$.

6. B
The vertex form of the equation is
$$y = (x-1)^2 + k - 1$$, minimum of y is $k-1$.
If $k - 1 = 3$. They have only one solution.
Or, you can use discriminant.
$$x^2 - 2x + k = 3 \rightarrow x^2 - 2x + k - 3 = 0$$
In order to have only one solution, discriminant should be 0.
$$D = (-2)^2 - 4(1)(k-3) = 0 \rightarrow k = 4$$

7. B
Let x = length of an edge.
Volume is $a = x^3$. $\rightarrow x = a^{\frac{1}{3}}$
Surface area is $6x^2 = 6\left(a^{\frac{1}{3}}\right)^2 = 6a^{\frac{2}{3}} = 6\sqrt[3]{a^2}$.

8. C

Answer Explanations

Sum of the roots: $a+b=4$
Product of the roots: $ab=2$
$(a+1)(b+1)=ab+(a+b)+1=2+4+1=7$

9. A
Add the equations. $8x-8y=40 \rightarrow x-y=5$

10. C
Zeros at $x=1$ and $3 \rightarrow f(x)=a(x-1)(x-3)$
For $(0,6)$, $6=a(-1)(-3) \rightarrow a=2$
Therefore, $f(x)=2(x-1)(x-3)$.

11. C
$a^7=\dfrac{2}{p} \rightarrow p=\dfrac{2}{a^7}=\dfrac{2}{a^5 a^2}=\dfrac{2}{10a^2}=\dfrac{1}{5a^2}$

12. D

	price	#	AMT
Adult	12	$n-c$	$12n-12c$
child	8	c	$8c$

Total amount is $2000.
$12n-12c+8c=2000 \rightarrow 12n=4c+2000$
$n=\dfrac{4c+2000}{12}=\dfrac{c+500}{3}$

13. D
When $x=0$, y-intercept is $f(0)=-6$.

14. D
The data are not spread farther from the mean than any other data.

15. C
$f(x)=x^2(x-2)=x^3-2x^2$
The graph is translated 2 units to the right and 2 units down. $g(x)=f(x-2)-2$: Therefore,
$g(x)=(x-2)^3-2(x-2)^2-2$.

16. $\dfrac{3}{2}$ or 1.5
$(x+y)^2=x^2+y^2+2xy=10$
$(x-y)^2=x^2+y^2-2xy=4$

Subtract. $4xy=10-4=6. \rightarrow xy=\dfrac{3}{2}$

17. 44
Input
$\dfrac{x}{3}=2 \rightarrow x=6 \rightarrow f(2)=6^2-2(6)=k=20$
$k=-4.\quad f\left(\dfrac{x}{3}\right)=x^2-2x-4$
For $f(-2) \rightarrow \dfrac{x}{3}=-2 \rightarrow x=-6$,
Therefore, $f(-2)=(-6)^2-2(6)-4=44$

18. 6

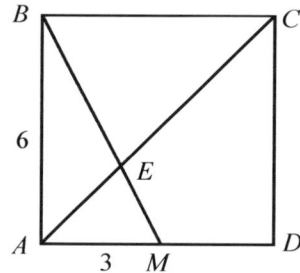

Since $\triangle BCE \sim \triangle AEM$, the ratio of any corresponding sides is 2:1.
Area of $\triangle ABM=\dfrac{6\times3}{2}=9$ and $\dfrac{BE}{EM}=\dfrac{2}{1}$
Therefore, the area of $\triangle ABE$ is $9\times\dfrac{2}{3}=6$.

19. 9
Line ℓ: $y=2x+b \rightarrow$ using $(2,6) \rightarrow y=2x+2$
Line m: $y=x+b \rightarrow$ using $(2,6) \rightarrow y=x+4$
Two x-intercepts are $Q(-4,0)$ and $R(-1,0)$.
$QR=3$ and the height of $\triangle PQR=6$.
Therefore, the area of $\triangle PQR=\dfrac{3\times6}{2}=9$.

20. $\dfrac{1}{2}$

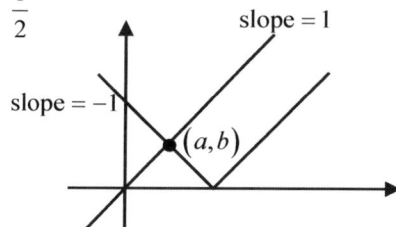

554

Since slope of the line is 1, they have only one solution.

For $x < 3$, $y = |x - 3| \rightarrow y = -x + 3$.

$x = -x + 3 \rightarrow 2x = 3 \rightarrow x = \dfrac{1}{2}$

For $x \geq 3$, $y = |x - 3| \rightarrow y = x - 3$

Now solve: $x = x - 3 \rightarrow 0 \neq -3$. (No solution)

Because -3 is not greater than or equal to 3.

SECTION 4

1. C

k cent $= \dfrac{k}{100}$ dollars

Total cost is $m + \left(\dfrac{k}{100} \right) x$.

2. D

$x > y \rightarrow$

% of increase $= \dfrac{x - y}{y} \times 100 = \dfrac{100(x - y)}{y}$

3. C

$\triangle ABC$ is isosceles.

Since $AB = 3\sqrt{3}$, $AC = BC = 3$. Therefore,

$B(x, y) \rightarrow x = 1 + 3 = 4$ and $y = 2 + 3 = 5$

4. B

Cofunction: $3x + 10 + x - 20 = 90 \rightarrow x = 25$

5. C

The function could be $f(x) = 5x^5 + \cdots + 3$.

y-intercept is 3.

As $x \rightarrow \infty$, $y \rightarrow \infty$: Graph rises up to the right.

As $x \rightarrow -\infty$, $y \rightarrow -\infty$: Graph falls down to the left.

6. C

$y = 2x^2 - 8x - 3 \rightarrow y = 2(x - 2)^2 - 11$

Range: $y \geq -11$

7. C

$P = 20000(1 - 0.07)^{10} \rightarrow 7\%$ decay

8. C

9. C

midpoint $= \dfrac{-10 + 16}{2} = 3$

distance from mid point to the end point is $16 - 9 = 7$.

Thereore, $|x - 3| \geq 9$.

10. C

Axis of symmetry: $x = \dfrac{a}{2} = 2 \rightarrow a = 4$

$y = x^2 - 4x + b$ Putting $(2, -5)$ in the

equation. $f(2) = 4 - 8 + b = 5 \rightarrow b = -1$

y-intercept $\rightarrow (0, b) = (0, -1)$

11. D

Think about their graphs as follows.

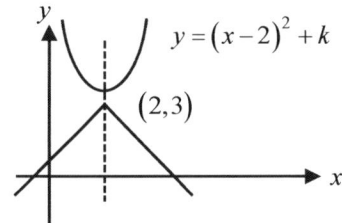

If $k = 3$, one intersection . If $k < 3$, two intersections. If $k > 3$, no intersection.

12. B

From the vertex form, axis of symmetry is $x = b$. From the zeros, axis of symmetry is

$x = \dfrac{-2 + 6}{2} = 2$. Therefore $b = 2$.

Now $f(x) = a(x - 2)^2 + 10$. Using

$(6, 0)$ or $(0, -4)$, $0 = a(4)^2 + 10 \rightarrow a = -\dfrac{5}{8}$

13. D

In order to have only one solution, $\dfrac{2}{a} \neq \dfrac{-3}{b}$.

D: $\dfrac{2}{10} \neq \dfrac{-3}{10}$

14. C

$\triangle OAM \sim \triangle MDN$, $DN = 8$ and $OA = 4$

$\dfrac{DN}{OA} = \dfrac{MN}{OM} = \dfrac{1}{2}$

Therefore, $OM = 12 \times \dfrac{1}{3} = 4 \rightarrow M(0,4)$

15. **B**

$$\text{slope} = \frac{\Delta F}{\Delta C} = \frac{9}{5} \rightarrow \frac{27}{\Delta C} = \frac{9}{5} \rightarrow$$

$$\Delta C = \frac{27 \times 5}{9} = 15$$

16. **B**

Actual profit $= 700$ and predicted profit $= 500$ for ticket price of $40.

$$\% \text{ decrease} = \frac{|500 - 700|}{700} \times 100 = 28.57 \cdots \approx 28\%$$

17. **D**

$$\frac{25}{40} \times 100 = 62.5\%$$

18. **B**

Multiply by $(x-1)(x-2)$.

$x+1 = a(x-2) + b(x-1) \rightarrow$

$x+1 = (a+b)x - 2a - b$, from the equation

$a+b=1$ and $-2a-b=1 \rightarrow a=-2, b=3$

19. **B**

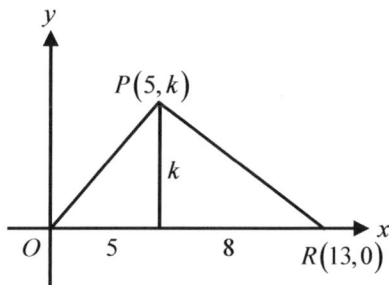

$k^2 = 5 \times 8 = 40 \rightarrow k = 2\sqrt{10}$

20. **B**

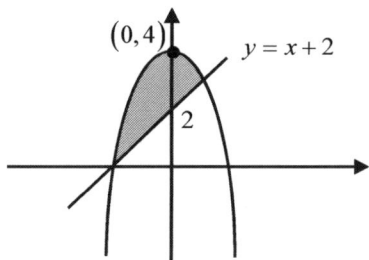

At point $(0,4)$, b has a maximum value.

21. **C**

Since machine A is available 20 hours,

$1(x) + 2(y) \le 20$

Machine B is available 15 hours,

$2(x) + 3(y) \le 15$

22. **C**

For original triangle, $A = \dfrac{bh}{2}$

For altered triangle, $A' = \dfrac{(1.2b)h'}{2}$

Since area is decreased by 16%,

$A' = (1 - 0.16)A = 0.84A$.

$$\frac{(1.2b)h'}{2} = 0.84\left(\frac{bh}{2}\right) \rightarrow h' = 0.7h$$

Since $h' = (1 - 0.3)h$, the height decreases by 30%.

23. **D**

Remainder theorem: $P(1) = 1 - 5 + k = 3 \rightarrow k = 7$

$P(x) = x^2 - 5x + 7$

When $P(x)$ is divided by $x+1$, the remainder is

$P(-1) = 1 + 5 + 7 = 13$.

24. **C**

$(2 - 3i)a + (3 - 2i)b = 5i \rightarrow$

$(2a + 3b) - (3a - 2b)i = 0 + 5i$

From the equation, $2a + 3b = 0$ and $3a + 2b = -5$.

Solution to the system of equations above is

$(-3, 2)$. $a = -3$ and $b = 2$.

25. **B**

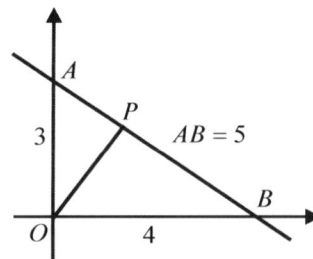

Area of $\triangle OAB = \dfrac{3 \times 4}{2}$, or $\dfrac{5 \times OP}{2}$.

Therefore, $\dfrac{3 \times 4}{2} = \dfrac{5 \times OP}{2} \rightarrow OP = 2.4$

26. B

$f(x-1) = 2^{x-1}$ and $f(x) = 2^x$

$2^{x-1} = 2^x - 8 \rightarrow \dfrac{2^x}{2} = 2^x - 8 \rightarrow$ Multiply by 2

$2^x = 2(2^x) - 16 \rightarrow 2^x = 2(2^x) - 2^x = 16$

Finally, $(2-1)2^x = 16 \rightarrow 2^x = 2^4 \rightarrow x = 4$

27. C

Change the variable. $f\left(\dfrac{k-2}{2}\right) = 5k - 10$

$\dfrac{k-2}{2} = x \rightarrow k = 2x + 2$. Now replace k with x.

$f(x) = 5(2x+2) - 10 \rightarrow f(x) = 10x$

28. B

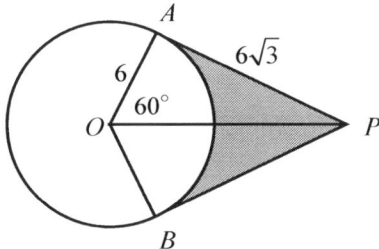

$\angle AOB = 180 - 60 = 120$

Area of sector AOB is $\pi(6)^2 \times \dfrac{1}{3} = 12\pi$

$AP = 6\sqrt{3}$ (Special triangle)

Area of $\triangle AOP = \dfrac{6 \times 6\sqrt{3}}{2} = 18\sqrt{3}$

Area of $APBO = 18\sqrt{3} \times 2 = 36\sqrt{3}$

Therefore, the area of shaded region is

$36\sqrt{3} - 12\pi = 24.6547\cdots \approx 25$

29. D

The graph of $f(x) = x^2 + mx + 4 > 0$ for all x is as follows.

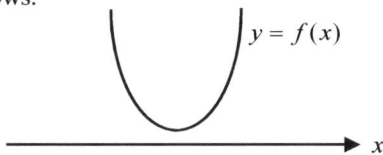

The graph doesn't intersect the x-axis.

Discriminant must be less than 0.

$D = m^2 - 4(1)(4) < 0 \rightarrow (m+4)(m-4) < 0$

Solution is $-4 < m < 4$.

30. C

A: neither B: even D: odd

31. 200

$\dfrac{1}{10}x + \dfrac{1}{20}y = 5 \rightarrow 2x + y = 100$

$4x + 2y = 2(2x+y) = 2(100) = 200$

32. 16

$2\sqrt{x} + 8 = x \rightarrow 2\sqrt{x} = x - 8 \rightarrow$ squaring

$4x = x^2 - 16x + 64 \rightarrow 0 = x^2 - 20x + 64$

Factor: $(x-4)(x-16) = 0 \rightarrow x = 4, 16$

Check: $x = 4$ doesn't work.

33. $\dfrac{12}{5}$ or 2.4

$x^2 - y^2 = 12$ and $x + y = 5$

$(x+y)(x-y) = 12 \rightarrow 5(x-y) = 12 \rightarrow$

$x - y = \dfrac{12}{5}$

34. 8

Since the parabola has zeros at $x = 0$ and $x = 16$, the quadratic equation in factored form is

$y = ax(x-16)$. Putting $(12,6)$ in the equation.

$6 = a(12)(12-16) \rightarrow a = -\dfrac{1}{8}$

Therefore, $y = -\dfrac{1}{8}x(x-16)$. When $x = 8$, the

graph has a maximum.

Maximum of $y = -\dfrac{1}{8}(8)(8-16) = 8$ meters.

35. 8

Solve for the intersections.

$(x-5)^2 - 10 = 6 \rightarrow (x-5)^2 = 16 \rightarrow$

$(x-5) = 4, -4 \rightarrow x = 9, 1$

Therefore, $AB = 9 - 1 = 8$.

36. 12

ACSP is a rectangle and $\triangle ABC$ is a right triangle. $PS = AC$ (Opposite sides are equal)

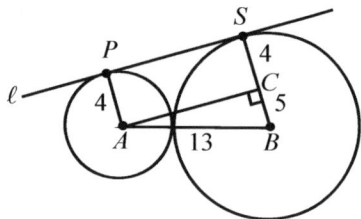

$$PS = AC = \sqrt{13^2 - 5^2} = 12$$

37. 800

Number of units produced is n.

Revenue $= 6 \times n = 6n$

Cost $= 3400 - n$

Profit $=$ revenue $-$ cost $= 6n - (3400 - n)$

$\qquad = 7n - 3400$

Profit is $2200. Therefore,

$7n - 3400 = 2200 \ \rightarrow \ 7n = 5600 \ \rightarrow \ n = 800$

38. 3

Area of $\triangle AOB = \dfrac{8 \times 12}{2} = 48$

Area of $\triangle APO = \dfrac{1}{4}$ of $\triangle AOB = \dfrac{1}{4}(48) = 12$

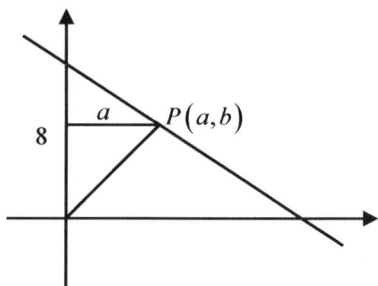

Also the area of $\triangle AOB$ is $\dfrac{8 \times a}{2} = 4a$.

Therefore, $4a = 12 \ \rightarrow \ a = 3$.

PRACTICE TEST 15

Dr. John Chung's SAT Math

Math Test - No Calculator

25 MINUTES, 20 QUESTIONS

Turn to Section 3 of your answer sheet to answer the questions in this section.

DIRECTIONS

For questions 1–15, solve each problem, choose the best answer from the choices provided, and fill in the corresponding circle on your answer sheet. **For questions 16–20,** solve the problem and enter your answer in the grid on your answer sheet. Please refer to the directions before question 16 on how to enter your answers in the grid. You may use any available space in your test booklet for scratch work.

NOTE

1. The use of a calculator **is not permitted**.

2. All variables and expressions used represent real numbers unless otherwise indicated.

3. Figures provided in this test are drawn to scale unless otherwise indicated.

4. All figures lie in a plane unless otherwise indicated.

5. Unless otherwise indicated, the domain of a given function f is the set of all real numbers x for which $f(x)$ is a real number.

REFERENCE

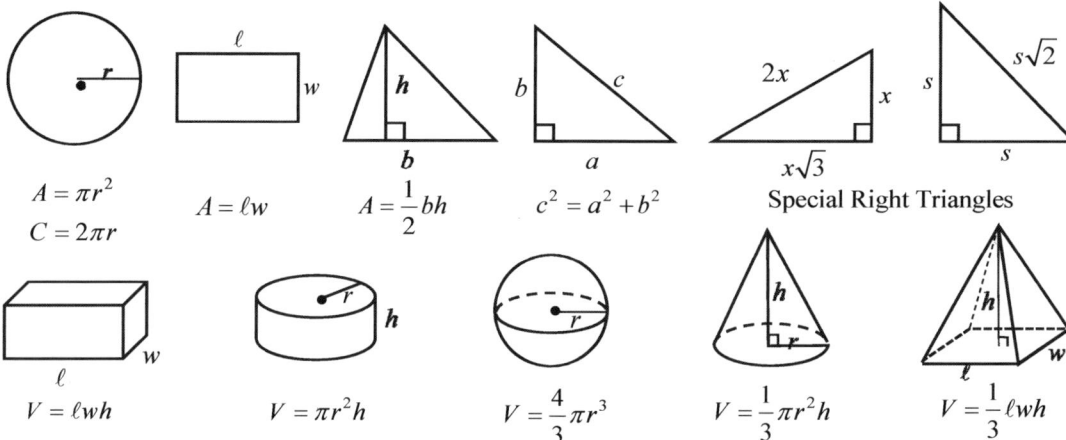

$A = \pi r^2$
$C = 2\pi r$

$A = \ell w$

$A = \frac{1}{2}bh$

$c^2 = a^2 + b^2$

Special Right Triangles

$V = \ell wh$

$V = \pi r^2 h$

$V = \frac{4}{3}\pi r^3$

$V = \frac{1}{3}\pi r^2 h$

$V = \frac{1}{3}\ell wh$

The number of degrees of arc in a circle is 360.
The number of radians of arc in a circle is 2π.
The number of the measures in degrees of the angles of a triangle is 180.

CONTINUE

1

What are the solutions to the equation
$$\frac{1}{10}x^2 - \frac{1}{5}x = \frac{4}{5}?$$

A) -2 and -4

B) -2 and 4

C) 4 and 6

D) 4 and 8

2

If $\dfrac{a}{2b} = 5$, what is the value of $\dfrac{15b}{2a}$?

A) $\dfrac{2}{3}$

B) $\dfrac{3}{4}$

C) $\dfrac{7}{2}$

D) $\dfrac{15}{2}$

3

Which of the following is equal to $\dfrac{a^{\frac{1}{2}}}{a^{-\frac{1}{3}}}$?

A) $\sqrt[6]{a}$

B) $\sqrt{6^5}$

C) $\sqrt[6]{a^5}$

D) $\sqrt[5]{a^6}$

4

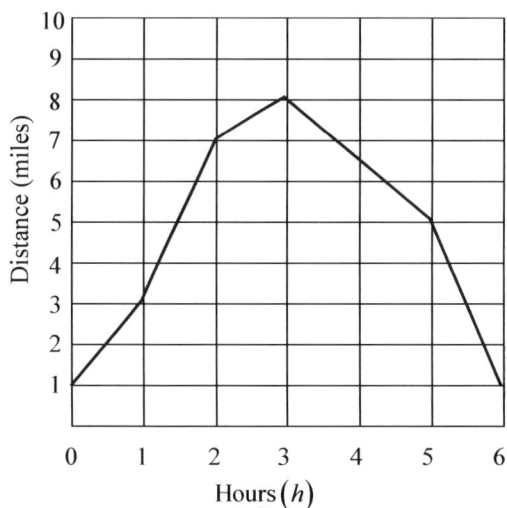

The graph above shows Peter's distance from his campsite in h hours. Which hourly interval had the greatest rate of change?

A) hour 0 to hour 1

B) hour 1 to hour 2

C) hour 4 to hour 5

D) hour 5 to hour 6

CONTINUE

5

If $\dfrac{8}{x} - \dfrac{12}{x+10} = 0$, what is the value of $\dfrac{x}{4}$?

A) 20

B) 10

C) 5

D) 2

6

$$2x - 5y = -9$$
$$5x - 2y = -5$$

If (a, b) is a solution to the system of equations above, what is the value of $a - b$?

A) -14

B) -4

C) -2

D) 2

7

The line $y = mx - 5$, where m is a constant, is graphed in the xy-plane. If the line passes through the points $(a, 0)$, where $a \neq 0$, what is the slope of the line in terms of a ?

A) $-\dfrac{a}{5}$

B) $-\dfrac{5}{a}$

C) $\dfrac{a}{5}$

D) $\dfrac{5}{a}$

8

In the xy-plane, a line with equation $y = -26$ intersects a parabola with the equation $y = -(x-10)^2 + 10$ at points P and Q. What is the length of $\overline{PQ}$?

A) 6

B) 8

C) 12

D) 20

CONTINUE

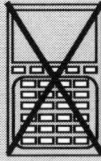

9

The equation $3x^3 - 5x + 5 = (3x^2 + 3x - 2)(x-1) + a$ is true for all values of x, where a is a constant. What is the value of a?

A) 3

B) 5

C) 7

D) 8

10

A pharmaceutical salesperson receives a monthly salary of $3000 plus a commission of k percent of his sales. If the salesperson's monthly wage is $7200, what are his sales in terms of k?

A) $7200k$

B) $4200k$

C) $\dfrac{4200}{k}$

D) $\dfrac{420000}{k}$

11

On a trip downtown, Peter drove his car at an average rate of 50 miles per hour. On the return trip due to bad weather, he drove his car at an average rate of 30 miles per hour and the return trip took 1 hour longer. How much time he spend on the returning trip?

A) 1.5 hours

B) 2 hours

C) 2.5 hours

D) 3 hours

12

What are the zeros of $f(x) = 2(x-4)^2 - 10$?

A) $x = -4 \pm \sqrt{5}$

B) $x = \dfrac{-4 \pm \sqrt{5}}{2}$

C) $x = 4 \pm \sqrt{5}$

D) $x = \dfrac{4 \pm \sqrt{5}}{2}$

CONTINUE

Questions 13-14 refer to the following information.

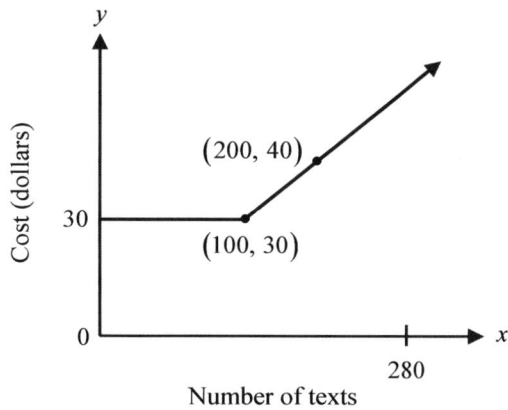

Cost (dollars)

30

(200, 40)

(100, 30)

0

280

Number of texts

The domestic texting plan of a mobile telephone company is modeled by the graph in the xy-plane above.

13

If Jackson uses 280 texts this month, what is his cost in dollars?

A) 42

B) 44

C) 46

D) 48

14

If $x > 100$, which of the following equations represents the cost?

A) $y = \dfrac{1}{10}x$

B) $y = \dfrac{1}{10}x + 30$

C) $y = \dfrac{1}{10}(x - 100) + 30$

D) $y = \dfrac{1}{10}(x - 200) + 30$

15

Age	For	Against	No opinion
21 – 30	80	90	10
31 – 40	80	100	20
Over 40	40	70	10

The data in the table above show a public opinion poll exploring the relationship between age and support for a candidate in an election in a town. If a person is chosen at random from those who are over 30 years old, what is the probability that the person belongs to the 31 – 40 age group and does not support the candidate?

A) $\dfrac{1}{5}$

B) $\dfrac{5}{16}$

C) $\dfrac{2}{5}$

D) $\dfrac{9}{24}$

DIRECTIONS

For questions 16–20, solve the problem and enter your answer in the grid, as described below, on the answer sheet.

1. Although not required, it is suggested that you write your answer in the boxes at the top of the columns to help you fill in the circles accurately. You will receive credit only if the circles are filled in correctly.

2. Mark no more than one circle in any column.

3. No question has a negative answer.

4. Some problems may have more than one answer.

5. **Mixed numbers** such as $3\frac{1}{2}$ must be gridded as 3.5 or 7/2. (If $\boxed{3\,|\,1\,|\,/\,|\,2}$ is entered into the grid, it will be interpreted as $\frac{31}{2}$, not $3\frac{1}{2}$.)

6. **Decimal answers:** If you obtain a decimal answer with more digits than the grid can accommodate, it may be either rounded or truncated, but it must fill the entire grid.

Answer: $\frac{7}{12}$

Write answer in boxes. → Fraction line

Grid in result.

Answer: 2.5 — Decimal point

Acceptable ways to grid $\frac{2}{3}$ are:

Answer: 201
Either position is correct.

Note: You may start your answers in any column, space permitting. Columns you don't need to use should be left blank.

CONTINUE

16

If one of the zeros of $f(x) = 2x^2 - 8x + k$ is -1, what is the other zero of the function?

17

If $f(x+1) = 5x + k$ and $f(3) = 8$, what is the value of $f(5)$?

18

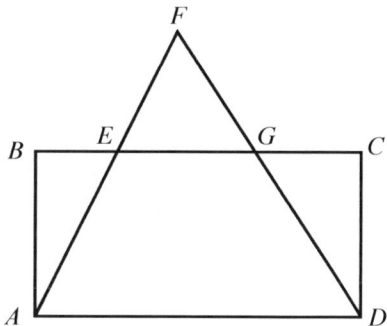

In rectangle $ABCD$ above, $AD = 12$, $AB = 6$, and $EG = 4$. What is the area of $\triangle EFG$?

19

$$3x - ay = 3$$
$$ax - by = 4$$

In the system of equations above, a and b are nonzero constants. If the system has infinitely many solutions, what is the value of b?

20

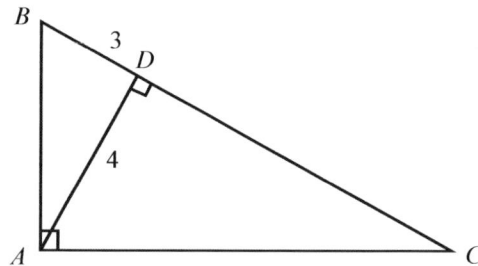

In the figure above, $AD = 4$ and $BD = 3$. What is the value of $\sin C$?

STOP

If you finish before time is called, you may check your work on this section only.
Do not turn to any other section in the test.

No Test Material on This Page

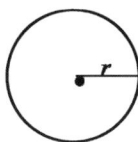
Math Test - Calculator

55 MINUTES, 38 QUESTIONS

Turn to Section 4 of your answer sheet to answer the questions in this section.

DIRECTIONS

For questions 1-30, solve each problem, choose the best answer from the choices provided, and fill in the corresponding circle on your answer sheet. **For questions 31-38**, solve the problem and enter your answer in the grid on your answer sheet. Please refer to the directions before question 31 on how to enter your answers in the grid. You may use any available space in your test booklet for scratch work.

NOTES

1. The use of a calculator **is permitted**.

2. All variables and expressions used represent real numbers unless otherwise indicated.

3. Figures provided in this test are drawn to scale unless otherwise indicated.

4. All figures lie in a plane unless otherwise indicated.

5. Unless otherwise indicated, the domain of a given function f is the set of all real numbers x for which $f(x)$ is a real number.

REFERENCE

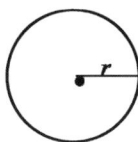 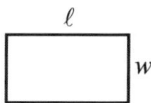 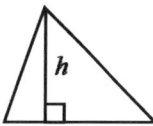 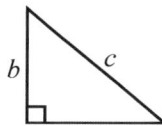 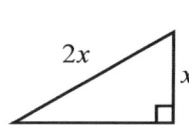 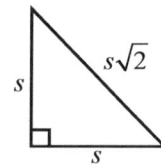

$A = \pi r^2$
$C = 2\pi r$

$A = \ell w$

$A = \frac{1}{2}bh$

$c^2 = a^2 + b^2$

Special Right Triangles

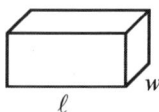 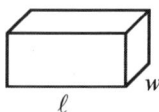   

$V = \ell wh$

$V = \pi r^2 h$

$V = \frac{4}{3}\pi r^3$

$V = \frac{1}{3}\pi r^2 h$

$V = \frac{1}{3}\ell wh$

The number of degrees of arc in a circle is 360.
The number of radians of arc in a circle is 2π.
The number of the measures in degrees of the angles of a triangle is 180.

CONTINUE ➤

4   **4**

1

If $\dfrac{3}{2}\left(x+\dfrac{3}{4}\right)=3$, what is the value of x ?

A) $\dfrac{5}{3}$

B) $\dfrac{5}{4}$

C) $\dfrac{4}{3}$

D) $\dfrac{3}{2}$

2

Mina can do a job in 2 hours, and Ruth can do the same job in 4 hours. If they work together, how long will it take them to do the job?

A) 50 minutes

B) 60 minutes

C) 80 minutes

D) 100 minutes

3

Which of the following equations has the same solution as $2x^2-8x-90=0$?

A) $(x-2)^2=98$

B) $(x-2)^2=49$

C) $(x-2)^2=45$

D) $(x-4)^2=45$

4

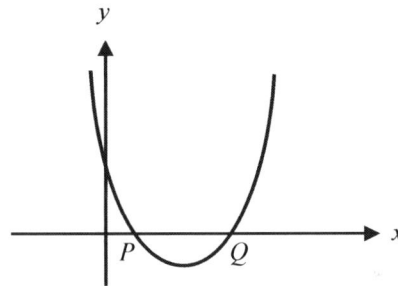

The graph of $y=2x^2-9x+4$ shown in the xy-plane above intersects the x-axis at points P and Q. What is the length of $\overline{PQ}$?

A) 1.5

B) 2.5

C) 3.5

D) 4.5

CONTINUE

Dr. John Chung's SAT Math Practice Test 15

5

For function f, $f(-1)=5$ and $f(3)=10$. For function g, $g(-1)=3$ and $g(3)=-1$. What is the value of $f(g(-1))$?

A) 3

B) 5

C) 8

D) 10

6

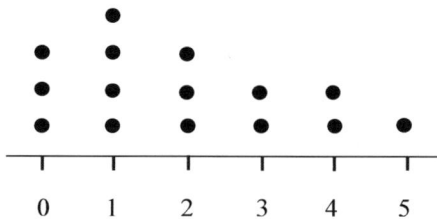

The dot plot shown above represents the number of pets owned by students in a class. What is the median number of pets?

A) 1

B) 1.5

C) 2

D) 2.5

7

$$(3-4i)^2 = a+bi$$

In the equation above, a and b are constants. What is the value of a? $\left(\text{Note: } i = \sqrt{-1}\right)$

A) −10

B) −7

C) 16

D) 25

8

$$\left(x+\frac{1}{x}\right)^2$$

Which of the following is equivalent to the expression above?

A) $x^2 + \dfrac{1}{x^2}$

B) $\left(x - \dfrac{1}{x}\right)^2 + 2$

C) $\left(x - \dfrac{1}{x}\right)^2 + 4$

D) $x^2 - \dfrac{1}{x^2}$

CONTINUE

9

Which of the following equations has a graph in the xy-plane for which y is always greater than 0 ?

A) $y = x + 5$

B) $y = x^2 - 2$

C) $y = (x-1)(x-3)$

D) $y = |x-10| + 1$

10

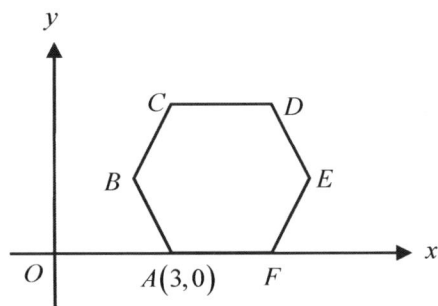

$A(3,0)$

In the xy-plane above, $ABCDEF$ is a regular hexagon with a side of 2. What is the x-coordinate of E?

A) 5

B) 6

C) 7

D) 8

11

If $k^2 - 16 = 16 - k^2$, what are all values of k ?

A) 0 only

B) 4 only

C) −4 and 4 only

D) −4, 0, and 4

12

$$\sin\left(2x - \frac{\pi}{12}\right) = \cos\left(x + \frac{\pi}{12}\right)$$

In the equation above, the angle measures are in radians. Which of the following could be the value of x ?

A) $\dfrac{\pi}{12}$

B) $\dfrac{\pi}{8}$

C) $\dfrac{\pi}{6}$

D) $\dfrac{\pi}{4}$

CONTINUE

Questions 13-14 refer to the following information.

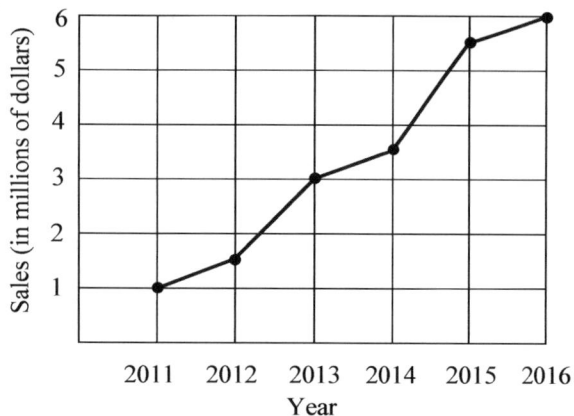

The graph above shows the sales for a smartphone company in the years 2011 through 2016.

13

Which interval has the greatest rate of change?

A) year 2012 to 2013
B) year 2013 to 2014
C) year 2014 to 2015
D) year 2015 to 2016

14

What is the average rate of change, in dollars per year, from year 2011 to 2016?

A) 1
B) 5
C) 1,000,000
D) 5,000,000

15

For a polynomial $p(x)$, the value of $p(2)$ and $p(-1)$ is 0. Which of the following could not be true about $p(x)$?

A) $x-2$ is a factor of $p(x)$.
B) $x+1$ is a factor of $p(x)$.
C) $x^2 - x - 2$ is a factor of $p(x)$.
D) $(x-2)+(x+1)$ is a factor of $p(x)$.

16

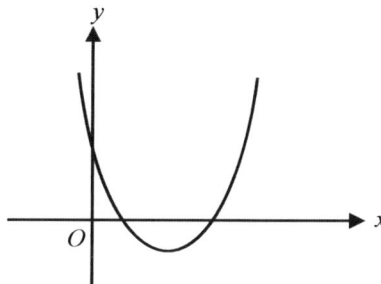

The graph of $f(x) = (x-a)(x-3)$, where a is constant, is shown in the xy-plane above. If the y-intercept of the graph is 3, what is the minimum value of f?

A) -1
B) -1.5
C) -2
D) -2.5

CONTINUE

17

If $ax^2 + bx + c = a(x-h)^2 + k$, which of the following is equal to h?

A) $-\dfrac{b}{2}$

B) $-\dfrac{b}{2a}$

C) $\dfrac{b}{2a}$

D) $\dfrac{b}{2}$

18

The speed s, in meters per second, of sound in air is given by the formula $s = 331 + 0.6C$, where C is the Celsius temperature. If the Celsius temperature increases by 10, what would be the increase in the speed of sound?

A) 0.6 meters per second

B) 6 meters per second

C) 331 meters per second

D) 331.6 meters per second

19

$$g(x) = x^2 + ax + b$$

For the function g defined above, a and b are constants. If $g(-2) = g(10)$, what is the value of a?

A)   8

B)   4

C)   −6

D)   −8

20

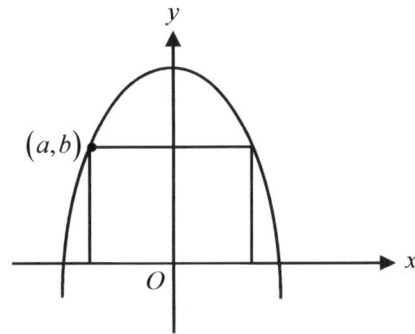

In the xy-plane above, the base of a rectangle is on the x-axis and the upper vertices are on the parabola $y = 10 - 2x^2$. Which of the following represents the area of the rectangle?

A) $10a - 2a^3$

B) $2a^3 - 10a$

C) $20a - 4a^3$

D) $4a^3 - 20a$

CONTINUE

21

$$x^2 - 2ax + y^2 - 4by = 0$$

The equation of a circle in the xy-plane with center $(2, 4)$ is shown above. What is the radius of the circle?

A) 4

B) $2\sqrt{5}$

C) 5

D) $3\sqrt{5}$

22

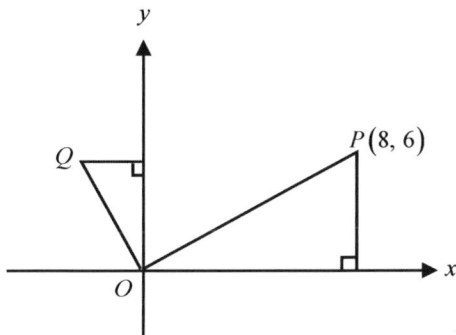

In the xy-plane above, $\overline{OQ}$ is perpendicular to $\overline{OP}$ and the length of $\overline{OQ}$ is 5. What are the coordinates of point Q?

A) $(-4, 3)$

B) $(-4, 5)$

C) $(-3, 4)$

D) $(-3, 5)$

23

The graph of the linear function g intersects points $(p, 0)$ and $(0, q)$ in the xy-plane. If $pq < 0$, which of the following is true about the graph of g ?

A) The graph of g has a positive y-intercept .

B) The graph of g has a negative y-intercept .

C) The graph of g has a positive slope.

D) The graph of g has a negative slope.

24

If $5x + y = 13$ and $\dfrac{8^x}{2^y} = 32$, What is the value of x ?

A) 2.25

B) 4.5

C) 6.5

D) 7.75

CONTINUE

Questions 25-26 refer to the following information.

$$h(t) = -4.9t^2 + v_o t$$

An object is thrown into the air from the ground with an initial velocity of v_o meters per second. Then, t seconds later, its height h meters above the ground is modeled by the function above.

25

If an object is thrown into the air with an upward velocity of 49 meters per second from the ground, after how many seconds will the object hit the ground?

A) 2.5

B) 5.0

C) 7.5

D) 10.0

26

Approximately what is the maximum height of the object?

A) 122.5

B) 245

C) 284

D) 320

27

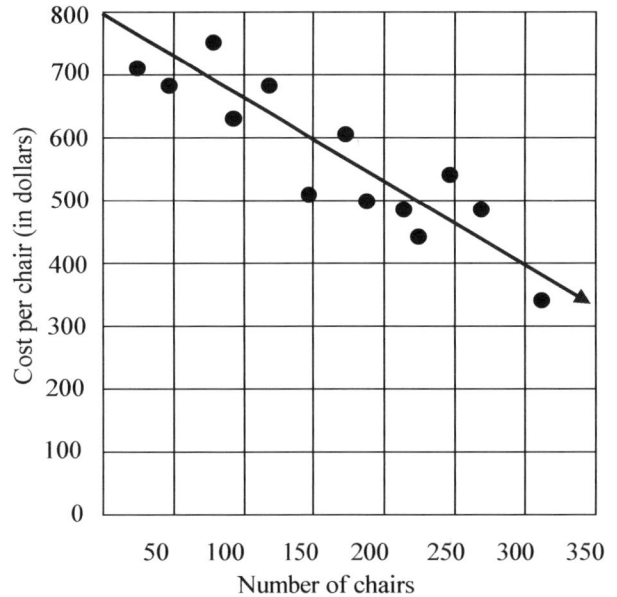

The scatterplot above shows the relationship between the number of chairs and the production cost per chair. The line of best fit is also shown. If 400 chairs have been produced, which of the following is closest to the cost per chair predicted by the line of best fit?

A) $280

B) $276

C) $270

D) $267

CONTINUE

28

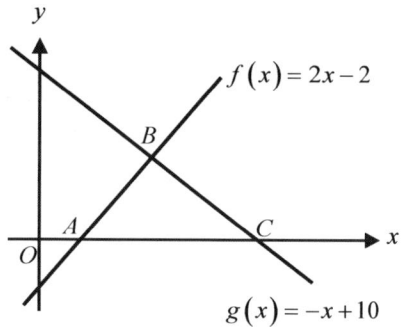

Note: Figure not drawn to scale.

The graphs of f and g are shown in the xy-plane above. What is the area of triangle ABC?

A) 20

B) 27

C) 40

D) 54

29

If the graph of $y = 2x - k$, where k is a positive constant, is tangent to the graph of $y = x^2$, what is the value of k?

A) -2

B) -1

C) 1

D) 2

30

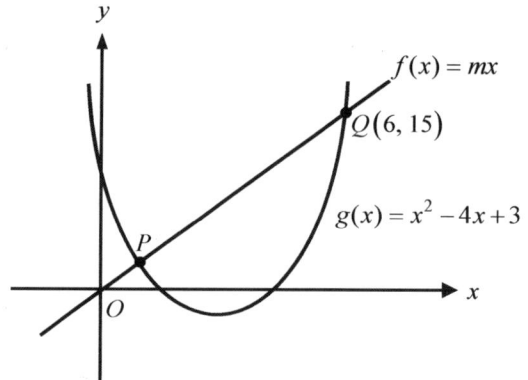

Note: Figure not drawn to scale.

The graphs of functions f and g intersect at points P and Q in the xy-plane above. What is the x-coordinate of point P?

A) $\dfrac{1}{4}$

B) $\dfrac{1}{2}$

C) 1

D) 2

CONTINUE

DIRECTIONS

For questions 31-38, solve the problem and enter your answer in the grid, as described below, on the answer sheet.

1. Although not required, it is suggested that you write your answer in the boxes at the top of the columns to help you fill in the circles accurately. You will receive credit only if the circles are filled in correctly.
2. Mark no more than one circle in any column.
3. No question has a negative answer.
4. Some problems may have more than one answer.
5. **Mixed numbers** such as $3\frac{1}{2}$ must be gridded as 3.5 or 7/2. (If $\boxed{3\;1\;/\;2}$ is entered into the grid, it will be interpreted as $\frac{31}{2}$, not $3\frac{1}{2}$.)
6. **Decimal answers:** If you obtain a decimal answer with more digits than the grid can accommodate, it may be either rounded or truncated, but it must fill the entire grid.

Answer: $\frac{7}{12}$

Write answer in boxes.

Fraction line

Grid in result.

Answer: 2.5

Decimal point

Acceptable ways to grid $\frac{2}{3}$ are:

Answer: 201
Either position is correct.

Note: You may start your answers in any column, space permitting. Columns you don't need to use should be left blank.

CONTINUE

31

If $9^{x-1} = \sqrt{27}$, what is the value of x ?

32

Let f be a linear function such that $f(-1) = -3$ and $f(3) = 12$. What is the value of $f(0)$?

33

$$x - \sqrt{x+11} = 1$$

What is the solution to the equation above?

34

Annie currently has \$200 in her bank account and deposits \$8 each week. Becky has \$440 in her account and withdraws \$4 each week. After how many weeks will they have the same balance in their accounts?

35

$$f(x) = \frac{x-10}{(x-10)^2 + 6(x-10) + 9}$$

For what value of x is the function f above undefined?

CONTINUE

Questions 36-37 refer to the following information.

Lee opened a bank account that earns 2 percent interest compounded quarterly. His initial deposit was $2000, and the value of the account after t years is given by the expression $2000(1+c)^{4t}$.

36

What is the value of c in the expression?

37

After 10 years, how much money will Lee have in his account? (Round your answer to the nearest dollar and ignore the dollar sign when gridding your answer.)

38

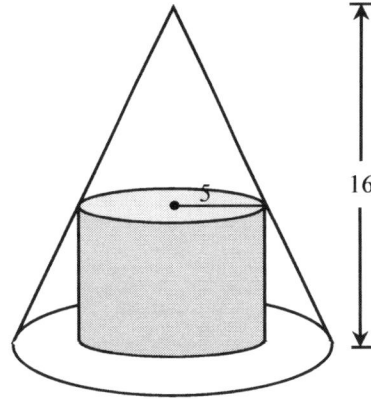

In the figure above, a cone has a height 16 and radius 8. A cylinder whose base has a radius of 5 is inscribed in the cone. What is the height of the cylinder?

STOP

If you finish before time is called, you may check your work on this section only.
Do not turn to any other section in the test.

Math Conversion Table

Raw Score	Scaled Score	Raw Score	Scaled Score
58	800	27	500
57	800	26	490
56	800	25	480
55	800	24	470
54	790	23	460
53	780	22	460
52	770	21	450
51	750	20	440
50	740	19	430
49	730	18	430
48	720	17	420
47	710	16	420
46	700	15	410
45	690	14	400
44	670	13	390
43	680	12	380
42	670	11	370
41	660	10	360
40	650	9	450
39	640	8	340
38	630	7	330
37	620	6	310
36	610	5	290
35	600	4	280
34	590	3	270
33	580	2	260
32	560	1	240
31	550	0	200
30	540		
29	530		
28	520		

Test 15				**Answers and Explanations**					

<table>
<tr><td rowspan="4">SECTION
3</td><td>1</td><td>2</td><td>3</td><td>4</td><td>5</td><td>6</td><td>7</td><td>8</td><td>9</td><td>10</td></tr>
<tr><td>B</td><td>B</td><td>C</td><td>B</td><td>C</td><td>C</td><td>D</td><td>C</td><td>A</td><td>D</td></tr>
<tr><td>11</td><td>12</td><td>13</td><td>14</td><td>15</td><td>16</td><td>17</td><td>18</td><td>19</td><td>20</td></tr>
<tr><td>C</td><td>C</td><td>D</td><td>C</td><td>B</td><td>5</td><td>18</td><td>6</td><td>16/3</td><td>3/5</td></tr>
<tr><td rowspan="6">SECTION
4</td><td>1</td><td>2</td><td>3</td><td>4</td><td>5</td><td>6</td><td>7</td><td>8</td><td>9</td><td>10</td></tr>
<tr><td>B</td><td>C</td><td>B</td><td>C</td><td>D</td><td>C</td><td>B</td><td>C</td><td>D</td><td>B</td></tr>
<tr><td>11</td><td>12</td><td>13</td><td>14</td><td>15</td><td>16</td><td>17</td><td>18</td><td>19</td><td>20</td></tr>
<tr><td>C</td><td>C</td><td>C</td><td>C</td><td>D</td><td>A</td><td>B</td><td>B</td><td>D</td><td>C</td></tr>
<tr><td>21</td><td>22</td><td>23</td><td>24</td><td>25</td><td>26</td><td>27</td><td>28</td><td>29</td><td>30</td></tr>
<tr><td>B</td><td>C</td><td>C</td><td>A</td><td>D</td><td>A</td><td>D</td><td>B</td><td>C</td><td>B</td></tr>
<tr><td></td><td>31</td><td>32</td><td>33</td><td>34</td><td>35</td><td>36</td><td>37</td><td>38</td><td></td></tr>
<tr><td></td><td>7/4</td><td>3/4</td><td>5</td><td>20</td><td>7</td><td>.005</td><td>2442</td><td>6</td><td></td></tr>
</table>

SECTION 3

1. **B**

$$\frac{1}{10}x^2 - \frac{1}{5}x = \frac{4}{5} \rightarrow x^2 - 2x = 8 \rightarrow x^2 - 2x - 8 = 0$$
$$\rightarrow (x-4)(x+2) = 0 \rightarrow x = 4, -2$$

2. **B**

Since $a = 10b$, $\dfrac{15b}{2a} = \dfrac{15b}{2(10b)} = \dfrac{15b}{20b} = \dfrac{3}{4}$.

3. **C**

$$\frac{a^{\frac{1}{2}}}{a^{-\frac{1}{3}}} = a^{\frac{1}{2}-\left(-\frac{1}{3}\right)} = a^{\frac{5}{6}} = \sqrt[6]{a^5}$$

4. **B**

A: rate of change $= \dfrac{2}{1} = 2$

B: rate of change $= \dfrac{6-2}{1} = 4$

C: rate of change $= \dfrac{4-5.5}{1} = -1.5$

D: rate of change $= \dfrac{0-4}{1} = -4$

5. **C**

$$\frac{8}{x} - \frac{12}{x+10} = 0 \rightarrow \frac{8}{x} = \frac{12}{x+10} \rightarrow 12x = 8x + 80$$
$$4x = 80 \rightarrow x = 20$$

Therefore, $\dfrac{x}{4} = \dfrac{20}{4} = 5$.

6. **C**

Add the equations: $7a - 7b = -14$
$$\rightarrow a - b = -2$$

7. **D**

Line $y = mx - 5$ contains $(a, 0)$.

$$0 = ma - 5 \rightarrow m = \frac{5}{a}$$

8. **C**

$$-(x-10)^2 + 10 = -26 \rightarrow (x-10)^2 = 36$$
$$x - 10 = 6, -6 \rightarrow x = 16, 4$$
Therefore, $PQ = 16 - 4 = 12$.

9. A

    Remainder Theorem;

    $P(1) = 3 - 5 + 5 = a \rightarrow a = 3$

10. D

    If x = his sales, then $3000 + \dfrac{k}{100}(x) = 7200$.

    $\dfrac{kx}{100} = 4200 \rightarrow x = \dfrac{420000}{k}$

11. C

	speed	hour	distance
Forward	50	x	$50x$
Backward	30	$x+1$	$30x+30$

    Same distance: $50x = 30x + 30 \rightarrow x = 1.5$ hours

    Returning trip: $x + 1 = 1.5 + 1 = 2.5$ hours

12. C

    For zeros: $2(x-4)^2 - 10 = 0 \rightarrow (x-4)^2 = 5$

    $x - 4 = \pm\sqrt{5} \rightarrow x = 4 \pm \sqrt{5}$

13. D

    Slope of the line $= \dfrac{40 - 30}{200 - 100} = \dfrac{1}{10}$

    Total cost $C = 30 + \dfrac{1}{10}(280 - 100) = 48$ dollars

14. C

    $y = \dfrac{1}{10}(x - 100) + 30$

15. B

    Over 40; $(80 + 40) + (100 + 70) + (10 + 20) = 320$

    $31 - 40$ with against:; 100

    $P = \dfrac{100}{320} = \dfrac{5}{16}$

16. 5

    Since $x = -1$ is a zero,

    $f(-1) = 2(-1)^2 - 8(-1) + k = 0 \rightarrow k = -10$

    Therefore, $f(x) = 2x^2 - 8x - 10$.

    Now solve.

    $2(x^2 - 4x - 5) = 0 \rightarrow 2(x - 5)(x + 1) = 0$

    You can see the other zero is $x = 5$.

17. 18

    $f(x+1) = 5x + k : x + 1 = 3 \rightarrow x = 2$

    Put this number in the equation.

    $f(2+1) = 5(2) + k \rightarrow 10 + k = 8 \rightarrow k = -2$

    Now we got $f(x+1) = 5x - 2$. Putting $x = 4$,

    $f(1+4) = 5(4) - 2 \rightarrow f(5) = 18$

18. 6

    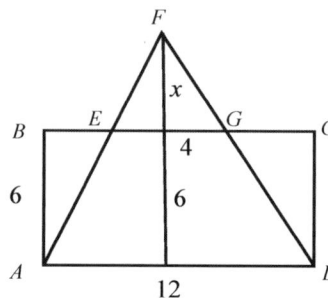

    Since $\triangle AFD \sim \triangle EFG$,

    $\dfrac{x}{x+6} = \dfrac{4}{12} \rightarrow 12x = 4x + 24 \rightarrow x = 3$.

    Therefore, the area of $\triangle EFG$ is $\dfrac{4 \times 3}{2} = 6$.

19. $\dfrac{16}{3}$

    In order to have infinitely many solutions,

    $\dfrac{3}{a} = \dfrac{-a}{-b} = \dfrac{3}{4}$. From the equation, $a = 4$,

    and $\dfrac{-4}{-b} = \dfrac{3}{4} \rightarrow b = \dfrac{6}{3}$.

20. $\dfrac{3}{5}$

    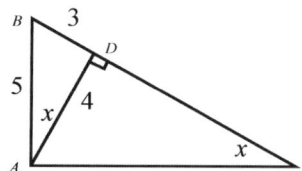

    The value of $\sin C$ is equal to the value of

    $\sin \angle BAD$. Therefore, $\sin \angle C = \sin \angle BAD = \dfrac{3}{5}$.

Answer Explanations

1. B

$$\frac{3}{2}\left(x+\frac{3}{4}\right)=3 \rightarrow x+\frac{3}{4}=2 \rightarrow x=\frac{5}{4}$$

2. C

	time	rate	combined
Mina	2	$\frac{1}{2}$	$\frac{1}{2}+\frac{1}{4}$
Ruth	4	$\frac{1}{4}$	$=\frac{3}{4}$

$$\text{Time}=1\div\frac{3}{4}=\frac{4}{3}\text{ hours} \rightarrow \text{ I hour 20 minutes}$$

3. B

$$2x^2-8x-90=0 \rightarrow x^2-4x-45=0$$
$$x^2-4x+4=49 \rightarrow (x-2)^2=49$$

4. C

$$0=2x^2-9x+4 \rightarrow (2x-1)(x-4)=0$$
$$x=4,\frac{1}{2} : \text{ Therefore, } PQ=4-0.5=3.5 .$$

5. D

$$g(-1)=3 \text{ and } f(3)=10$$

6. C

Two is in the middle.

7. B

$$(3-4i)^2=a+bi \rightarrow -7-24i=a+bi \rightarrow a=-7$$

8. C

$$\left(x+\frac{1}{x}\right)^2=\left(x-\frac{1}{x}\right)^2+4$$

9. D

10. B

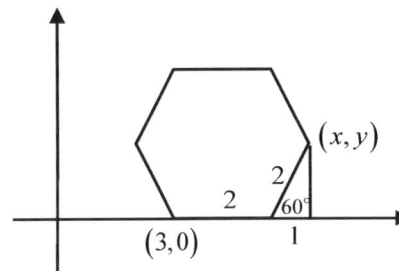

$$x=3+2+1=6$$

11. C

$$k^2-16=16-k^2 \rightarrow 2k^2=32 \rightarrow k^2=16$$
Therefore, $k=\pm4$.

12. C

$$\sin\left(2x-\frac{\pi}{12}\right)=\cos\left(x+\frac{\pi}{12}\right): \text{ Cofunction}$$
$$2x-\frac{\pi}{12}+x+\frac{\pi}{12}=\frac{\pi}{2} \rightarrow 3x=\frac{\pi}{2} \rightarrow x=\frac{\pi}{6}$$

13. C

$$\frac{5.5M-3.5M}{1}=2M \text{ per year}$$

14. C

$$\frac{6M-1M}{5}=1,000,000$$

16. C

Since $(0,3)$ lies on the graph,
$$3=(0-a)(0-3) \rightarrow 3=3a \rightarrow a=1$$
$$f(x)=(x-1)(x-3) \rightarrow$$
$$\text{Axis of symmetry}: x=\frac{1+3}{2}=2$$
Therefore, minimum of f is $f(2)=-1$.

17. B

$$ax^2+bx+c=a\left(x^2+bx\right)+c \rightarrow$$
$$a\left(x+\frac{b}{2a}\right)^2+\frac{4ac-b^2}{4a}$$
Therefore, $h=-\frac{b}{2a}$.

18. B

$$\Delta s=0.6\times10=6 \text{ meters per second}$$

Answer Explanations

19. D

$g(2)=4-2a+b$ and $g(10)=100+10a+b$

$g(2)=g(10) \rightarrow a=-8$

20. C

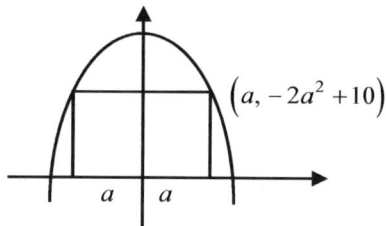

$\left(a,\,-2a^2+10\right)$

Area of the rectangle is $2a\left(10-2a^2\right)=20a-4a^3$.

21. B

$x^2-2ax+y^2-4by=0 \rightarrow$ Standard form is

$(x-a)^2+(y-2b)^2=a^2+4b^2$.

Center $(2,4)=(a,2b) \rightarrow a=2,b=2$

Therefore, $r=\sqrt{a^2+4b^2}=\sqrt{20}=2\sqrt{5}$.

22. C

$OP=10$ and $OQ=5$

Since the triangles are similar, corresponding sides are half.

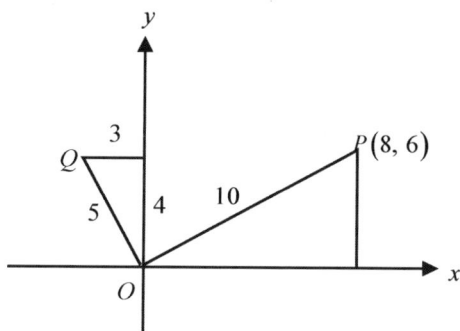

Therefore, coordinates of point Q is $(-3,4)$.

23. C

$pg<0 \rightarrow \begin{cases} p>0 \text{ and } q<0 \\ p<0 \text{ and } q>0 \end{cases}$

Therefore, the graph g has always a positive slope.

24. A

$\dfrac{8^x}{2^y}=32 \rightarrow \dfrac{2^{3x}}{2^y}=2^5 \rightarrow 2^{3x-2y}=2^5$

From the equation: $3x-y=5$

Solve the system of equations.

$5x+y=13$ and $3x-y=5 \rightarrow x=2.25$

25. D

$h(t)=-4.9t^2+49t \rightarrow -4.9t(t-10)=0$

$t=0$ and 10 Therefore, $10-0=10$.

26. A

At $t=5$, it has a maximum.

Therefore, $h(5)=-4.9(5)^2+49(5)=122.5$.

27. D

Slope of the line is $\dfrac{600-800}{150-0}=-\dfrac{4}{3}$.

Now, $y=-\dfrac{4}{3}x+800$.

For $x=400$,

$y=-\dfrac{4}{3}(400)+800=266.66\cdots\approx 267$.

28. B

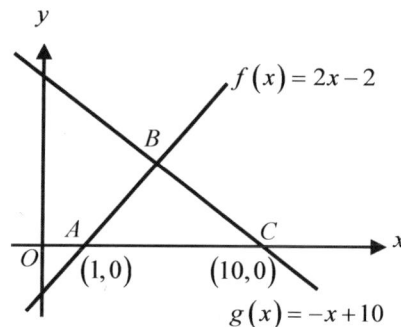

The x-intercepts of the graphs are $(1,0)$ and $(10,0)$. The intersection of the graphs: $\rightarrow 2x-2=-x+10 \rightarrow x=4$ and $y=6$

Since $AC=9$ and the height is 6, the area of $\triangle ABC$ is $\dfrac{9\times6}{2}=27$.

29. C

Discriminant should be zero.

$x^2=2x-k \rightarrow x^2-2x+k=0$

$D=(-2)^2-4(1)(k)=0 \rightarrow k=1$

30. B

Answer Explanations

Slope of f is $\dfrac{15}{6} = \dfrac{5}{2}$. $\rightarrow$ $f(x) = \dfrac{5}{2}x$

Find the points of intersection.

$x^2 - 4x + 3 = \dfrac{5}{2}x \rightarrow 2x^2 - 13x + 6 = 0$

$(2x-1)(x-6) = 0 \rightarrow x = \dfrac{1}{2},\ 6$

Or, you can use sum of the roots.

31. 5/4

$9^{x-1} = \sqrt{27} \rightarrow 3^{2(x-1)} = 3^{\frac{3}{2}} \rightarrow 2x - 1 = \dfrac{3}{2}$

$x = \dfrac{5}{4}$

32. 3/4

$f(-1) = -3 \rightarrow (-1,-3)$, $f(3) = 12 \rightarrow (3,12)$

$f(0) = k \rightarrow (0,k)$

Since all these points lie on the line, slopes between ant two points are equal.

$\dfrac{12-(-3)}{3-(-1)} = \dfrac{k-12}{0-3} \rightarrow k = \dfrac{3}{4}$

33. 5

$x - \sqrt{x+11} = 1 \rightarrow x - 1 = \sqrt{x+11}$

$x^2 - 2x + 1 = x + 11 \rightarrow x^2 - 3x - 10 = 0 \rightarrow$

$(x-5)(x-2) = 0 \rightarrow x = 5,\ 2\,(\text{not working})$

34. 20

$y = 200 + 8x$ and $y = 440 - 4x$

Therefore, $200 + 8x = 440 - 4x \rightarrow x = 20$

35. 7

Let $X = x - 10$, then $X^2 + 6X + 9 = 0$

$(X+3)^2 = 0 \rightarrow X = -3 \rightarrow X - 10 = -3 \rightarrow x = 7$

36. .005

$A = 2000\left(1 + \dfrac{0.02}{4}\right)^{4t} \rightarrow c = .005$

37. 2442

$A = 2000(1 + 0.005)^{40} = 2441.588 \cdots \approx 2442$

38. 6

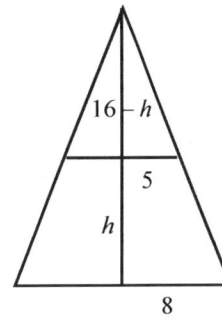

Similar triangles:

$\dfrac{16-h}{5} = \dfrac{16}{8} \rightarrow 128 - 8h = 80 \rightarrow 8h = 48$

$h = 6$

No Test Material on This Page

PRACTICE TEST 16

Dr. John Chung's SAT Math

Math Test - No Calculator

25 MINUTES, 20 QUESTIONS

Turn to Section 3 of your answer sheet to answer the questions in this section.

DIRECTIONS

For questions 1–15, solve each problem, choose the best answer from the choices provided, and fill in the corresponding circle on your answer sheet. **For questions 16–20**, solve the problem and enter your answer in the grid on your answer sheet. Please refer to the directions before question 16 on how to enter your answers in the grid. You may use any available space in your test booklet for scratch work.

NOTE

1. The use of a calculator **is not permitted**.

2. All variables and expressions used represent real numbers unless otherwise indicated.

3. Figures provided in this test are drawn to scale unless otherwise indicated.

4. All figures lie in a plane unless otherwise indicated.

5. Unless otherwise indicated, the domain of a given function f is the set of all real numbers x for which $f(x)$ is a real number.

REFERENCE

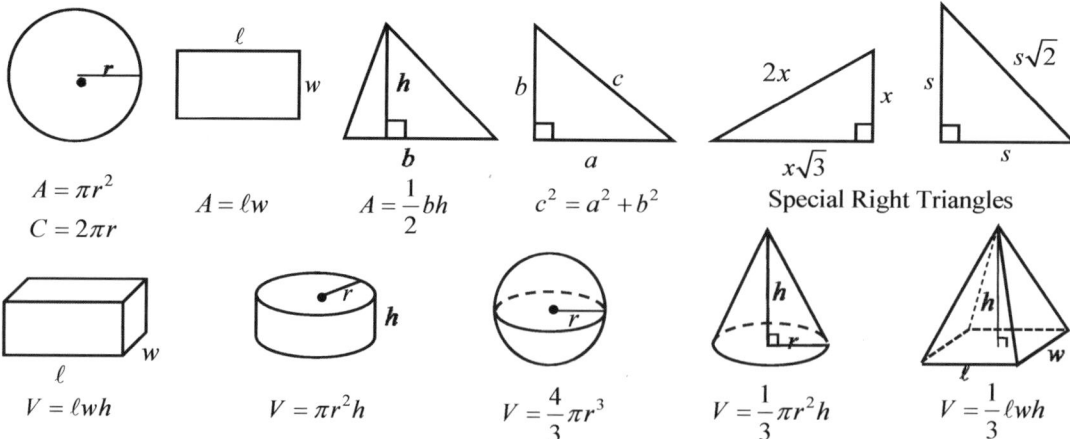

$A = \pi r^2$
$C = 2\pi r$

$A = \ell w$

$A = \dfrac{1}{2}bh$

$c^2 = a^2 + b^2$

Special Right Triangles

$V = \ell w h$

$V = \pi r^2 h$

$V = \dfrac{4}{3}\pi r^3$

$V = \dfrac{1}{3}\pi r^2 h$

$V = \dfrac{1}{3}\ell w h$

The number of degrees of arc in a circle is 360.

The number of radians of arc in a circle is 2π.

The number of the measures in degrees of the angles of a triangle is 180.

CONTINUE

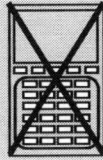

1

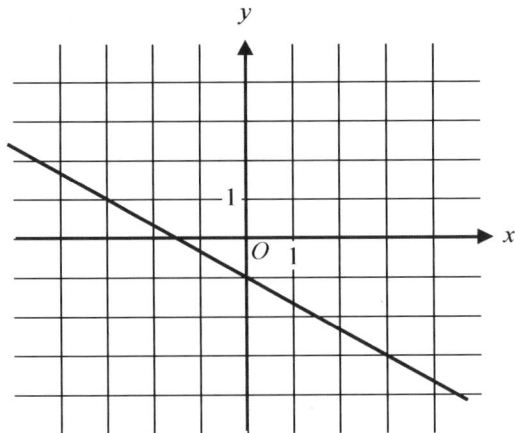

Which of the following is an equation of line ℓ in the xy-plane above?

A) $y = 3x - 1$

B) $y = \dfrac{1}{3}x - 1$

C) $y = -\dfrac{1}{3}x - 1$

D) $y = -\dfrac{2}{3}x - 1$

2

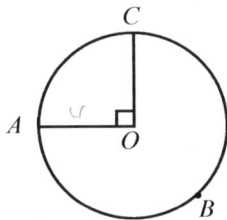

The circle above with center O has an area of 36π. What is the length of major arc $\overparen{ABC}$?

A) 3π
B) 9
C) 9π
D) 27

3

What are the solutions of the quadratic equation

$$\frac{1}{10}x^2 - \frac{1}{2}x + \frac{3}{5} = 0 ?$$

A) $x = -2$ and $x = -3$

B) $x = -2$ and $x = 3$

C) $x = 2$ and $x = -3$

D) $x = 2$ and $x = 3$

4

Which of the following is a function whose graph in the xy-plane has no x-intercepts ?

A) $y = 2x - 1$

B) $y = x^2 - 3x - 10$

C) $y = -x^2 - 4$

D) $y = x^3 - 1$

CONTINUE

5

$$\sqrt{k+11} + 2 = x + k$$

In the equation above, k is a constant. If $x = 1$, what is the value of k?

A) -2

B) 2

C) 5

D) 14

(handwritten work):
$\sqrt{k+11} + 2 = 1 + K$
$\sqrt{k+1} = 1 + K - 2$
$\sqrt{k+1} = K - 1$
$k+1 = k^2 + 1$
$3 + 2 = 1 - 2$
$\sqrt{13}$ $\sqrt{16} + 2 = 5 + 1$

6

Which of the following is equivalent to the sum of $a^3 - 2a$ and $a^2 - 2$?

A) $a^3 - 2$

B) $a^2(a+1)$

C) $(a^2 - 2)(a+1)$

D) $a(a+1)(a-1)$

(handwritten work):
$a^3 - 2a + a^2 - 2$
$a^3 + a^2 - 2a - 2$

7

Peter has two weekly jobs. He works in a town library, which pays $12 per hour, and he works as a cashier at a grocery mart, which pays $10.5 per hour. He can work no more than 15 hours per week, but he want to earn at least $120 per week. Which of the following systems of inequalities represents this situation in terms of x and y, where x is the number of hours he works in the town library and y is the number of hours he works as a cashier at the grocery mart?

A) $\begin{cases} 12x + 10.5y \geq 120 \\ x + y \geq 15 \end{cases}$

(handwritten): $12x + 10.5x \geq 120$ $x + y \leq 15$

B) $\begin{cases} 12x + 10.5y \leq 120 \\ x + y \geq 15 \end{cases}$

C) $\begin{cases} 12x + 10.5y \leq 120 \\ x + y \leq 15 \end{cases}$

D) $\begin{cases} 12x + 10.5y \geq 120 \\ x + y \leq 15 \end{cases}$

8

A kitchen appliance manufacturing company determines that the total cost C, in dollars, of producing n units of a blender is given by $C(n) = 30n + 2500$. Which of the following statements is the best interpretation of the number 2500 in this context?

A) The cost of producing each unit.

B) The total cost of producing n units

C) The cost that must be paid regardless of the number of units produced.

D) The cost of producing 30 units

CONTINUE

9

$$2x + 12 = 2(y - 6)$$
$$x^2 = y$$

If (x, y) is a solution of the system of equations above. If $x > 0$, what is the value of $x + y$?

A) 20
B) 16
C) 15
D) 9

(handwritten work)
$2x + 12 = 2(x^2 - 6)$
$2x + 12 = 2x^2 - 12$
$2x^2 - 24 - 2x$
$2x^2 - 2x - 24$
$x^2 - x - 48$
$\frac{(x - 8)(x + 6)}{2}$
$2(x - 4)(x + 3)$
$x = 4 \quad y = 16$
$16 + 4 = 20$

10

If $a^2 - b^2 = x$ and $y = 2ab$, which of the following is equivalent to $x^2 + y^2$?

A) $\left(a^2 + b^2\right)^2$

B) $\left(a^2 - b^2\right)^2$

C) $a^4 - b^4$

D) $a^4 + b^4$

(handwritten work)
$(a^2 - b^2)^2 (2ab)^2$
$a^4 - b^4 + 4a^2b^2$

11

The volume of a right circular cone is 33 cubic feet. What is the volume, in cubic feet, of a right circular cone with half the radius and twice the height of the cone?

(handwritten) $33 = \pi r^2 h$

A) 16.5
B) 33
C) 49.6
D) 66

12

Which of the following is equivalent to $2\left(\sqrt[3]{16}\right)$?

A) $4\sqrt{2}$

B) $2\sqrt[3]{4}$

C) $2\sqrt[3]{2}$

D) $4\sqrt[3]{2}$

CONTINUE

13

At a coffee shop, the coffee urn has a certain amount of coffee in it at the start of the afternoon rush. It is filled by adding k coffee bags to hot water. If a cups of coffee are made by k coffee bags is given by the equation $a = 10k + 5$, how many additional coffee bags are needed to make each additional 20 cups of coffee?

$20 = 10k + 5$
$10k = 15$
$k = 1.5$

A) One

B) Two

C) Three

D) Four

14

Jackson drives an average of 300miles each week. His car can travel an average of 25 miles per gallon of gasoline. Jackson would like to reduce his weekly expenditure on gasoline by $ 8. If gasoline costs $2.50 per gallon, how many average miles should he drive each week?

$25 \overline{)300}$ 12 gal.

2.50×12

500
2500
20.00 $30/8 = 22$

22

A) 200 miles

B) 220 miles

C) 240 miles

D) 250 miles

15

$$f(x) = \sqrt{x - 1}$$

The function f is defined by the function above. Which of the following is the graph of $y = f(-x)$ in the xy-plane ?

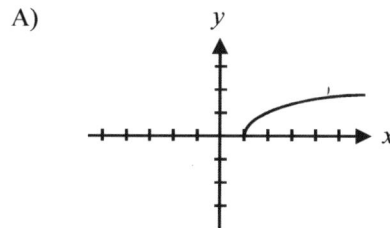

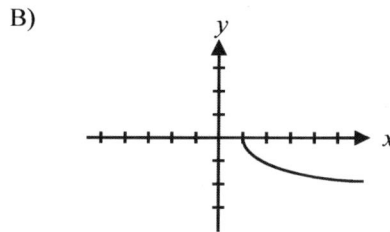

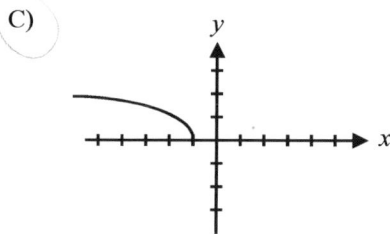

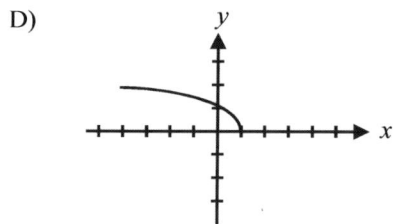

A)

B)

C)

D)

CONTINUE

Answer: $\frac{7}{12}$

Answer: 2.5

DIRECTIONS

For questions 16–20, solve the problem and enter your answer in the grid, as described below, on the answer sheet.

Write answer in boxes. → Fraction line

Grid in result. → Decimal point

1. Although not required, it is suggested that you write your answer in the boxes at the top of the columns to help you fill in the circles accurately. You will receive credit only if the circles are filled in correctly.

2. Mark no more than one circle in any column.

3. No question has a negative answer.

4. Some problems may have more than one answer.

5. **Mixed numbers** such as $3\frac{1}{2}$ must be gridded as 3.5 or 7/2. (If $3|1|/|2|$ is entered into the grid, it will be interpreted as $\frac{31}{2}$, not $3\frac{1}{2}$.)

6. **Decimal answers:** If you obtain a decimal answer with more digits than the grid can accommodate, it may be either rounded or truncated, but it must fill the entire grid.

Acceptable ways to grid $\frac{2}{3}$ are:

Answer: 201
Either position is correct.

Note: You may start your answers in any column, space permitting. Columns you don't need to use should be left blank.

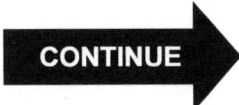

CONTINUE

16

Claire plans to go an amused park The entrance fee costs $8 and each ride cost $4. Claire wants to spend no more than $35 in the park. What is the maximum number of rides for which Claire can take in the park?

$8 + 4x = 35$
$4x = 27$
$x = 6$

6

17

$$8(p-1) = 5(p^2 - 1) - 6(p-1)$$

If $p > 1$, what is the solution of the equation above?

$8p - 8 = 5p^2 - 5 - 6p + 6$
$5p^2 - 6p - 8p + 6 - 5 + 8$
$5p^2 - 14p + 9$ $p^2 - 14p + 45$
$(x-9)(x-5)$
9 or 5

18

$$\frac{1}{4}(3x + y) = \frac{15}{8}$$

$$\frac{1}{3}y = 9x$$

$\frac{3}{4}x + \frac{1}{4}y = \frac{15}{3}$
$\frac{6}{8}x + \frac{2}{8}y = \frac{15}{8}$

$\boxed{\frac{27}{9}?}$

The system of equations above has solution (a, b).

What is the value of b?

$y = 3x$
$x = \frac{1}{27}y$

$\frac{6}{8}x + \frac{2}{8}(3x) - \frac{15}{8}$
$\frac{6}{8}x + \frac{6}{8}x = \frac{15}{8}$
$\frac{12}{8}x = \frac{15}{8}$
$x = \frac{15}{12}$

$\frac{6}{8}(\frac{1}{27}y) + \frac{2}{8}y = \frac{15}{8}$
$\frac{6}{216}y + \frac{2}{8}y = \frac{15}{8}$
$\frac{1}{36}y + \frac{9}{36}y = \frac{15}{8}$ $\frac{10}{36}y = \frac{15}{8}$

19

$$\frac{A}{(x+2)^2} + \frac{B}{x+2}$$

$\boxed{4}$

The expression above is equivalent to $\dfrac{3x+10}{(x+2)^2}$, where

A and B are constants and $x \neq 2$. What is the value of A?

$A + B(x+2) = 3x + 10$
$A + Bx + 2B = 3x + 10$
$\quad 5 \qquad 5$
$B = 5$
$A + 5x + 10 = 3x + 10$
$A = -2$

20

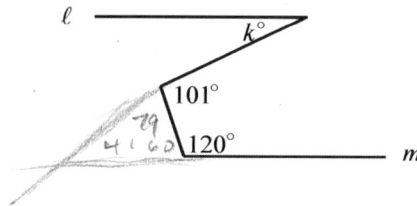

In the figure above, line ℓ is parallel to line m. What is the value of k?

180
-101
79

180
$+66$
126
-79
41

$\boxed{41}$

STOP

If you finish before time is called, you may check your work on this section only.
Do not turn to any other section in the test.

No Test Material on This Page

Math Test - Calculator

55 MINUTES, 38 QUESTIONS

Turn to Section 4 of your answer sheet to answer the questions in this section.

DIRECTIONS

For questions 1-30, solve each problem, choose the best answer from the choices provided, and fill in the corresponding circle on your answer sheet. **For questions 31-38**, solve the problem and enter your answer in the grid on your answer sheet. Please refer to the directions before question 31 on how to enter your answers in the grid. You may use any available space in your test booklet for scratch work.

NOTES

1. The use of a calculator **is permitted**.

2. All variables and expressions used represent real numbers unless otherwise indicated.

3. Figures provided in this test are drawn to scale unless otherwise indicated.

4. All figures lie in a plane unless otherwise indicated.

5. Unless otherwise indicated, the domain of a given function f is the set of all real numbers x for which $f(x)$ is a real number.

REFERENCE

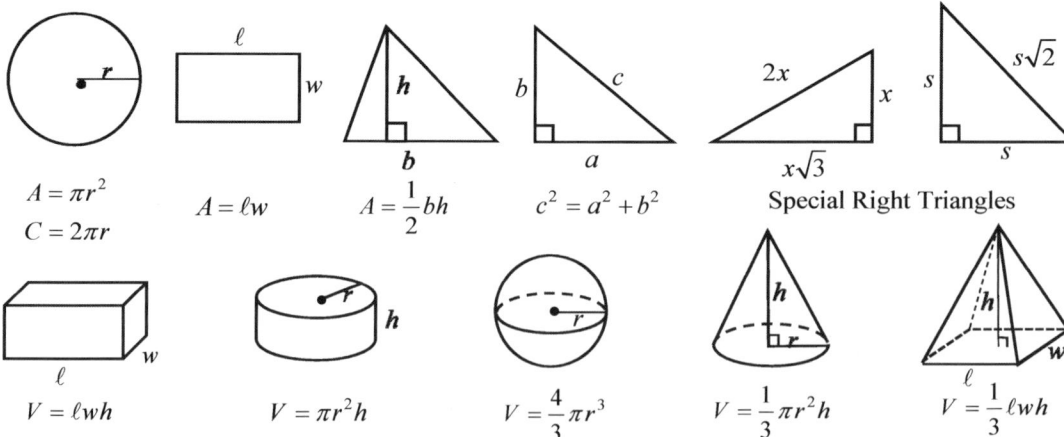

$A = \pi r^2$
$C = 2\pi r$

$A = \ell w$

$A = \frac{1}{2}bh$

$c^2 = a^2 + b^2$

Special Right Triangles

$V = \ell w h$

$V = \pi r^2 h$

$V = \frac{4}{3}\pi r^3$

$V = \frac{1}{3}\pi r^2 h$

$V = \frac{1}{3}\ell w h$

The number of degrees of arc in a circle is 360.

The number of radians of arc in a circle is 2π.

The number of the measures in degrees of the angles of a triangle is 180.

CONTINUE

1

Spring-Field County Population

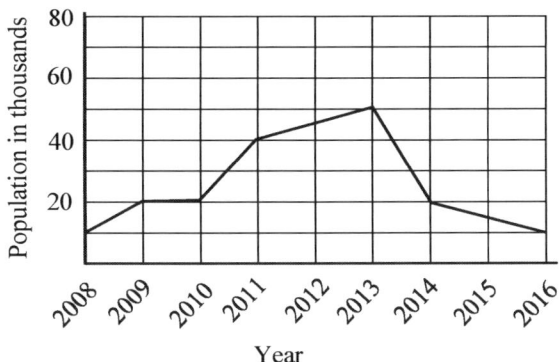

According to the graph above, which of the following two consecutive years has the greatest change of rate in the population of the county?

A) 2008-2009

B) 2010-2011

C) 2013-2014

D) 2015-2016

2

$$C = (10 + 0.1m)(1 + r)$$

The Sunny Information and Telephone Company (SITC) calculates a customer's total monthly cell phone charge using the formula above, where C is the total cell phone charge, r is the tax rate, and m is the number of minutes the customer used that month. Last month a customer used 6 hours 40 minutes and paid $54. If the customer uses 8 hour this month, which of the following will be the customer's cost this month?

A) $56.25

B) $62.64

C) $70.56

D) $85.25

3

x	$g(x)$
-3	0
1	6
4	0

Some values of the quadratic function g are shown in the table above. Which of the following defines g ?

A) $y = -\dfrac{3}{2}(x+3)(x-4)$

B) $y = -\dfrac{3}{2}(x-3)(x+4)$

C) $y = -\dfrac{1}{2}(x-3)(x+4)$

D) $y = -\dfrac{1}{2}(x+3)(x-4)$

4

$2\dfrac{1}{2}$ cups of flour

3 eggs

4 table spoons butter

$1\dfrac{3}{4}$ cups of sugar

The recipe, which makes a pan of 24 cookies, lists these items above. Peter only can buy eggs by the dozen. If he wants to make a pan of 125 cookies for his birthday party, what is the minimum number of dozens of eggs he should buy to make enough cookies?

A) 2   B) 3   C) 10   D) 27

CONTINUE

5

If $5\left(a^2 - b^2\right) = 13$ and $a = b + 7$, what is the value of $a + b$?

A) 8

B) 2

C) $\dfrac{23}{33}$

D) $\dfrac{13}{35}$

Handwritten work:
$a^2 - b^2 = \dfrac{13}{5}$
$(b+7)^2$
$b^2 + 14b + 49 - b^2 \; \dfrac{13}{5}$
$14b + 49 = \dfrac{13}{5}$
$14b = \dfrac{13}{5} - 49$
$b = -3.314$
0.37171429

6

Members of a reading club want to buy a science fiction book for their spnosors. If each member of the club contributes $4.00 towards the purchase of the book, they will still be short $7.00. If each member contributes $6.00, they will have a surplus of $9.00 after the purchase of the book. What is the price of the book in dollars?

A) 8

B) 24

C) 39

D) 48

Handwritten work:
$4x + 7 = 6x - 9$
$16 = 2x$
$x = 8$

7

A new online bookstore sells algebra and geometry workbooks. Each algebra workbook sells for $8, and each geometry workbook sells for $12. If Peter purchased a total of 9 algebra and geometry workbooks that have a combined selling price of $88, how many geometry workbooks did he purchase?

A) 4

B) 5

C) 6

D) 7

Handwritten work:
$8x + 12y = 88$
$x + y = 9$
$x = 9 - y$
$8(9-y) + 12y = 88$
$72 - 8y + 12y = 88$
$72 + 4y = 88$
$4y = 16$
$y = 4$

8

Which of the following is an equivalent form of $\left(1.2a - 3.6\right)^2 - \left(2.5a^2 - 1.2\right)$?

A) $-1.06a^2 + 14.16$

B) $-1.06a^2 + 11.76$

C) $-1.06a^2 - 8.64a + 14.16$

D) $-1.06a^2 - 8.64a + 11.76$

Handwritten work:
$1.44a^2 \; 12.96 - 2.5a^2 + 1.2$
$-1.06a^2 - 1176$

CONTINUE

9

A group of men agreed to share equally the cost of a hunting trip which they estimated to be $720. At the last minute, two of the men were unable to go, with the result that each of the others had to pay an additional $12. How many men actually went on the hunting trip?

A) 10

B) 12

C) 13

D) 14

handwritten: 24
handwritten: $12\overline{)720}$, 72, $10\overline{)720}$

10

$$h = v_o t - 0.8t^2$$

An object is shot upward with an initial velocity of 60 meters per second and it returns to the surface. The height is given by the formula above, where v_o is the initial velocity. How long does the object stay in the air?

A) 75

B) 65

C) 60

D) 50

handwritten: $60t\ 0.8t^2$, $t(60 - 0.8t)$, $t = 0$ or 75

11

The average of m numbers is a and the average of n numbers is b. If the average of all the $(m+n)$ numbers is k, what is the value of m?

A) $\dfrac{b}{a}(nk)$

B) $\left(\dfrac{k-b}{k-a}\right)n$

C) $\left(\dfrac{k-a}{k-b}\right)n$

D) $-\left(\dfrac{k-b}{k-a}\right)n$

handwritten:
$$\frac{ma+nb}{m+n} = k$$
$$ma+nb = mk+nk$$
$$ma - mk = nk - nb$$
$$m(a-k) = n(k+b)$$
$$\frac{}{a-k} \quad \overline{a-k}$$

12

$$x + y = 8$$
$$x^2 + y^2 = 48$$

If (x, y) is a solution of the system of equations above, what is the value of xy?

A) 4

B) 8

C) 12

D) 16

CONTINUE

13

$$4x + y \leq 3$$
$$4x \geq y + 1$$

What are the coordinates of the point with the maximum value of y in the solution of the system of inequalities above?

A) $\left(-\dfrac{1}{2}, \dfrac{1}{2}\right)$

B) $\left(-\dfrac{1}{2}, 1\right)$

C) $\left(\dfrac{1}{2}, 1\right)$

D) $\left(1, \dfrac{3}{2}\right)$

Handwritten work:
$4x \geq 1 + 1$
$4x \geq 2$
$x \geq \frac{1}{2}$
$4 \geq \frac{5}{2}$ $2+1 \leq 2$
$4 \geq \frac{2}{2}$

14

$$x - y = 1$$
$$x^2 + y^2 < 16$$

Which of the following are the points that satisfy the system above.

A) A point

B) An arc of the circle

C) A straight line segment not including the end-points

D) Points in the semicircle

15

A polling agency recently surveyed 300 students who were selected at random from a certain high school to see if they are for or against purchasing a water filtration system for the school water fountains. Of these surveyed, 80 percent responded that they favor the water filtration system and 20 percent responded that they oppose the system. Based on the results of the survey, which of the following statements must be true?

   I. Eighty percent of all students in the school favor the water filtration system.

   II. If 300 students selected at random from a different high school were surveyed, 20 percent of them would oppose the water filtration system.

   III. If another 300 students selected at random from the school were surveyed, 80 percent of them would favor the water filtration system

A) I only

B) I and III only

C) I and III only

D) None

Handwritten work:
$300 \times 0.8 = 240$
2400
240 for
60 against

CONTINUE

Questions 16- 17 refer to the following information.

Planets	Gravity $\left(m/s^2\right)$
Earth	9.8
Mercury	3.7
Venus	9.1
Mars	3.8
Jupiter	2.4
Pluto	0.7

The weight of an object is the force of gravity exerted on that object. The mass of an object is the amount of matter it has. The method of calculating the approximate weight (in newtons) from mass on planets is given by the equation $F = mg$, where m is the mass of an object (in kilograms), and g is the gravity on a planet (in meters per square second).

16

According to the information in the table, what is the approximate weight of an object with a mass of 85 kilograms on the Mercury?

$85 \cdot 3.7$

A) 314.5 newton

B) 256.4 newton

C) 215.5 newton

D) 200.4 newton

17

Weight versus Mass

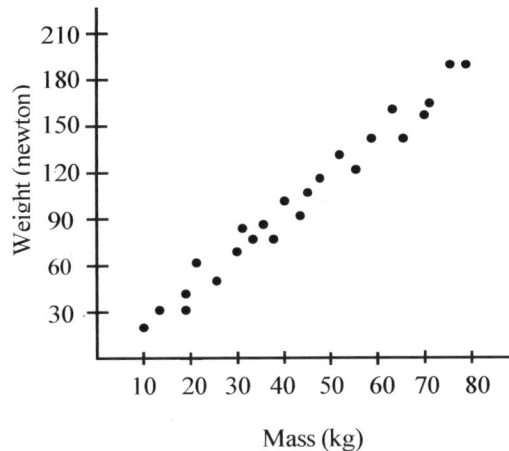

The scatter plot gives the mass of an object plotted against the weight of the object. The gravity of a planet is closest to that of which of the following planets?

A) Mercury

B) Venus

C) Mars

D) Jupiter

CONTINUE

18

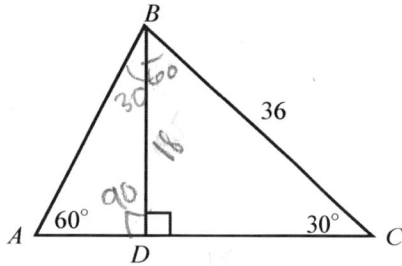

Note: Figure not drawn to scale

In $\triangle ABC$ above, what is the length of $\overline{AB}$?

A) $6\sqrt{2}$

B) $6\sqrt{3}$

C) $12\sqrt{2}$

D) $12\sqrt{3}$

(handwritten:) $18 = \chi\sqrt{3}$

$\dfrac{18}{\sqrt{3}}\ \dfrac{\sqrt{3}}{\sqrt{2}}\quad \dfrac{18\sqrt{3}}{3}$

$6\sqrt{3}$

19

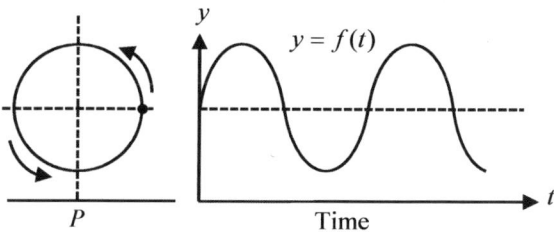

The figure on the left above shows a Ferris wheel with the mark of a passenger car above the ground. The Ferris wheel rotates counterclockwise at a constant rate. The graph of $y = f(t)$ on the right could represent which of the following as a function of time when the Ferris wheel starts immediately afterward.

A) The length of the radius of the Ferris wheel

B) The distance of the Ferris wheel from the passenger car

C) The distance of the passenger car from point P on the ground

D) The height the passenger car from the ground

20

$$k = \frac{a^2 - b}{a}$$

In the equation above, if a is positive and b is negative, which of the following must be true?

A) $k > 1$

B) $k > a$

C) $k < b$

D) $k < 1$

(handwritten:) $\dfrac{2}{2}\ \dfrac{\sqrt{3}}{}\ \dfrac{5}{2}$

21

In City Spring Lake, Mr. Lee's tenth-grade class, consisting of 30 students, was surveyed and 25.6 percent of the students reported that they spent at least two hours on their school homework. The average class size in the state is 30. If there are 800 tenth-grade classes in the state, which of the following best estimates the number of tenth-grade students in the state who spend fewer than two hours for their school homework?

A) 6,100

B) 10,000

C) 18,000

D) 25,000

(handwritten:) $\dfrac{768}{30} = \dfrac{1}{24000}$

CONTINUE

Question 22 and 23 refer to the following information.

World Electric Car Group		
Styles	Purchase price	Monthly lease price
Model 3	$43,000	$375
Model B	$84,000	$990
Model S	$64,000	$690
Model E	$92,000	$1,110
Model 5	$88,000	$1,050

The World Electric Cart Group offers the purchase and lease price for five different models of electric cars next year is as shown in the table above. The table shows the purchase prices and the corresponding monthly lease prices for the cars.

22

The relationship between the monthly lease price m, in dollars, and the purchase price p, in thousands of dollars, can be represented by a linear function. Which of the following functions represents the relationship?

A) $m(p) = 40p - 1345$

B) $m(p) = 15p - 270$

C) $m(p) = 12.5p - 20$

D) $m(p) = 10.5p - 76.5$

23

In the New Year, the company will offer an 18% discount off the original price for purchasing the all models with an additional 8% off the discounted price. What is the total percent of discount of the cars?

A) 21.35%

B) 24.56%

C) 26%

D) 26.5%

24

A teacher set up a survey to study the tendency of a student to select a favorite subject when presented with five subjects. In the survey, 100 students were selected and each student was asked to choose a favorite subject. Of the first 80 students, 15 students choice mathematics as their favorite subject. Among the remaining 20 students, k students chose mathematics. If more than 25% of all participants chose mathematics, which of the following could be the value of k?

A) $k = 5$

B) $k = 10$

C) $k = 15$

D) $k = 25$

25

If the volume of a cube is $27\left(\dfrac{a^3}{8}\right)$, where a is a positive constant, which of the following gives the surface area of the cube?

A) $27\left(\dfrac{a^2}{4}\right)$

B) $27\left(\dfrac{a^2}{2}\right)$

C) $9\left(\dfrac{a^2}{4}\right)$

D) $3\left(\dfrac{a^2}{2}\right)$

CONTINUE

26

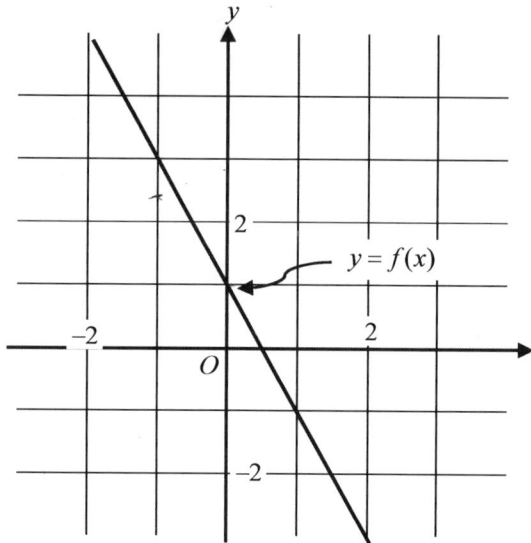

The graph of the linear function f is shown in the xy-plane above. If the graph of g, defined by $g(x) = 4f(x-2) + k$, where k is a constant, passes through the point $(1, -4)$, what is the value of $g(3)$?

A) -20
B) -17
C) 12
D) 17

27

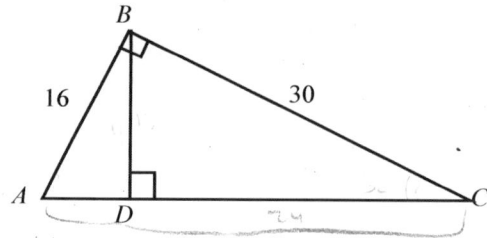

Note: Figure not drawn to scale.

In the figure above, an altitude of a right triangle from the right angle splits the hypotenuse into segments of lengths AD and DC. Which of the following is closest to the value of AD?

A) 7.5
B) 8.5
C) 10
D) 12

28

$$x^2 + y^2 + 10x - 16y = p$$

The equation above defines a circle in the xy-plane and p is a constant. If the area of the circle is 81π, what is the value of p?

A) 11
B) 8
C) -8
D) -11

CONTINUE

29

$$y = 2x^2 - ax + b$$

In the equation above, a and b are constants and the graph of the equation has a vertex at point $(3, -6)$. What is the value of b?

A) 3
B) 9
C) 12
D) 15

$$\frac{a}{4} = 3$$

$$a = 12$$

30

A high school library spends $2000 a year on algebra workbooks for the students. The average price for algebra workbooks bought this year was 2 dollars more than the average price last year. Because of the price increase, the school was forced to buy 50 fewer math workbooks. Which of the following equations best describes the value of p, the number of algebra workbooks that the school bought last year?

A) $\dfrac{2000}{p} = \dfrac{2000}{p-2} - 50$

B) $\dfrac{2000}{p} = \dfrac{2000}{p+2} - 50$

C) $\dfrac{2000}{p-2} = \dfrac{2000}{p} - 50$

D) $\dfrac{2000}{p+2} = \dfrac{2000}{p} - 50$

CONTINUE

For questions 31-38, solve the problem and enter your answer in the grid, as described below, on the answer sheet.

1. Although not required, it is suggested that you write your answer in the boxes at the top of the columns to help you fill in the circles accurately. You will receive credit only if the circles are filled in correctly.

2. Mark no more than one circle in any column.

3. No question has a negative answer.

4. Some problems may have more than one answer.

5. **Mixed numbers** such as $3\frac{1}{2}$ must be gridded as 3.5 or 7/2. (If $\boxed{3\,|\,1\,|\,/\,|\,2}$ is entered into the grid, it will be interpreted as $\frac{31}{2}$, not $3\frac{1}{2}$.)

6. **Decimal answers:** If you obtain a decimal answer with more digits than the grid can accommodate, it may be either rounded or truncated, but it must fill the entire grid.

Answer: $\frac{7}{12}$

Write answer in boxes. →

Fraction line

Grid in result.

Answer: 2.5

Decimal point

Acceptable ways to grid $\frac{2}{3}$ are:

Answer: 201
Either position is correct.

Note: You may start your answers in any column, space permitting. Columns you don't need to use should be left blank.

CONTINUE

31

The elastic potential energy in joules is the energy stored in a spring when it is stretched. The elastic potential energy is directly proportional to the square of the distance that the spring has been stretched from its original length. When a spring is stretched a distance of 0.5 meter, the elastic potential energy is 12 joules. If the spring is stretched 0.25 meter, what is the elastic potential energy, in joules, stored in the spring?

$$\frac{0.5}{12} = \frac{0.25}{x}$$ 3 joules

32

$$\begin{array}{c} \xleftarrow{\hspace{1.5cm}} x+6 \xrightarrow{\hspace{1.5cm}} \\ A \qquad B \qquad\quad C \qquad D \\ \xleftarrow{\hspace{1cm}} 3x-10 \xrightarrow{\hspace{1cm}} \end{array}$$

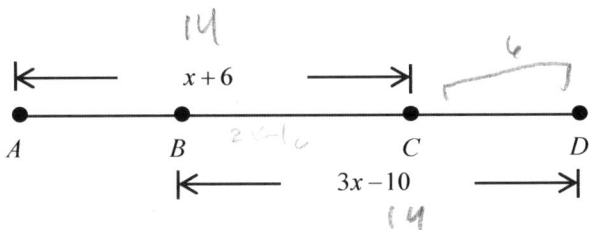

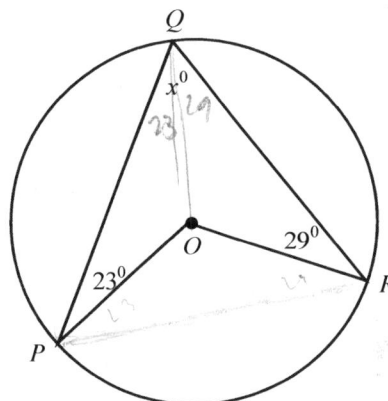

Note: Figure not drawn to scale.

On $\overline{AD}$ above, $AC = x+6$, $BD = 3x-10$, and $AD = 20$. If $AC = BD$, what is the length of $\overline{BC}$?

$3x-10-x-6$ $x+6 = 3x-10$
$2x-16 =$ $\boxed{BC = 8}$ $16 = 2x$
                              $x = 8$

33

In the xy-plane, the point $(1,3)$ is the vertex of the graph of the function f. If $f(x) = x^2 + ax + b$, where a and b are constants, what is the value of b?

$$\frac{-a}{2} = 1$$

$$a = -2$$

$b =$ $\boxed{4}$ $x^2 - 2x + b$

34

A landscaper is designing a regular hexagonal garden. If the area of the garden will be $150\sqrt{3}$ square feet, what will be the perimeter, in feet, of the garden?

$\frac{150\sqrt{3}}{6}$

60

35

Note: Figure not drawn to scale.

Point O is the center of the circle in the figure above.

What is the value of x?

$\boxed{52^\circ}$

Questions 36 and 37 refer to the following information.

Lidia lives on a street that runs west and east. Her office is to the east and the town library is to the west of her house. Both are on the same street as her house. At 9:00 a.m., she left her home and drove at an average speed of 30 mph for 6 minutes to her office. When she arrived at the office, she realized she has forgotten to pick up a book at the library. Immediately, she drove to the library at an average speed of 36 mph for 8 minutes. After 3 minutes at the library, she drove directly to her office.

36

What is the distance, in miles, from her house to the town library?

$$\frac{30}{6} = \frac{36}{8}$$

37

If she arrived at the office at 9:27 a.m., what was her average speed, in miles per hour, from the town library to her office?

38

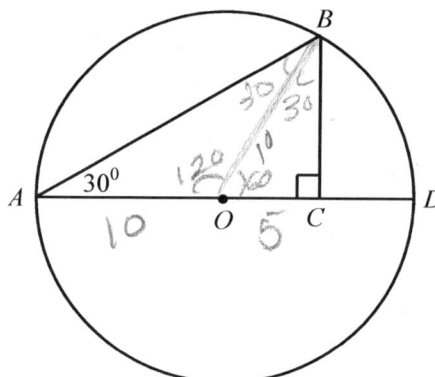

The circle in the figure above has center O and radius 10. If $\overline{BC}$ is perpendicular to $\overline{AD}$, what is the length of $\overline{AC}$?

$\boxed{15}$

STOP

If you finish before time is called, you may check your work on this section only.
Do not turn to any other section in the test.

No Test Material on This Page

Math Conversion Table

Raw Score	Scaled Score	Raw Score	Scaled Score
58	800	27	500
57	800	26	490
56	800	25	480
55	800	24	470
54	790	23	460
53	780	22	460
52	770	21	450
51	750	20	440
50	740	19	430
49	730	18	430
48	720	17	420
47	710	16	420
46	700	15	410
45	690	14	400
44	670	13	390
43	680	12	380
42	670	11	370
41	660	10	360
40	650	9	450
39	640	8	340
38	630	7	330
37	620	6	310
36	610	5	290
35	600	4	280
34	590	3	270
33	580	2	260
32	560	1	240
31	550	0	200
30	540		
29	530		
28	520		

Answer Explanations

Test 16 — Answers and Explanations

SECTION 3

	1	2	3	4	5	6	7	8	9	10
	D	C	D	C	C	C	D	C	A	A
	11	12	13	14	15	16	17	18	19	20
	A	D	B	B	C	6	1.8	6.75	4	39

ASECTION 4

	1	2	3	4	5	6	7	8	9	10
	B	B	D	A	D	C	A	C	A	A
	11	12	13	14	15	16	17	18	19	20
	D	B	C	C	D	A	D	D	D	B
	21	22	23	24	25	26	27	28	29	30
	C	B	B	C	B	A	A	C	C	D
	31	32	33	34	35	36	37	38		
	3	8	4	60	52	1.8	28.8	15		

SECTION 3

1. **D**

   Slope of the line is $\dfrac{-2}{3}$ and its y-intercept is -1.

2. **C**

   Since $\pi r^2 = 36\pi$, $r = 6$. Circumference $= 12\pi$

   Length of arc $\overparen{ABC} = 12\pi \times \dfrac{3}{4} = 9\pi$

3. **D**

   $\dfrac{1}{10}x^2 - \dfrac{1}{2}x + \dfrac{3}{5} = 0 \xrightarrow{\times 10} x^2 - 5x + 6 = 0$

   $(x-2)(x-3) = 0 \rightarrow x = 2, 3$

4. **C**

   Since the graph of $y = -x^2 - 4$ is below the x-axis, it doesn't have x-intercept.

5. **C**

   $\sqrt{k+11} + 2 = 1 + k \rightarrow \sqrt{k+11} = k - 1 \rightarrow$

   $k + 11 = k^2 - 2k + 1 \rightarrow k^2 - 3k - 10 = 0 \rightarrow$

   $(k-5)(k+2) = 0 \rightarrow k = 5, -2$ (not working)

6. **C**

   $a^3 - 2a + a^2 - 2 \rightarrow a^2(a+1) - 2(a+1) \rightarrow$

   $(a^2 - 2)(a+1)$

7. **D**

   Total number of hours he can work is less than or equal to 15 hours and the total amount of money should be greater or equal to $120.

8. **C**

9. **A**

   Substitute $y = x^2$ in the equation.

   $2x + 12 = 2(y - 6) \rightarrow x + 6 = x^2 - 6 \rightarrow$

   $x^2 - x - 12 = 0 \rightarrow (x+3)(x-4) = 0 \rightarrow$

   $x = \cancel{-3}, 4 \rightarrow y = 4^2 = 16$

   Therefore, $x + y = 4 + 16 = 20$.

10. **A**

    $x^2 + y^2 = (a^2 - b^2)^2 + (2ab)^2 = a^4 + 2a^2b^2 + b^4$

    $= (a^2 + b^2)^2$

11. **A**

$\dfrac{\pi r^2 h}{3} = 33$ and the volume of the new circular

cone is $\dfrac{\pi\left(\dfrac{r}{2}\right)^2 (2h)}{3} = \dfrac{1}{2}\left(\dfrac{\pi r^2}{3}\right) = \dfrac{1}{2}(33) = 16.5$.

12. D

$2\left(\sqrt[3]{16}\right) = 2\left(\sqrt[3]{2^3 \times 2}\right) = 2\left(2\sqrt[3]{2}\right) = 4\sqrt[3]{2}$

13. B

$\dfrac{\triangle a}{\triangle k} = 10 \ \rightarrow \ \dfrac{20}{\triangle k} = 10 \ \rightarrow \ \triangle k = 2$

14. B

$\dfrac{300}{25} = 12 \ \rightarrow$ He used 12 gallons to drive

300 miles. Amount of money is $12 + 2.5 = \$30$.
He wants to spend $22 on gasoline.

Therefore, $\dfrac{300}{30} = \dfrac{x}{22} \ \rightarrow \ x = \$220$

15. C

The graph of $y = f(-x)$ is symmetric with the

graph of $y = f(x)$ in the y-axis.

16. 6

If n is the number of rides,

$8 + 4n \le 35 \ \rightarrow \ n \le 6.25 \ \rightarrow$ maximum number of

rides is 6.

17. $\dfrac{9}{5}$ or 1.8

Since $p - 1 > 0$, divide both side of the equation by

$p - 1$. $8(p-1) = 5(p^2 - 1) - 6(p - 1) \ \rightarrow$

$8 = 5(p+1) - 6 \ \rightarrow \ 14 = 5p + 5 \ \rightarrow \ p = \dfrac{9}{5}$

18. $\dfrac{27}{4}$ or 6.75

$\dfrac{1}{3}y = 9x \ \rightarrow \ y = 27x$ Putting in the first equation.

$\dfrac{1}{4}(3x + 27x) = \dfrac{15}{8} \ \rightarrow \ \dfrac{30x}{4} = \dfrac{15}{8} \ \rightarrow$

$x = \dfrac{4 \times 15}{30 \times 8} = \dfrac{1}{4} \ \rightarrow \ y = 27\left(\dfrac{1}{4}\right) = \dfrac{27}{4}$

19. 4

$\dfrac{A}{(x+2)^2} + \dfrac{B}{x+2} = \dfrac{3x+10}{(x+2)^2} \ \rightarrow$ Multiply by $(x+2)^2$

$A + B(x+2) = 3x + 10 \ \rightarrow \ Bx + A + 2B = 3x + 10$

Expressions of both sides are equal.
$B = 3$ and $A + 2B = 10 \ \rightarrow \ A + 2(3) = 10 \ \rightarrow$
$A = 4$

20. 39

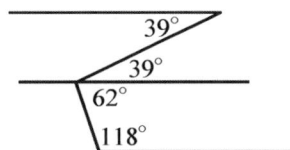

1. B
2010~2011 : 40 thousands/year (positive)
2013~2014 : −60 thousands/year (negative)

2. B
For 6 hour 40 minutes $\rightarrow$ 400 minutes,

$(10 + 0.1 \times 400)(1 + r) = 54 \ \rightarrow \ (1+r) = \dfrac{54}{50}$

For 8 hours $\rightarrow$ 480 minutes,

$(10 + 0.1 \times 480)(1+r) = 58\left(\dfrac{54}{50}\right) = 62.64$

3. D
Two zeros at $x = -3$ and $x = 4$
$y = a(x+3)(x-4)$: factored form
Since the graph passes through the point $(1,6)$,

$6 = a(1+3)(1-4) \ \rightarrow \ 6 = -12a \ \rightarrow \ a = -\dfrac{1}{2}$.

Therefore, $y = -\dfrac{1}{2}(x+3)(x-4)$.

4. A

Proportion: $\dfrac{24}{3} = \dfrac{125}{x} \ \rightarrow \ 15.625 \, \text{eggs} \ \rightarrow$

$\dfrac{5.625}{12} \approx 1.3 \, \text{dozens} \ \rightarrow \ 2 \, \text{dozens}$.

5. D

$5(a^2 - b^3) = 13 \ \rightarrow \ 5(a-b)(a+b) = 13$

$a = b + 7 \ \rightarrow \ a - b = 7$

Thus, $5(7)(a+b) = 13 \ \rightarrow \ (a+b) = \dfrac{13}{35}$

6. C
For x members, the book price is $4x + 7$ or
$6x - 9$. $4x + 7 = 6x - 9 \ \rightarrow \ 2x = 16 \ \rightarrow \ x = 8$
Therefore, the price is $4(8) + 7 = \$39$

Answer Explanations

7. **A**

    The number of geometry book $= x$

Alg	\$8	$9 - x$
Geo	\$12	x

    $8(9 - x) + 12x = 88 \rightarrow 4x = 16 \rightarrow x = 4$

8. **C**

    $1.44a^2 - 8.64a + 12.96 - 2.5a^2 + 1.2 =$

    $-1.06a^2 - 8.64a + 14.16$

9. **A**

    x = number of men actually on the hunting trip

    $\dfrac{720}{x} - \dfrac{720}{x + 2} = 12 \rightarrow \dfrac{60}{x} - \dfrac{60}{x + 2} = 1$

    $60(x + 2) - 60x = x(x + 2)$

    $120 = x^2 + 2x \rightarrow x^2 + 2x - 120 = 0$

    $(x + 12)(x - 10) = 0 \rightarrow x = 10 \ (x \neq -12)$

10. **A**

    $0 = 60t - 0.8t^2 \rightarrow 0 = t(60 - 0.8t)$

    Therefore, $t = 0$ or $t = \dfrac{60}{0.8} = 75. \rightarrow 75 - 0 = 75$

11. **B**

    $\dfrac{ma + nb}{m + n} = k \rightarrow ma + nb = mk + nk$

    $ma - mk = nk - nb \rightarrow m(a - k) = n(k - b)$

    $m = \left(\dfrac{k - b}{a - k}\right)n = -\left(\dfrac{k - b}{k - a}\right)n$

12. **B**

    $x^2 + y^2 = 48 \rightarrow (x + y)^2 - 2xy = 48$

    $8^2 - 2xy = 48 \rightarrow 2xy = 16 \rightarrow xy = 8$

13. **C**

    At the intersection of the graphs, the solution has a maximum value of y.

    $4x + y = 3$ and $4x = y + 1 \rightarrow 4x = 3 - y$

    $y + 1 = 3 - y \rightarrow 2y = 2 \rightarrow y = 1$

    and $4x = 3 - 1 = 2 \rightarrow x = \dfrac{1}{2}$. Therefore, $\left(\dfrac{1}{2}, 1\right)$.

14. **C**

    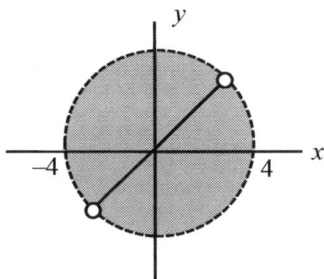

15. **D**

    The result of the survey cannot guarantee anything about other group.

16. **A**

    $85 \times 3.7 = 314.5$

17. **D**

    Slope of the scatter points is approximately 2.4.

18. **D**

    Special triangles. $BD = 18$ and $AB = 12\sqrt{3}$

19. **D**

20. **B**

    $k = \dfrac{a^2 - b}{a} \rightarrow k = a + \dfrac{(-b)}{a} \rightarrow \dfrac{(-b)}{a} = \text{positive}$

    Therefore, $k > a$.

21. **C**

    $100 - 25.6 = 74.4$

    $30(0.744) = 22.32 \rightarrow 22.32 \times 800 = 17856$

    18,000 is closest to the number.

22. **B**

    You can use a graphing calculator to find the equation of a linear regression.

    Or check each equation from the data given.

23. **B**

    $p(1 - 0.18)(1 - 0.08) = 0.7544p \rightarrow (1 - 0.2456)p$

    Total percent of D.C is 24.56%.

24. **C**

    The inequality is

    $15 + k > 0.25(100)$ and $k < 20$

    $k > 10$ and $k < 20 \rightarrow 10 < k < 20$

25. **B**

    $V = \left(\dfrac{3a}{2}\right)^3 \rightarrow$ The length of the cube is $\dfrac{3a}{2}$.

    Surface area is $6\left(\dfrac{3a}{2}\right)^2 = 27\left(\dfrac{a^2}{2}\right)$.

26. **A**

    $(1, -4)$ lies on the graph of g.

    $-4 = 4f(1 - 2) + k \rightarrow -4 = 4f(-1) + k$

    Since $f(-1) = 3$, $-4 = 4(3) + k \rightarrow k = -16$.

    $g(x) = 4f(x - 2) - 16$

    $g(3) = 4f(3 - 2) - 16 \rightarrow g(3) = 4f(1) - 16$

    Since $f(1) = -1$, $g(3) = -4 - 16 = -20$.

27. **A**

    $AD = \sqrt{AB^2 + BC^2} = \sqrt{16^2 + 30^2} = 34$ A

    From the formula: $AB^2 = AD \times AC$

Answer Explanations

$16^2 = AD \times 34 \ \rightarrow \ AD \approx 7.5$

28. C

$x^2 + 10x + y^2 - 16y = p \ \rightarrow$

$x^2 + 10x + 25 + y^2 - 16y + 64 = p + 25 + 64$

$(x+5)^2 + (y-8)^2 = p + 89 = r^2$

Since the radius of the circle is 9, $p + 89 = 81$.

$p = -8$

29. C

Using Axis of symmetry

$\dfrac{-(-a)}{2(2)} = 3 \ \rightarrow \ a = 12, \ y = 2x^2 - 12x + b$

$f(3) = 2(3)^2 - 12(3) + b = -6 \ \rightarrow \ b = 12$

30. D

The number of books last year is $\dfrac{2000}{p}$ and the

number of this year is $\dfrac{2000}{p+2}$. Since their difference

is 50, $\dfrac{2000}{p+2} = \dfrac{2000}{p} - 50$.

31. 3

From direct proportion,

$E = kx^2 \ \rightarrow \ \dfrac{E}{x^2} = k(\text{constant})$

Therefore, $\dfrac{12}{0.5^2} = \dfrac{E}{0.25^2} \ \rightarrow \ E = \dfrac{12(0.25^2)}{0.5^2} = 3$

32. 8

Since $\overline{BC}$ is common, $AC = BC$.

$x + 6 = 3x - 10 \ \rightarrow \ x = 8 \ \rightarrow \ AC = 14$ and $CD = 6$

Therefore, $BC = BD - CD = 14 - 6 = 8$.

33. 4

Axis of symmetry: $x = \dfrac{-a}{2} = 1 \ \rightarrow \ a = -2$

$f(x) = x^2 - 2x + b \ \rightarrow \ f(1) = 1 - 2 + b = 3$

$b = 4$

34. 60

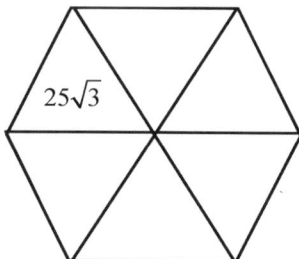

Hexagon has six equilateral triangles.

Each triangle has a area of $25\sqrt{3}$.

Remember: If the length of a equilateral triangle is

a, it's area is $\dfrac{a^2 \sqrt{3}}{4}$.

Therefore, $\dfrac{a^2 \sqrt{3}}{4} = 25\sqrt{3} \ \rightarrow \ a^2 = 100 \ \rightarrow \ a = 10$.

Now the perimeter of the hexagon is $10(6) = 60$.

35. 52

$x = 23 + 29 = 52$. $\triangle OPQ$ and $\triangle ORQ$ are isosceles.

36. 1.8

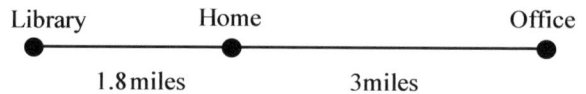

Distance from office to library:

$36 \times \dfrac{8}{60} = 4.8$ miles

Distance from home to office:

$30 \times \dfrac{6}{60} = 3$ miles

Therefore, $4.8 - 3 = 1.8$ miles.

37. 28.8

She spent 37 minutes on that morning. The time
she spent from the library to her office is

$27 - (6 + 8 + 3) = 10$ minutes.

Average speed is $\dfrac{4.8}{\frac{10}{60}} = 28.8$ mph.

38. 65

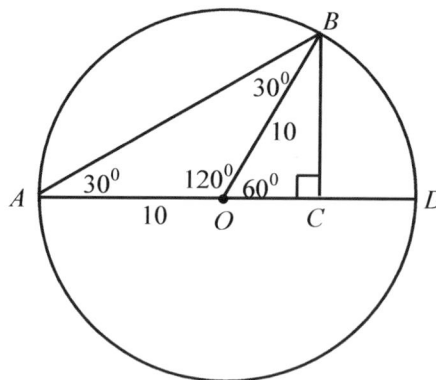

$AO = BO = 10$ and $OC = \dfrac{1}{2}(10) = 5$ (Special

triangle). Therefore, $AC = 15$.

Made in the USA
Lexington, KY
04 June 2018